Emulsion Science

Second Edition

Fernando Leal-Calderon
Véronique Schmitt
Jérôme Bibette

Emulsion Science

Basic Principles

Second Edition

Fernando Leal-Calderon
ISTAB, Université Bordeaux 1
Avenue des facultés
33405 Talence
FRANCE
f.leal@istab.u-bordeaux1.fr

Véronique Schmitt
Centre de Recherche Paul Pascal, CNRS
33600 Pessac
FRANCE
schmitt@crpp-bordeaux.cnrs.fr

Jérôme Bibette
ESPCI, Laboratoire Colloïdes
et Matériaux Divisés
10 rue Vauquelin
75231 Paris
FRANCE
jerome.bibette@espci.fr

Library of Congress Control Number: 2007921974

ISBN-10: 0-387-39682-9 e-ISBN-10: 0-387-39683-7
ISBN-13: 978-0-387-39682-8 e-ISBN-13: 978-0-387-39683-5

Printed on acid-free paper.

9 8 7 6 5 4 3 2 1

springer.com

Contents

Preface

Emulsions and Common Sense

Droplets of oil in water allow us to transport oil soluble materials (in a non viscous form) and ultimately to release them on a chosen target. Cosmetics, paints, foods, are often based on these "emulsions".

But the droplets are fragile, and must be lucidly protected. Formulating an industrial emulsion implies numerous conditions: stability, efficiency, easy delivery, price, ... This is an art, and like all forms of art it requires experience and imagination. The present book provides both. It describes basic experiments on realistic model systems. I like this matter of fact approach. For instance, instead of beginning by formal discussions on interaction energies, the book starts with *methods of fabrication*. And, all along the text, the theoretical aspects are restricted to basic needs.

Of course (as always in a delicate subject like the present one) I have my own critiques on certain points in the presentation: for instance, in Section 5.3.1, coalescence is attributed to the thermal nucleation of a pore between two adjacent droplets. For me, discussing this channel is like discussing the sex of angels. Nucleation, in most physical systems, does not occur via plain thermal fluctuations. It involves external defects: a cosmic ray in a bubble chamber, or a dust particle in a condenser. I believe that the same holds for emulsions: dust particles (or small surfactant aggregates) control coalescence.

But these byzantine discussions on mechanism are beyond the point. This book is based on experiments. It does not claim to solve all problems (e.g: what is the real origin of Bancroft's rule) but it presents them with common sense and precision. I am convinced that it will be of great help.

P.-G. de Gennes
January 2007

Acknowledgments

The authors thank S. Abramson, S. Arditty, P. Aronson, J. Baudry, E. Bertrand, L. Bressy, M. Chamerois, L. Cohen-Tannoudji, A. Colin, B. Deminière, T. Dimitrova, R. Dreyfus, J. Fattaciolli, M.-F. Ficheux, P.-G. de Gennes, J. Giermanska-Kahn, P. Gorria, C. Goubault, A. Griffith, J. Guery, F. Guimberteau, A. Koenig, C. Keichinger, C. Mabille, T. Mason, O. Mondain-Monval, P. Omarjee-Rivière, P. Pays, J. Philip, F. Placin, B. Pouligny, P. Poulin, V. Ravaine, A. Royère, T. Stora, J. Sylvestre, F. Thivilliers, G. Vetter, D.A. Weitz, and D. Zerrouki for their contributions.

Introduction

Colloids comprise a very broad class of materials. Their basic structure consists of a dispersion of one phase into another one, in which the dispersed phase possesses a typical length scale ranging from a few molecular sizes up to several microns. Some colloids are thermodynamically stable and generally form spontaneously, whereas others are metastable, requiring energy for preparation and specific properties to persist. Metastable colloids are obtained by two main distinct routes: one is nucleation and growth, including precipitation, and the other is fragmentation. In both cases, as a consequence of the intrinsic off-equilibrium nature of this class of colloids, specific surface properties are required to prevent recombination. Surface-active species are generally employed to stabilize freshly formed fragments or growing nuclei, as they are expected to provide sufficient colloidal repulsive forces.

Emulsions are one example of metastable colloids. They are generally made out of two immiscible fluids, one being dispersed in the other, in the presence of surface-active agents. They are obtained by shearing two immiscible fluids, leading to the fragmentation of one phase into the other. Emulsion droplets exhibit all classical behaviors of metastable colloids: Brownian motion, reversible phase transitions as a result of droplet interactions, and irreversible transitions that generally involve the destruction of the emulsion. The droplet volume fraction may vary from zero to almost one: dense emulsions are sometimes called biliquid foams since their structure is very similar to the cellular structure of air–liquid foams for which the continuous phase is very minor. From dilute to highly concentrated, emulsions exhibit very different internal dynamics and mechanical properties. When the emulsion is strongly diluted, droplets are agitated by Brownian motion [1,2], and the emulsion behaves as a viscous Newtonian fluid. When the emulsion is concentrated, namely above the random-close-packing volume fraction, which is 64% for monodisperse droplets, the internal dynamics are severely restricted and the emulsion behaves as a viscoelastic solid [3,4]. Simple direct emulsions are composed of oil droplets dispersed in water (O/W), while inverse emulsions are composed of water droplets dispersed in an oil continuous phase (W/O). Emulsions are in general made of two immiscible phases for which the surface tension is nonzero, and may in principle involve other hydrophilic-like or lipophilic-like fluids in the presence of suitable surface-active species, each phase being possibly composed of numerous components. As an example, simple emulsions may also contain smaller droplets of the continuous phase dispersed within each droplet of the dispersed phase. Such systems are called double emulsions or multiple emulsions [5]. Simple emulsions may also contain solid dispersion within each droplet, as a possible route to produce magnetic colloids [6].

The destruction of emulsions may proceed through two distinct mechanisms. One, called Ostwald ripening, is due to the diffusion of the dispersed phase through the continuous phase. This mechanism does not involve any film rupture; instead, there is a continuous exchange of matter through the continuous phase, which increases the average droplets diameter while reducing their number. The other mechanism, called coalescence, consists of the rupture of the thin liquid film that forms between two adjacent droplets. This rupture requires the formation of a hole within the thin film which then grows, resulting in the fusion of two adjacent droplets. This ultimately leads to a total destruction of the dispersed system, since two macroscopic immiscible phases are recovered. The lifetime of emulsions is increased by the presence of surface-active species, which are known to cover the interfaces and to delay both coalescence and Ostwald ripening. As a matter of fact, the metastability of emulsions is strongly correlated to the presence of these surface-active species at their interfaces. Because the lifetime of these materials may become significant (longer than one year) they become good candidates for various commercial applications.

Emulsions are widely used in a variety of applications because of their ability to transport or solubilize hydrophobic substances in a water continuous phase. All kinds of surface treatments will take advantage of emulsion technology: painting, paper coating, road surfacing, and lubricating. Because homogeneous mixtures of two immiscible fluids may be obtained, organic solvents may be avoided when solubilizing hydrophobic substances into water. When the mixture is applied, water evaporates and is safely released into the atmosphere, while the dispersed phase concentrates and ultimately leads to the formation of a hydrophobic film (painting, paper coating, lubricating). Moreover, emulsion technology drastically simplifies the pourability of many hydrophobic substances. Indeed, at ambient temperature some material may be almost solid whereas by dispersing it within small droplets in water it remains fluid at room temperature. One famous example is bitumen used for road surfacing. Emulsions are also involved in the food and cosmetic industries because of their rheological properties which may vary from an essentially Newtonian liquid to an elastic solid. Moreover, they are also efficient drug carriers (medicines, food, and pesticides) for various types of targets. Indeed, double direct emulsions will allow transporting a water soluble molecule within the internal water droplets throughout a water continuous phase.

All these applications have already led to an important empirical control of these materials, from their formation to their destruction. Besides this empirical background which is considerably widespread among the various specific applications, the basic science of emulsions is certainly progressing and we aim within this book to give an overview of the most recent advances.

The production of emulsions is certainly one of the most important aspects related to the industrial use of these materials. After a review of the available and currently employed techniques, this book aims to present advances in making controlled size emulsions at a large scale and rate production. It will be shown that controlled shear applied to a polydisperse emulsion can transform it into a monodispersed one through Rayleigh instability (Chapter 1). Interdroplet forces, from long range to very short range, are of prime importance in understanding the collective behavior of emulsion droplets. A variety of interactions that occur between these liquid colloids are described. Repulsive interactions between droplets are directly measured by using the magnetic chaining technique, providing detailed descriptions of steric and electrostatic forces (Chapter 2). Soft attractions and particularly the depletion induced interactions are described, as well as the resulting equilibrium phase transitions that can also be used to fractionate polydisperse emulsions. Strong adhesion is also explored through the measurement of contact angles, for a variety of interfaces and compositions, as well as the very characteristic gelation transition that takes place in the regime of deep attractive interaction quench (Chapter 3). Because emulsion droplets are deformable they can span droplet volume fraction from zero to almost one. We present the basic physics that governs both compressibility and shear elasticity of dense emulsions as a function of droplet packing and the nature of adsorbed species from short surfactants to macromolecules and solid particles. When droplets are still capable to slip under stress, the role of disorder has been revealed to be of most importance and to dictate the subtle scaling of the shear elastic modulus (Chapter 4). Understanding the lifetime and destruction of emulsions is obviously a crucial aspect. The various scenarios of destruction are reviewed and correlated to the two well accepted limiting mechanisms: coalescence and diffusion or permeation. The basic understanding of thermally activated hole nucleation, which is responsible for coalescence, is presented, in close relation with the nature of the adsorbed species (Chapter 5). The very rich domain of double emulsion is discussed. A detailed description of these materials is presented owing to their very promising potential in various applications, in addition to their remarkable contribution in understanding the metastability of thin films (Chapter 6). At that stage of understanding, the scientific background about emulsions can direct their potential use to new fields of applications: droplets can act either as minute substrates or reservoirs. They can also be manipulated by applying external forces to selectively sort desirable products, or to create local stress-controlled conditions. Liquid droplets can compartmentalize minute amounts of defined reactants, either to screen a large compound library or parallelize a directed process imposed by confinement. We present in Chapter 7 some particularly promising examples of new applications of emulsions in nano or microtechnologies, related to biotechnologies, biophysics, and processing of high-tech materials for micro-optics.

References

[1] J. Perrin: " La Loi de Stockes et le Mouvement Brownien." C. R. Acad. Sci. **147**, 475 (1908).

[2] J. Perrin: In: Felix Alean (ed): "Mouvement Brownien—Emulsions." Les Atomes. Gallimard PUF, Paris (1913).

[3] H.M. Princen: "Rheology of Foams and Highly Concentrated Emulsions I. Elastic Properties and Yield Stress of a Cylindrical Model System." J. Colloid Interface Sci. **91**, 160 (1983).

[4] T.G. Mason, J. Bibette, and D.A. Weitz: "Elasticity of Compressed Emulsions." Phys. Rev. Lett. **75**, 2051 (1995).

[5] S.S. Davis, J. Hadgraft, and K.J. Palin: In: P. Becher (ed): "Medical and Pharmaceutical Applications of Emulsions." Encyclopedia of Emulsion Technology, Vol. 2. Marcel Dekker New York (1985).

[6] J. Bibette: "Monodisperse Ferrofluid Emulsions." J. Magn. Magn. Mater. **122**, 37 (1993).

1 Emulsification

1.1. Introduction

Emulsification consists of dispersing one fluid into another, nonmiscible one, via creation of an interface. Properties of emulsions (e.g., stability, rheological properties) and their industrial uses are governed not only by variables such as temperature and composition but also by the droplet size distribution. The highest level of control consists of producing "monodisperse," that is, narrow size distributed emulsions with a tunable mean size. From a fundamental perspective, monodispersity has allowed significant progress in emulsion science as will be shown throughout this book. Monodispersity also opens perspectives for new technological applications that are reviewed in Chapter 7. Usually, industrial emulsification is empirically controlled and the purpose of this chapter is to provide fundamental concepts that support such empirical knowledge.

In the first part, we briefly review some possible routes to fabricate emulsions such as high-pressure homogenization and membrane, microchannel and spontaneous emulsification. Then, the basic principles of the phase inversion temperature (PIT) method are presented and the influence of different parameters such as surfactant concentration and stirring intensity is discussed. The following section is devoted to emulsification via application of a controlled shear. The mechanism of drop rupturing and the conditions leading to monodispersity are described. From fundamental studies on shear emulsification, some useful strategies for formulators can be proposed and we shall explain how they can be exploited to produce monodisperse materials of technological interest.

1.2. High-Pressure Homogenization

High-pressure homogenization (microfluidization) is widely used for producing dairy and food emulsions. It consists of forcing the two fluids or a coarse premix to flow through an inlet valve, into a mixing chamber, under the effect of a very high pressure. The fluids undergo a combination of elongation and shear flows, impacts, and cavitations. Despite the complexity of the mechanisms involved [1], the size distributions are usually reproducible with a mean size ranging from 50 nm to 5 μm.

Emulsification by high-pressure homogenization results from a dynamical equilibrium between breakup promoted by drop deformation resulting from the high speed flow and of recombination (coalescence) promoted by collisions. Numerous studies have been performed [2–6] to determine the effect of stabilizing agent concentration (protein or surfactant), applied pressure, number of cycles on the droplets size and emulsion stability. Taisne and Cabane [6] have developed a refractive index contrast matching technique allowing the determination of oil exchange between the droplets. They were able to distinguish two regimes of emulsification in a high-pressure homogenizer depending on the surfactant concentration C_{surf}. In the **"surfactant-poor" regime** ($C_{surf} < \text{CMC}/10$ where CMC is the critical micellar concentration), the average drop size, d, only weakly depends on the applied pressure. Lobo and co-workers [7] have elaborated a quantitative method based on the dilution of a fluorescent excimer signal during oil exchange to determine the number of coalescence events during emulsification. They showed [8] that a high level of coalescence leads to emulsions with average diameters ranging from 0.3 to 2 μm depending on the surfactant concentration. Drops are first fragmented at a low size and then coalesce because of insufficient interfacial coverage. In the **"surfactant-rich" regime** ($C_{surf} > 10$ CMC), the average droplet diameter d is lower, typically varying from 50 to 350 nm, and is almost independent of the surfactant concentration [6, 8, 9]. Even though coalescence can not be completely arrested in a high-pressure homogenizer, a low level of recombination is attained [6, 8]. Hence, the size is determined mainly by droplet fragmentation and scales with the applied pressure P_h as:

$$d \propto P_h^{-\alpha} \tag{1.1}$$

where the power law α typically varies between 0.6 and 0.9 [6,10,11]. Brösel and Schubert [4] showed that during the deformation and breakup of a single drop, almost no surfactant molecules adsorb at the new interface because the adsorbing time is larger than that of disruption. Surfactant adsorbs between two breakup events, thus lowering the interfacial tension and facilitating further rupturing. The existence of two regimes can be generalized to protein-stabilized emulsions: larger sizes are obtained by drop coalescence for low protein concentrations [2]. Other parameters may influence the final droplet size distribution: (1) an increasing number of passes reduces the size distribution width [12,13], (2) whatever the emulsifier (surfactant or proteins), large dispersed phase volume fractions favor collisions and recombinations but the droplet volume fraction ϕ has little influence on the average size for $\phi < 30\%$.

1.3. Membrane Emulsification

Membrane emulsification [14] consists of forcing the dispersed phase to permeate into the continuous phase through a membrane having a uniform pore

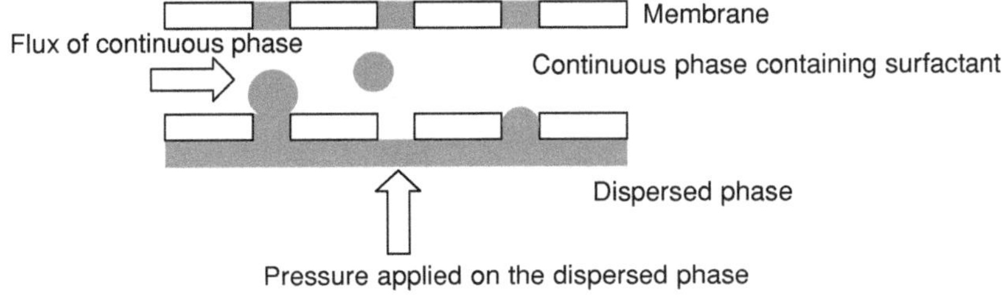

FIGURE 1.1. Schematic principle of membrane emulsification.

size distribution. The dispersed phase is pressed perpendicular to the membrane while the continuous phase is flowing tangential to the membrane (Fig. 1.1). Although easy in principle, membrane emulsification is dependent on many parameters such as membrane properties, fluxes, and formulation, all influencing the emulsion size distribution. To obtain a monodisperse emulsion, the membrane pores must themselves have a narrow size distribution [15]. Usually, the drop size is proportional to the pore size [16–18]. The choice of membrane porosity is the result of a compromise: if the pore density is too large, coalescence of freshly formed drops is likely to occur, increasing polydispersity; conversely, if the pore density is too low, the production rate is insufficient [19]. The dispersed phase should not wet the membrane coating and consequently a hydrophilic membrane should be used to produce an oil-in-water (O/W) emulsion [16,20]. High continuous phase velocity and low interfacial tension will promote small drops [19–27]. The pressure to be applied to the dispersed phase depends on both the interfacial tension [21] and the membrane pore size. A compromise between high pressures promoting either large drops or a dispersed phase jet and low pressures decreasing the production rate should be found. For a more detailed review on membrane emulsification of simple and double emulsions, the reader can refer to [28] and [29] and references therein.

1.4. Microchannel Emulsification

Microchannel technology allows fabrication of monodisperse emulsions with an average droplet diameter ranging from 10 to 100 μm [30,31]. The principle is reminiscent of membrane emulsification. The dispersed phase is forced into the continuous phase through microchannels manufactured via photolithography. A scheme of a microchannel device is shown in Fig. 1.2. The use of a high-speed camera and a microscope allows direct observation of the flow and of the emulsification process [32,33]. The phase to be dispersed is pushed through a hole in the center of the plate in such a way that it passes through the microchannels and inflates on the terrace in a disk-like shape. When it reaches the end of the terrace, the phase falls onto the well and a drop detaches. The spontaneous detachment and relaxation into the spherical drop are driven by interfacial tension. Outstanding

(a)

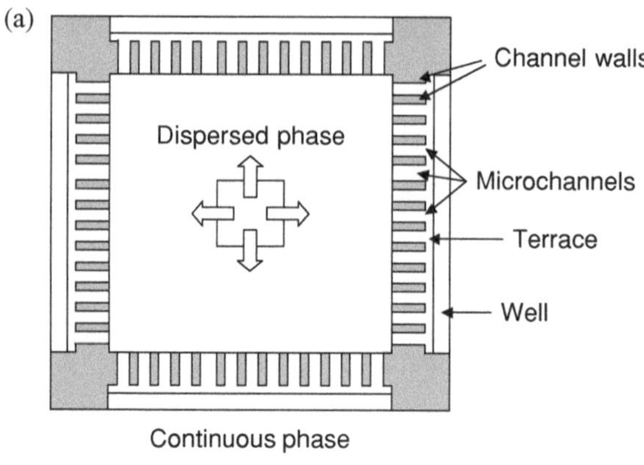

(b)

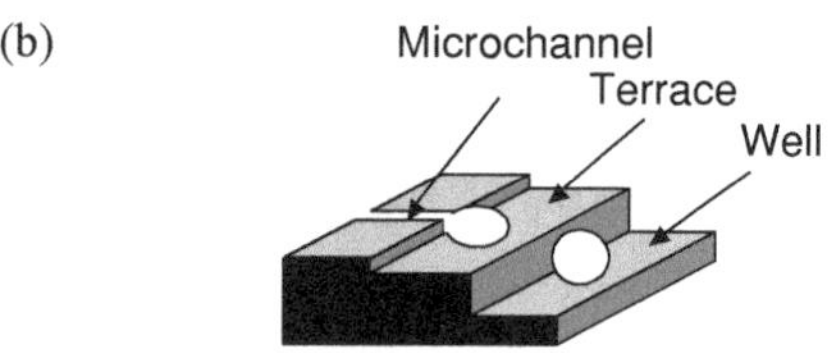

FIGURE 1.2. Schematic principle of microchannel emulsification. **(a)** Top view; **(b)** side view.

monodispersity is obtained by this process. Direct, reverse, and multiple emulsions [34] can be developed. As for membrane emulsification, an O/W emulsion is produced using hydrophilic microchannels, whereas producing a W/O emulsion requires a hydrophobic device [35]. The influence of various parameters on droplet size and monodispersity has been studied:

- The geometry of the device is important; the terrace length and microchannel depth are size-determining factors.
- At low flow velocity of the dispersed phase, the interfacial tension does not influence the droplet diameter but it affects the time-scale parameters for droplet formation [35–37]; the detachment time becomes shorter at high interfacial tension (low surfactant concentration) [38].
- The surfactant type (anionic, nonionic) is indifferent [39], but cationic surfactants should be avoided to produce O/W emulsions because they lead to complete wetting of the dispersed phase on the microchannel plate.

More complex geometries have been developed [40] and the influence of the geometrical structure has been examined. Although straight-through microchannel emulsification has been developed [39,41], the production rates are still low compared to those obtained with standard emulsification methods. However, the very high monodispersity makes this emulsification process very suitable for some specific technological applications such as polymeric microsphere synthesis [42,43], microencapsulation [44], sol–gel chemistry, and electro-optical materials.

Microchannel technology has also opened the route to microfluidics. Devices with different geometries have been reported. Emulsification can proceed through flow focusing [45,46] where geometries are composed of three inlet channels and a small orifice located downstream (Fig. 1.3a and b). The liquid that becomes the dispersed phase flows into the middle channel and the second immiscible fluid flows into the two outside channels. When passing through the orifice, the inner fluid breaks into drops of size comparable with the orifice width. As for microchannel emulsification, the geometrical and hydrodynamic parameters influence both droplet size and polydispersity [20,45]. Use of an expanding nozzle geometry (Fig. 1.3c) allows fixing the position of drop breakup at the orifice

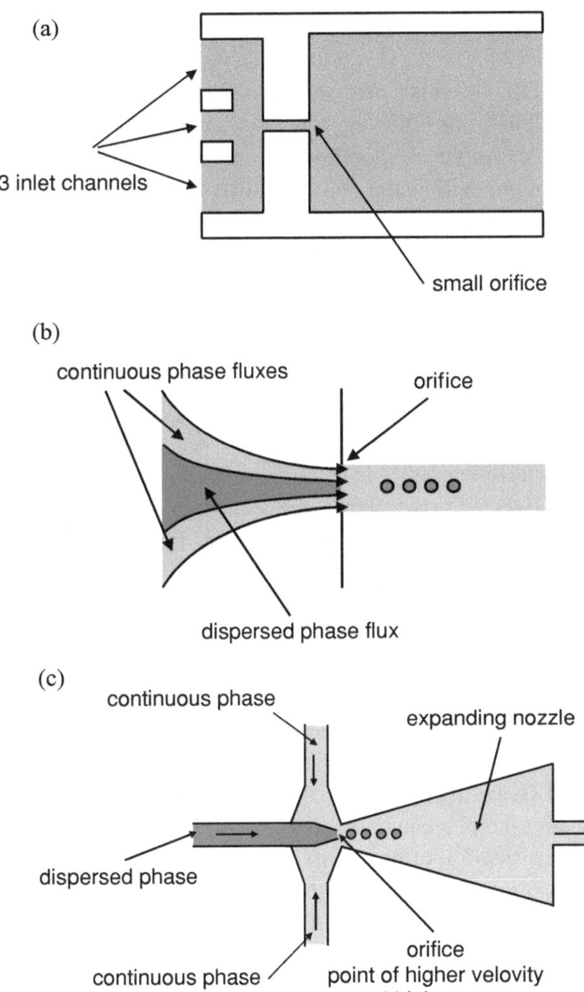

FIGURE 1.3. Schematic principle of "flow focusing" emulsification.

where the shear force is the highest [47]. Emulsification can also be induced by a junction [48–50] defined as the intersection of two microchannels, where the shear is locally highest [49]. Using a two-step method and both hydrophilic and hydrophobic junctions, double emulsions can be obtained [50]. Microfluidic emulsification is suitable for certain specific applications such as in microactuators, allowing rapid manipulation of microdroplets [51], or in microreactors, where it is useful for screening of protein crystallization conditions [52–54], glucose detection in clinical diagnostic [55], and controlling a reaction or mixing [56,57].

1.5. Spontaneous Emulsification

Spontaneous emulsification is a process that occurs without external energy supply when two immiscible fluids with very low interfacial tension are brought in contact. The most famous example of spontaneous emulsification is the famous "pastis" beverage put in contact with water. Without stirring, the blend becomes turbid. Because increasing the interfacial area generally requires energy input, spontaneous emulsification is an intriguing phenomenon, as revealed by the abundant literature devoted to it. It is worth noting that in industrial processes, the kinetics of this kind of emulsification, also termed self-emulsification, is accelerated by an energy supply [58]. Spontaneous emulsification was reported for the first time in 1878 by Joahnnes Gad [59]. Although observed a long time ago, this phenomenon is still not fully understood. So far, different mechanisms have been proposed and some of them are described hereafter.

Interfacial turbulence [60]: Due to a nonuniform distribution of surfactant molecules at the interface or to local convection currents close to the interface, interfacial tension gradients lead to a mechanical instability of the interface and therefore to production of small drops.

Negative interfacial tension [58,61–66]: Due to adsorption of surfactants or cosurfactant molecules, the interfacial tension can become extremely low (less than 1 mN/m) and eventually transiently negative. Therefore, the interface can increase and any fluctuation can break it.

The two aforementioned mechanisms involve a *mechanical* instability of the interface that breaks up and produces small droplets.

Diffusion and stranding [61,67–70]: In this case, emulsification has a *chemical* origin and can take place even for quite high interfacial tensions. This kind of spontaneous emulsification often occurs when a cosolvent, soluble in both phases, is present. For example, if a mixture of alcohol and oil is brought in contact with water, the alcohol diffuses from the oil to the water phase, carrying with it some oil that can be "stranded" in fine drops as soon as water becomes supersaturated in oil. Although emulsification is an out-of-equilibrium process, phase diagrams coupled with the diffusion path theory [67–69] can be used to predict the phases that are likely to form when the two fluids are brought in contact and to determine the phase in which spontaneous emulsification will take place. More recently, other mechanisms have been proposed to explain spontaneous emulsification owing to the development of new experimental techniques [71–77]. Among others, one can mention the formation and swelling of water/surfactant aggregates at the vicinity of the interface. The structural change and the swelling can be driven by temperature, osmotic, or concentration gradients. In addition to the amazing and spectacular nature of this phenomenon, spontaneous emulsification is attractive because of the numerous potential applications in various fields such as agriculture (emulsifiable concentrates for insecticides, pesticides, and herbicides), pharmaceutics, cosmetics, oil recovery, and all applications in which nanometric emulsions are required. For a more complete review of spontaneous emulsification, the reader can refer to the reviews of Miller [67] and Lopez-Montilla et al. [78]. Despite the ease of production (no or low energy input) and the diversity of applications, spontaneous emulsification applies only to moderate dispersed volume fractions (less than 10%).

1.6. Phase Inversion

Emulsification through "phase inversion" is also often considered as a spontaneous emulsification process because it requires low energy input. Its advantage resides in the possibility of producing concentrated emulsions. Emulsification through phase inversion is often used industrially, especially in cosmetics [79,80]. Its interest lies in the low energy input required and the emulsions obtained are usually fine (average diameter of the order or lower than 1 μm) and monodisperse. Understanding the mechanism of phase inversion still remains a challenge; however, several studies provide some insight into this amazing and sometimes spectacular phenomenon. Phase inversion is said to occur when the structure of the emulsion inverts, that is, when the continuous phase becomes the internal phase or vice versa. This may happen with a change of any variable such as temperature, pressure, salinity, use of a cosurfactant, or the proportion of oil and water.

1.6.1. PIT Method

The most frequent emulsification using phase inversion is known as the **PIT** (Phase Inversion Temperature) **method** [81–83] and occurs through a temperature quench. This method is based on the phase behavior of nonionic surfactants and the correlation existing between the so-called surfactant spontaneous curvature and the type of emulsion obtained.

The well-known empirical Bancroft's rule [84] states that the phase in which the surfactant is preferentially soluble tends to become the continuous phase. An analogous empirical correlation has been reported by Shinoda and Saito [85]. For a nonionic surfactant of the polyethoxylated type [$R\text{-}(CH_2\text{-}CH_2\text{-}O)_n\text{-}OH$, where R is an alkyl chain], as temperature increases, the surfactant head group becomes less hydrated and hence the surfactant becomes less soluble in water and more soluble in oil. Its phase diagram evolves as schematically shown in Fig. 1.4. At low

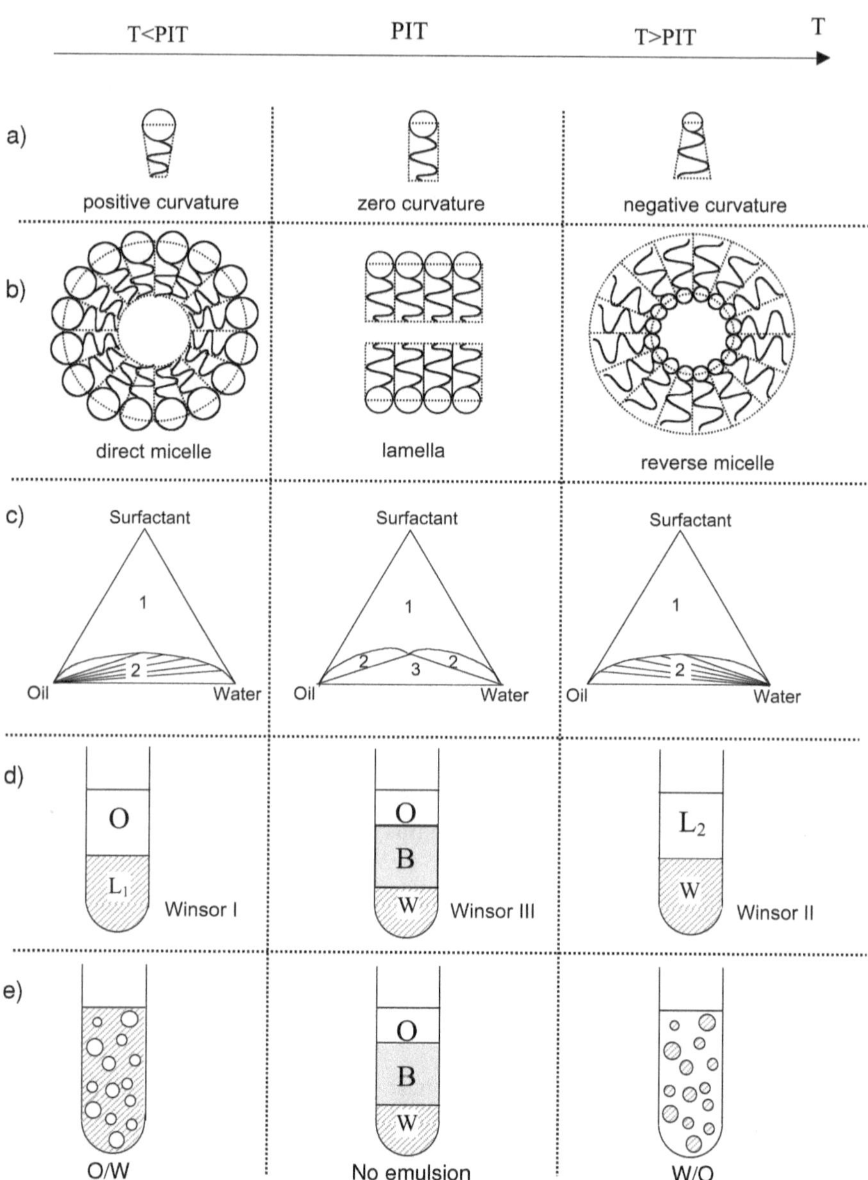

FIGURE 1.4. For a nonionic surfactant, influence of the temperature on **(a)** the surfactant morphology and hence the spontaneous curvature, **(b)** the type of self-assembly, **(c)** the phase diagram, the number of coexisting phases is indicated **(d)** the coexisting phases at equilibrium, and **(e)** the corresponding emulsions.

temperatures, the surfactant is preferentially soluble in water and a ternary mixture of oil, water, and surfactant will phase separate into a surfactant-rich aqueous phase and almost pure oil (Winsor I). The surfactant-rich aqueous phase, referred to as L_1, is made of direct micelles swollen by oil. The surfactant is said to have a positive spontaneous curvature. Under stirring, an O/W emulsion is obtained. At high temperatures, the surfactant polar head is dehydrated and the surfactant becomes preferentially oil-soluble. Its spontaneous curvature is negative and it self-assembles into reverse micelles. A ternary mixture of oil, water, and surfactant will phase separate into an oil phase containing the surfactant, referred to as L_2, in coexistence with almost pure water (Winsor II). On stirring, a W/O emulsion is obtained. At the PIT, the surfactant is "equilibrated" and its spontaneous curvature is close to zero. A ternary mixture of oil, water, and surfactant will separate into three phases: a bicontinuous or liquid crystalline phase (referred to as B in Figure 1.4d and e) in equilibrium with oil and water phases (Winsor III). At this temperature, both O/W and W/O emulsions are immediately destroyed through coalescence (for a better comprehension of the stability of emulsions related to the spontaneous curvature, the reader can refer to Chapter 5).

The PIT method exploits the fact that the surfactant affinity and hence the type of emulsion can be tuned by temperature. A W/O emulsion is first prepared at high temperature and then is rapidly cooled below the PIT to obtain an O/W emulsion without the need for stirring. Close to the PIT, the interfacial tension decreases drastically (three orders of magnitude; see Fig. 1.5) as reported by numerous authors [86–100], thus promoting droplet fragmentation. Hence, the resulting emulsions often have small droplet sizes (of the order or less than 1 μm). However, a low interfacial tension also facilitates coalescence. Therefore the PIT must be crossed fast enough to avoid destabilization and, to achieve good kinetic stability, the storage temperature must be well below the PIT [79,101–103]. In practice, a crude O/W emulsion is first produced at low temperature. Then, a temperature cycle is applied. The primary O/W emulsion converts into the W/O type by increasing the temperature above the PIT, and finally turns again into the O/W type on cooling. The main inconvenient of the PIT method is its restriction to nonionic surfactants.

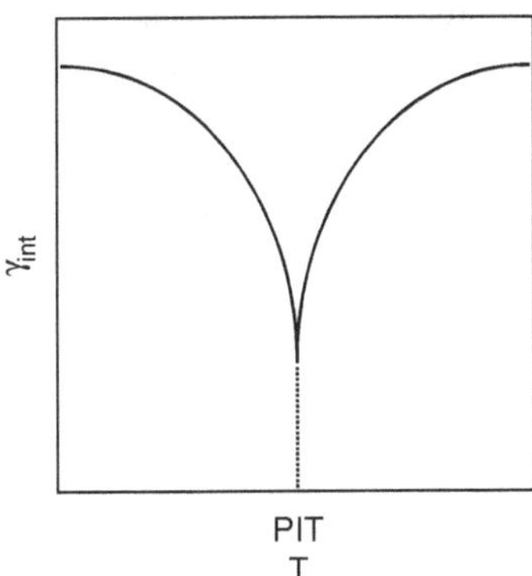

FIGURE 1.5. Influence of temperature on the interfacial tension between oil and water in the presence of a polyethoxylated surfactant.

1.6.2. Generalization

Emulsification through phase inversion is based on a change in the surfactant spontaneous curvature induced by temperature. This concept can be generalized considering any parameter influencing the spontaneous curvature of a surfactant, for example, salinity, pH, presence of a cosurfactant, and nature of the oil. The concept of inversion has often been reported in the literature by means of a formulation–composition map. In the following, we shall sum up this empirical approach which can be useful for formulators.

To take into account physicochemical parameters (salinity, temperature, pH) the Surfactant Affinity Difference (SAD), also named Hydrophilic–Lipophilic Difference (HLD), has been introduced [104–106]. It is an empirical number taking into account the surfactant type and its environment. The case SAD = 0 corresponds to a formulation in which the surfactant has the same affinity for both aqueous and oil phases. It corresponds to a surfactant spontaneous curvature close to zero and the phase behavior is of the Winsor III type. It is also called optimal formulation, the interfacial tension being close to zero. When SAD is negative (respectively positive), the surfactant has more affinity for the aqueous (respectively oil) phase, the phase behavior is of the Winsor I (respectively Winsor II) type, and the spontaneous curvature is positive (respectively negative). The emulsion is of the O/W type (respectively W/O) when SAD is negative (respectively positive). This phenomenology is expected if equivalent amounts of water and oil are present. The type of emulsion is also dependent on the composition defined by the water-to-oil volume ratio (WOR). Very high water (respectively oil) contents corresponding to diverging values of WOR (respectively close to zero) favor O/W (respectively W/O) emulsions. A schematic map is shown in Fig. 1.6a. Six different regions can be distinguished denoted by a letter and a sign. In − regions, SAD is negative and formulation favors O/W emulsions, whereas in + regions, SAD is positive and W/O emulsions are favored. The A letter corresponds to a WOR close to 1 and the type of emulsion is governed only by formulation. The B letter corresponds to samples in which the amount of oil exceeds that of water and where composition favors W/O emulsions. Finally, the C letter corresponds to samples in which water is more abundant than oil so that composition favors O/W emulsions. Among the six regions, four of them (A−, A+, B+ and C−) are referred to as normal regions because the composition preference agrees with formulation and two of them (B− and C+, in gray) are considered abnormal because they are in agreement with the volumetric requirements and in opposition with formulation ones. Because of the conflict between formulation and composition, close to the oblique lines in regions B− and C+, double emulsions can be present. The inner emulsion is thought to be governed by formulation whereas the outer emulsion is governed by composition; in the B− region, the double emulsion should be of the O/W/O type and of the W/O/W type in the C+ region. The bold black line represents the inversion locus.

Two kinds of transitions are generally distinguished in the literature. The emulsion can be inverted by crossing the bold line from A+ to A− or vice versa.

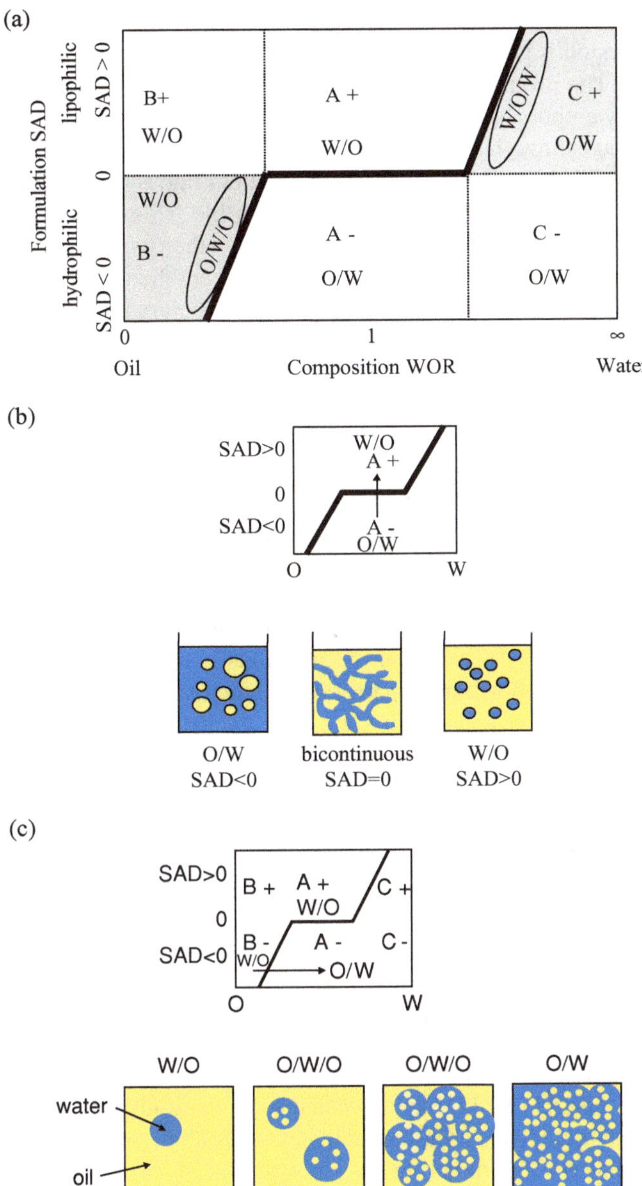

FIGURE 1.6. **(a)** Schematic formulation–composition map. SAD is the surfactant affinity difference; it is positive for a lipophilic surfactant and negative for a hydrophilic one. The gray zones are abnormal. **(b)** Schematic representation of the proposed mechanism for a transitional inversion. **(c)** Schematic representation of the proposed mechanism for a catastrophic phase inversion.

In this case, the inversion is said to be **transitional**, with an almost horizontal line; the inversion is governed by the formulation, with almost no dependence on composition. The inversion is characterized by a dramatic decrease of interfacial tension. Many authors have reported that emulsification via transitional inversion, involves passage through a bicontinuous microemulsion phase or a lamellar liquid crystalline phase [107–113] (Fig. 1.6b). If the bold line is crossed by a change in composition, the transition is said to be **catastrophic**. This happens when the type of emulsion is first governed by the composition and then by the formulation or vice versa. In catastrophic phase inversion, a region where the type of emulsion is not well defined is passed through and formation of temporary multiple emulsions has often been reported [102,114–116]. The proposed mechanism is as follows (Fig. 1.6c): during stirring, droplets of the continuous phase can be entrapped in the dispersed droplets, increasing the effective internal volume fraction until reaching the required value for inversion. Because the spontaneous curvature of the surfactant is frustrated at the interface of the external drops, coalescence occurs [117,118] with a high frequency leading to a rapid inversion. Once the required volume fraction is reached, the inversion phenomenon is rapid and often dramatic. For catastrophic inversion, no decrease of the interfacial tension is observed. The inversion locus can also be crossed at the intersections of the horizontal and oblique lines; both formulation and composition favor the inversion that is called "**combined inversion**" [119–123]. This allows production of emulsions with low average droplet size. The formulation–composition map is only a qualitative tool since it is well known that transient effects give rise to hysteresis in catastrophic inversion, depending on the emulsification protocol [116,123–126]. An increase of the surfactant concentration enlarges both the central region and the transition zones, whereas stirring has the opposite influence. An increase in the oil phase viscosity shrinks the A+ region, favoring the catastrophic inversion from A+ to C+. In a similar way, an increase of the aqueous phase viscosity makes the region A- narrower promoting inversion from the A− to B− region. As a consequence, catastrophic inversion is well adapted for emulsifying viscous phases (see example later).

To sum up, it is commonly accepted that transitional inversion likely happens through a bicontinuous or a liquid crystalline phase characterized by a very low interfacial tension, whereas catastrophic inversion proceeds through the formation of multiple emulsions. Although pedagogic, this picture is probably oversimplified since some papers [113,127] have revealed a correlation between the possibility of forming nano-emulsions via the emulsion inversion method and the presence of lamellar phases. Only direct investigations [128] and especially the promising observation using nonintrusive dye tracing techniques [129] will provide an answer to the strenuous question of inversion mechanism and dynamics.

1.6.3. Examples

1.6.3.1. Emulsification of Very Concentrated Emulsions Using the PIT Method

Phase inversion may be used as an emulsification method for systems with a very high internal phase content [130–133]. The process is illustrated in Fig. 1.7 for

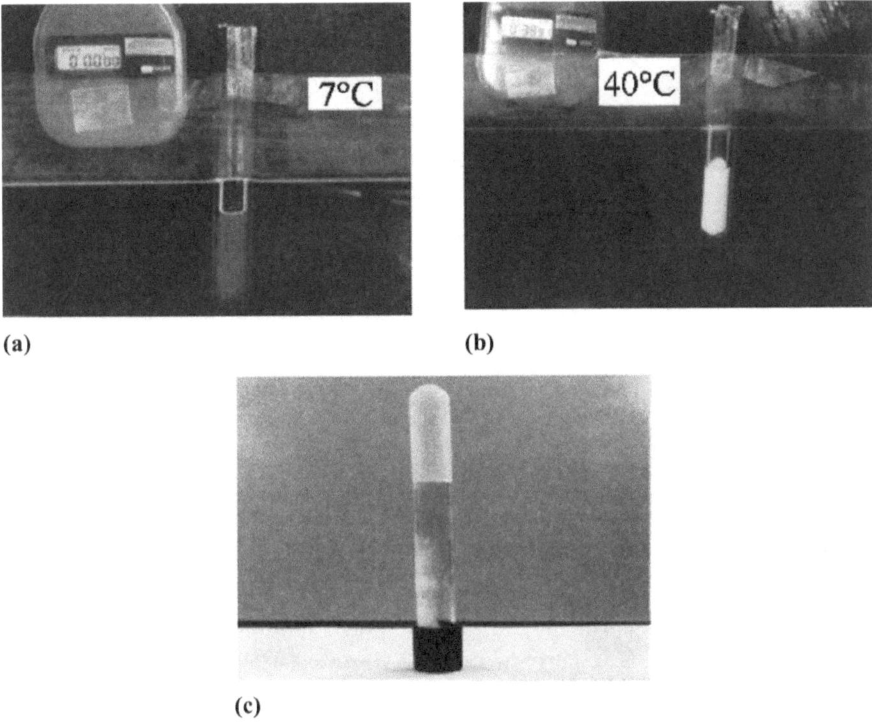

(a) (b) (c)

FIGURE 1.7. Process of emulsification of a very concentrated W/O emulsion. An oil-in-water microemulsion at 7°C **(a)** is rapidly heated to 40°C, the gel emulsion is formed after less than 40 s **(b)**, the emulsion does not flow if the tube is turned upside down **(c)**. (From [133], with permission.)

a W/O emulsion. At 7°C, the mixture of 90 wt% 0.1 M NaCl aqueous phase, 3 wt% of $C_{12}E_4$ surfactant (tetraethylene glycol monododecyl ether), and 7 wt% decane forms an isotropic L1 phase. As temperature is increased, the surfactant is dehydrated and becomes more lipophilic (its PIT is around 18°C). From 7°C to 40°C, different phases in the equilibrium phase diagram (liquid crystal lamellar phase, bicontinuous sponge phase) may be crossed [131]. At 40°C, an L2 phase is in equilibrium with excess water. When the sample is rapidly heated from 7°C to 40°C, it becomes milky in less than 1 minute (Fig. 1.7b), revealing the formation of an emulsion. This is confirmed by the microscope image of Fig. 1.8, showing small water droplets (≈1 μm). Because of the very high water content (90 wt%), the W/O emulsion exhibits the behavior of a solid (Fig. 1.7c).

1.6.3.2. Emulsification of Very Viscous Phases

Phase inversion may be used to emulsify highly viscous substances such as bitumen (the viscosity of which can be as high as 10^6 Pa·s at room temperature) [134]. The most frequent industrial technique used to prepare bitumen-in-water emulsions consists of mixing hot bitumen (130°C) with an aqueous phase (≈60°C at

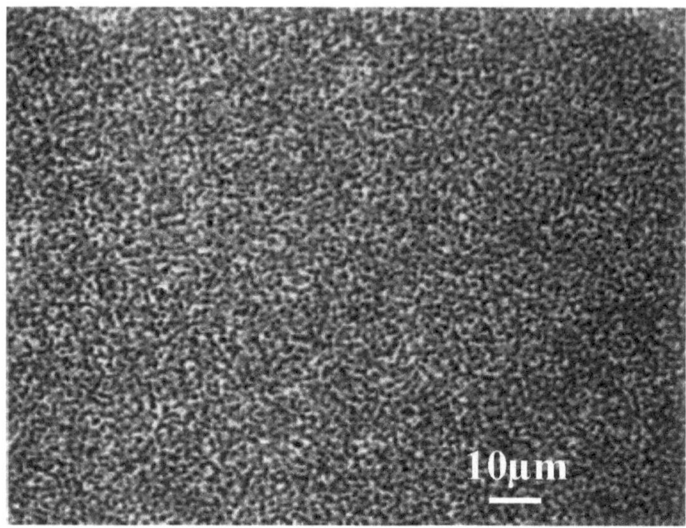

FIGURE 1.8. Microphotograph of the obtained gel emulsion. (From [133], with permission.)

1.5 atm) under turbulent stirring. Such fabrication conditions are very constraining and do not allow precise control of the drop size. Leal-Calderon et al. [134] have proposed directly mixing the two phases in a laminar regime. This emulsification process is based on a catastrophic inversion. It allows production of emulsions with bitumen volume fractions as high as 95%. The average droplet size is in the micron range and the size distribution is quite narrow compared to the same emulsion obtained via the classical technique (Fig. 1.9). This method is applicable to other very viscous components and for the preparation of both direct and inverse emulsions.

1.7. Application of a Controlled Shear

In industrial applications, emulsions are often obtained by exerting a crude stirring, made of a very complex combination of extensional and shearing flows. To study and understand the fragmentation process, the flow must be simplified before

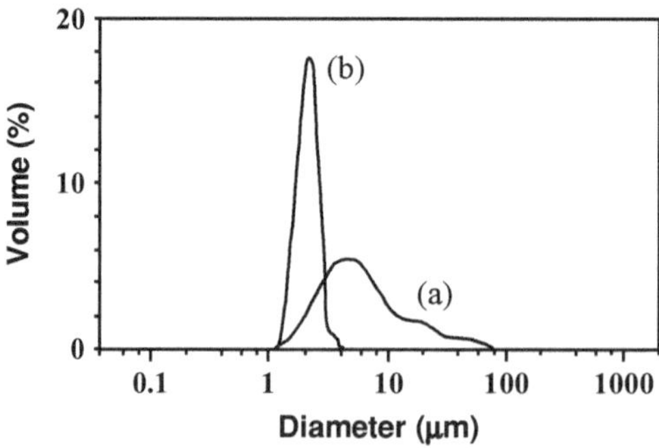

FIGURE 1.9. Comparison between two bitumen-in-water emulsions prepared following **(a)** the classical protocol, **(b)** the protocol described in [134]. (Adapted from [134].)

generalizing the results to uncontrolled stirring. Mason and Bibette [135] showed that a controlled high shear applied to crude polydisperse emulsions can lead to monodisperse systems. They took advantage of this observation to design a process allowing the production of monodisperse materials in large quantities [136]. Systematic work [137] performed on various systems provided a complete mapping of the conditions leading to monodisperse fragmentation. In this section, we discuss the mechanisms of shear rupturing and the origin of monodispersity.

Many authors have worked on drop deformation and breakup, beginning with Taylor. In 1934, he published an experimental work [138] in which a unique drop was submitted to a **quasi-static deformation**. Taylor provided the first experimental evidence that a drop submitted to a quasi-static flow deforms and bursts under well-defined conditions. The drop bursts if the capillary number Ca, defined as the ratio of the shear stress σ over the half Laplace pressure (excess of pressure in a drop of radius R: $P_L = \frac{2\gamma_{\text{int}}}{R}$ where γ_{int} is the interfacial tension):

$$Ca = \frac{2\sigma}{P_L} \tag{1.2}$$

exceeds some critical value, Ca_{cr}, that depends on the viscosity ratio, p, between the dispersed and continuous phases:

$$p = \frac{\eta_d}{\eta_c} \tag{1.3}$$

Several studies [139–141] led to a complete description giving Ca_{cr} as a function of p for flows spanning from simple shear to extensional flow (Fig. 1.10). Moreover, different mechanisms of rupturing have been observed, as for example (1) the development of a Rayleigh instability [142–145], in which an undulation develops and grows at the surface of the drop deformed into a cylinder until breakage; (2) tip streaming [146], in which very small droplets are expelled from pointed ends of the deformed initial drop; and (3) end pinching [147], in which droplets are progressively formed by pinching of both ends from the initial drop deformed into a cylinder. Again, all these results have been obtained in quasi-static conditions, where the flow is slowly increased in such a manner that either a stationary deformation is reached, or the drop breaks. Usually, emulsions are prepared under

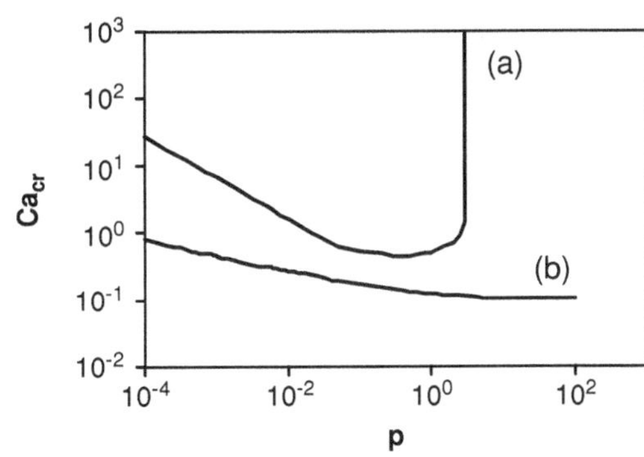

FIGURE 1.10. Condition of rupturing in quasi static-conditions for **(a)** a simple shear flow and **(b)** a pure elongational flow. (Adapted from [139].)

conditions far away from ideal quasi-static deformation, changing radically the accessible states and hence the rupturing conditions as showed by Hinch and Acrivos [148]. Hereafter, we describe in more detail the work of Mabille et al. [149,150] on fragmentation under **non quasi-static** conditions.

1.7.1. Emulsion Preparation and Characterization

To investigate the effect of a shear, the first step consists of preparing a crude polydisperse emulsion called "premix," which is obtained by gently incorporating the oil phase into the aqueous one. This allows one to obtain a macroscopically homogeneous sample. During this first step, the stirring must be soft enough to avoid the production of small droplets that could perturb the investigation of further fragmentation. Once this premix is obtained, different monodisperse emulsions are produced under shear and the droplet volume fraction is adjusted to the required value. An example of a premix and of the resulting monodisperse emulsion is shown in Fig. 1.11. The shear clearly reduces the average size and the distribution width.

The emulsions are characterized by the mean diameter d and the polydispersity P, defined by:

$$d = \frac{\sum_i N_i d_i^4}{\sum_i N_i d_i^3} \quad \text{and} \quad P = \frac{1}{\bar{d}} \frac{\sum_i N_i d_i^3 \left|\bar{d} - d_i\right|}{\sum_i N_i d_i^3} \tag{1.4}$$

where N_i is the total number of droplets with diameter d_i. $\bar{d}$ is the median diameter, that is, the diameter for which the cumulative undersized volume fraction is equal to 50%. Both d and P are obtained from static light-scattering measurements and the use of Mie theory. In the following, an emulsion is considered as monodisperse if $P \leq 25\%$ because below this limit value, concentrated drops organize into crystallized domains as can be observed in Fig. 1.11 (for the premix $d = 23$ μm and $P = 40\%$, while for the sheared emulsion $d = 1$ μm and $P = 12\%$).

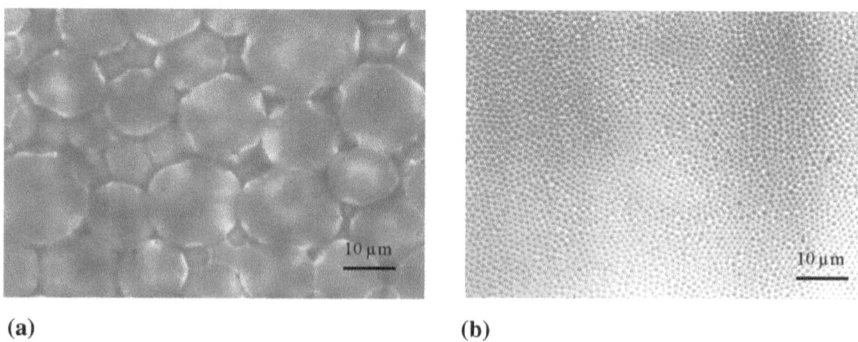

FIGURE 1.11. Microphotography of **(a)** the premix and **(b)** the emulsion after application of a shear. (Adapted from [149].)

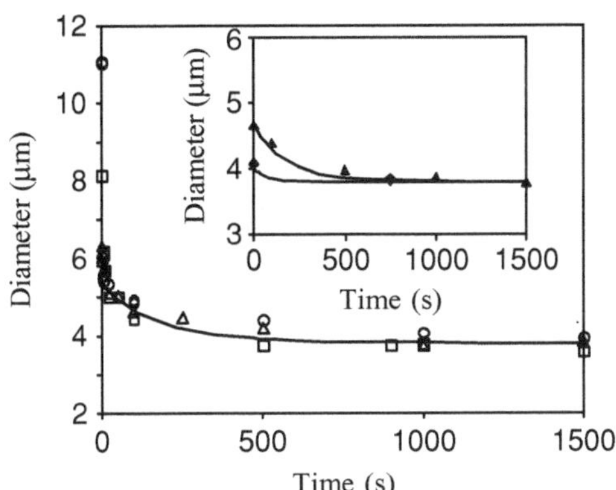

FIGURE 1.12. Emulsification kinetics. The emulsions are of the same composition and only the initial sizes are different: (○) 11 μm; (□) 8.2 μm; (△) 6 μm; (▲) 4.7 μm; and (♦) 4.1 μ m. The lines are fit to the experimental data in the second regime using Eq. 1.12. (Adapted from [149].)

1.7.2. Fragmentation Kinetics of Model Emulsions

The model system used by Mabille et al. [149, 150] was a set of monodisperse dilute (2.5 wt% of dispersed oil) emulsions of identical composition, whose mean size ranged from 4 μm to 11 μm. A sudden shear of 500 s^{-1} was applied by means of a strain-controlled rheometer for durations ranging from 1 to 1500 s. All the resulting emulsions were also monodisperse. At such low oil droplet fraction, the emulsion viscosity was mainly determined by that of the continuous phase (it was checked that the droplet size had no effect on the emulsion viscosity). The viscosity ratio $p = \eta_d/\eta_c = 0.4$ and the interfacial tension $\gamma_{\text{int}} = 6$ mN/m remained constant.

In Fig. 1.12, the evolution of the mean diameter as a function of the shearing time is plotted. For large initial sizes (8 μm and 11 μm), one clearly distinguishes two regimes. The diameter jumps from its initial value to about 6 μm in less than 1 s. The size reached after this first abrupt decrease is independent of the initial diameter. Then, a second slow decrease takes place and the size decreases from 6 to 3.8 μm. For emulsions with an initial diameter smaller than 6 μm, only this second regime is observed with variable characteristic times. Whatever the initial size, all the curves converge toward a unique asymptotic diameter, equal to 3.8 μm. If the resulting emulsions are sheared once again at the same shear rate, no further fragmentation occurs, showing that the value of 3.8 μm corresponds to an asymptotic diameter.

1.7.3. Breaking Mechanisms

1.7.3.1. The Fast Regime

i. Evidence of the Occurrence of a Rayleigh Instability Under Shear

A dilute emulsion comprising oil drops dispersed in an aqueous solution was sheared between two glass slides under an optical microscope [149]. Original drops were deformed into long threads (Fig. 1.13a) that broke in numerous identical aligned and regularly spaced droplets (Fig. 1.13b). One can distinguish some very

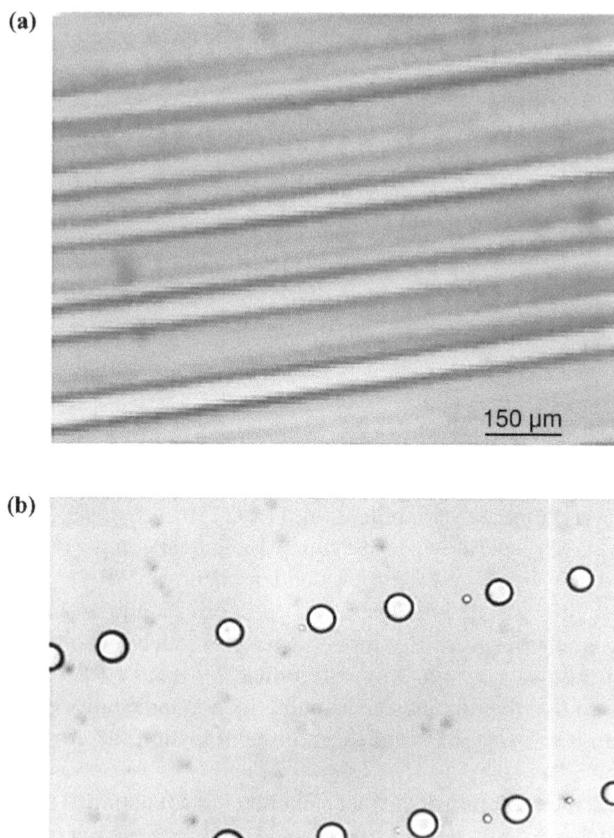

FIGURE 1.13. Mechanism of rupturing. Under shear, drops elongate into long cylinders **(a)** that undergo a Rayleigh instability leading to identical aligned droplets **(b)**. (Adapted from [149].)

small satellites in between the new formed droplets. The alignment of regularly spaced droplets is characteristic of a Rayleigh instability. It can hence be concluded that the first fast size decrease is due to a Rayleigh instability that develops on the elongated threads.

ii. Description of the Rayleigh Instability at Rest

The breakup of a cylindrical thread **at rest** has been studied [142,145,151]. A varicosity of wavelength λ (Fig. 1.14) at the surface of a liquid cylinder immersed in another liquid will be amplified if its development causes a decrease of the interfacial area. Such a condition is satisfied if λ is larger than $2\pi r_c$, r_c being the radius

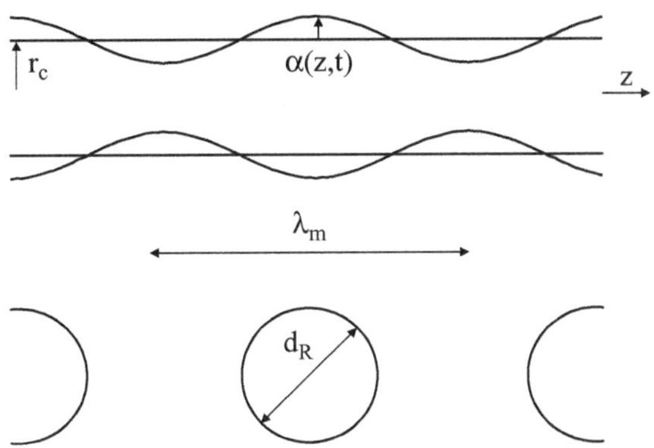

FIGURE 1.14. Scheme of an undulation perturbation of the interface of a thread with radius r_c. $\alpha(z,t)$ is the perturbation amplitude. The thread breaks with a characteristic wavelength λ_m into drops of diameter d_R.

of the cylinder. Tomotika [151] showed that the perturbation grows exponentially with a rate that depends on the interfacial tension γ_{int}, the external fluid viscosity η_c, and the internal to external viscosity ratio p. Instead, the shape fluctuation is damped if it causes an increase of the interfacial area, corresponding to λ smaller than $2\pi r_c$. Among all the amplified perturbations, the one leading to rupturing is characterized by the wavelength λ_m that depends on p. When the amplitude of the disturbance becomes equal to the thread radius r_c, the cylinder breaks up into a set of drops regularly spaced by λ_m and having a diameter d_R. The perturbation leading to the Rayleigh instability is characterized by the dimensionless parameter x_m defined as:

$$x_m = \frac{2\pi r_c}{\lambda_m} \tag{1.5}$$

The x_m values can be found in [151] or in [145]. These tabulated values and the two following volume conservation relations allow a complete determination of the characteristic lengths of the instability:

$$\frac{d_{init}^3}{6} = l r_c^2 \tag{1.6}$$

and:

$$\frac{d_R^3}{6} = \lambda_m r_c^2 \tag{1.7}$$

where l is the cylinder length just before rupture and d_{init} is the initial drop diameter. The number of formed droplets N is given by:

$$N = \left(\frac{d_{init}}{d_R}\right)^3 \tag{1.8}$$

The values of x_m are known only for **stationary threads**. The characteristic lengths are in good agreement with experimental observations [145] and were used in polymer melts in order to deduce the interfacial tension [152,153].

When a shear flow is applied, the situation becomes more complex because the stretching rate and the orientation of the thread are linked [148,154] and are continuously evolving. Moreover a comparison of an analytical description with experimental results requires knowledge of the cylinder radius and orientation just before the instability develops. Because these two quantities are not accessible, a complete description of the Rayleigh instability under shear is not possible.

iii. Rayleigh Instability in Non-stationary Conditions

In the experiments of Mabille et al. [149,150],only the resulting size d_R was accessible. Because d_R was independent of the initial size and because of the volume conservation condition, the product $r_c^2\lambda_m$ remained constant. By shearing viscous polymer blends, Rusu et al. [155] have directly examined the fragmentation of long cylinders. They observed that the larger the initial drop, the longer the cylinders but one and the same final diameter was obtained after rupturing. Moreover, they provided experimental evidence for the existence of a unique cylinder diameter before breakup. Such results suggest that the criterion for the Rayleigh instability to occur is linked only to the diameter of the threads and is independent of their lengths. Because nonmiscible polymer blends are quite similar to oil-water mixtures, it is likely that a unique critical value of the threads radius r_c and consequently a unique value of λ_m for the instability development [149,150] also exists during emulsion fragmentation. The uniqueness of r_c and λ_m with respect to the initial drop size explains the fact that a polydisperse emulsion, made of a mixture of different initial sizes, is fragmented into a monodisperse one.

1.7.3.2. The Slow Regime

In the second regime, by considering that one drop breaks into $\nu + 1$ droplets per unit time, the increase in the number of drops per unit time can be written as:

$$\frac{\partial N(t)}{\partial t} = \nu N(t) \tag{1.9}$$

Volume conservation provides a relation between the diameter $d(t)$ and the number of droplets $N(t)$ at time t. As a consequence, the differential equation governing the size evolution is:

$$\frac{\partial d(t)}{\partial t} = -\frac{\nu}{3}\, d(t) \tag{1.10}$$

The parameter ν can also be seen as a probability of rupture. Because the size saturates at a value d_f, ν must depend on the size. A linear dependence of ν with d can be assumed:

$$\begin{cases} \nu = \nu_0 \dfrac{d - d_f}{d_f} & if\ d > d_f \\ \nu = 0 & if\ d \le d_f \end{cases} \tag{1.11}$$

Taking into account that the initial size of this second regime, d_0, is either the emulsion initial size or the size after the first regime, that is, d_R, the expression of $d(t)$ can be derived:

$$d(t) = \frac{d_f}{1 - \dfrac{d_0 - d_f}{d_0} \exp(-\nu_0 t/3)} \tag{1.12}$$

The data of Fig. 1.12 were fitted using ν_0 as the unique fitting parameter. One and the same value was obtained for all sets of point, $\nu_0 = 0.015$ drops per second, indicating that the drop production rate is very low and that this second regime has only little efficiency in comparison with the first one.

1.7.4. Generalization

The observations of model systems made of monodisperse dilute emulsions can be generalized to polydisperse and concentrated emulsions.

1.7.4.1. Polydisperse Emulsion

The fragmentation kinetics obtained for a polydisperse dilute oil-in-water emulsion ($\eta_d = 815$ mPa·s, $p = 1.1$) characterized by $d = 19$ μm and $P = 55\%$, is shown in Fig. 1.15 [156]. As a 500 s^{-1} shear is applied, d decreases to 6.6 μm after 1 s and P approaches 25%. By further shearing, the diameter still decreases to 5.8 μm. The kinetic curve exhibits the same characteristic regimes as in Fig. 1.12. This experiment shows that there is no difference in fragmenting initially polydisperse or monodisperse emulsions, because, as explained earlier, the size resulting from the Rayleigh instability is independent of the initial size. A polydisperse emulsion contains drops of different sizes that undergo the same Rayleigh instability under shear. The drops deform into threads of different lengths and fragmentation occurs when the threads reach the same critical radius r_c.

1.7.4.2. Concentrated Emulsions

The previous experiments were all performed on dilute emulsions for which the dispersed phase represents 2.5 wt% of the emulsion. The results obtained for a concentrated emulsion with oil mass fraction equal to 75% sheared at 500 s^{-1} and 3000 s^{-1} are reported in Fig. 1.16 [156]. The primary emulsion was polydisperse with $d = 57$ μm. The two previously described regimes still exist. The first regime is particularly efficient in reducing the diameter because one drop breaks into 160 droplets through the Rayleigh instability for an applied shear rate of 500 s^{-1} ($d_R = 10.5$ μm) and into 6200 droplets for an applied shear rate of 3000 s^{-1}

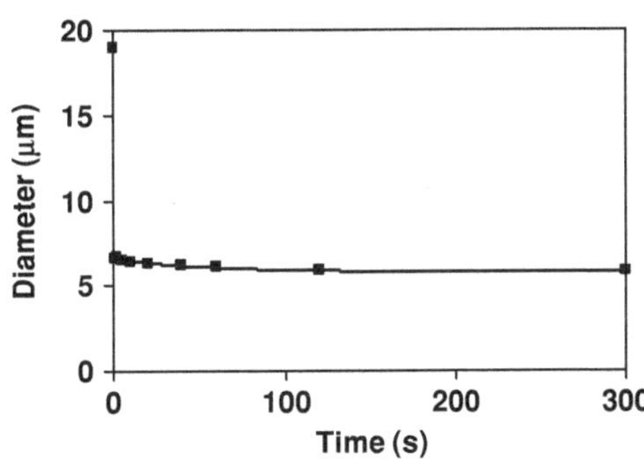

FIGURE 1.15. Fragmentation kinetics of a polydisperse emulsion. (Adapted from [156].)

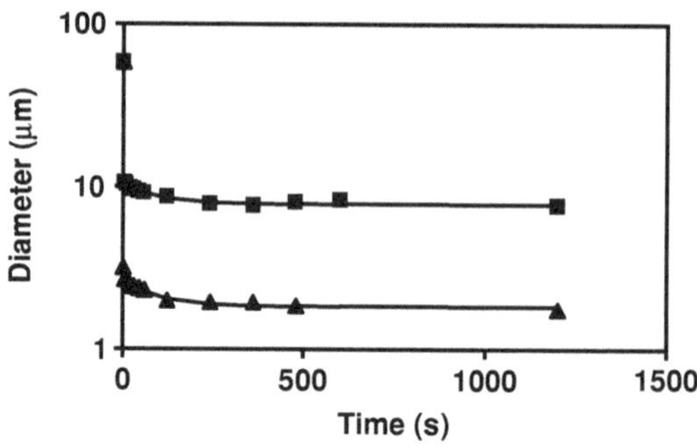

FIGURE 1.16. Fragmentation of a concentrated emulsion (75 wt%) sheared at (■) 500 s^{-1} and (▲) 3000 s^{-1} (Adapted from [156].)

($d_R = 3.1$ µm). After this first step, the emulsion was already monodisperse. The second slow regime is less efficient because one drop breaks only into three and six droplets, respectively. Between 2.5 wt% and 75 wt% of the dispersed phase, the same phenomenology was observed with d_R decreasing as the droplet mass fraction increased. This tendency merely reflects the fact that the emulsion viscosity and thus the applied stress increases with droplet concentration (at constant shear rate).

1.7.5. Parameters Governing the Rayleigh Instability

In this section, the parameters influencing the mean diameter d_R resulting from the Rayleigh instability are examined. Hereafter, the second fragmentation regime is not considered because a narrow size distribution is already obtained after the first one. The parameters that influence d_R are the applied stress σ, the viscosity ratio p, the rheological behavior, and the way the shear is applied.

1.7.5.1. Applied Stress

To explore the influence of σ, at constant shear rate $\dot{\gamma}$, a non adsorbing thickening polymer (sodium alginate) was added to the continuous phase at different concentrations (Fig. 1.17) [149]. The oil viscosity was adjusted to keep p unchanged and it was checked that the interfacial tension γ_{int} remained the same. That way, the effect of σ was investigated, all other parameters remaining constant ($\gamma_{\text{int}} =$ 6 mN/m, $p = 1$, $\dot{\gamma} = 500$ s^{-1} and 2.5 wt% of dispersed phase). The evolution of the daughter drop diameter d_R as a function of σ is plotted in Fig. 1.18a. The experimental dependence $d_R = f(\sigma)$ is valid for any mother emulsion consisting of large droplets with diameters larger than d_R. This curve can also be seen as a fragmentation limit: for a given stress, only drops with diameter larger than d_R will break. In Fig. 1.18b, the stress is plotted as a function of the Laplace pressure of the daughter drops:

$$P_D = 4\gamma_{\text{int}}/d_R \tag{1.13}$$

The straight line defines the transition between two regimes: in the upper half plane, fast fragmentation occurs (<1 s) under shear, whereas fragmentation does not take

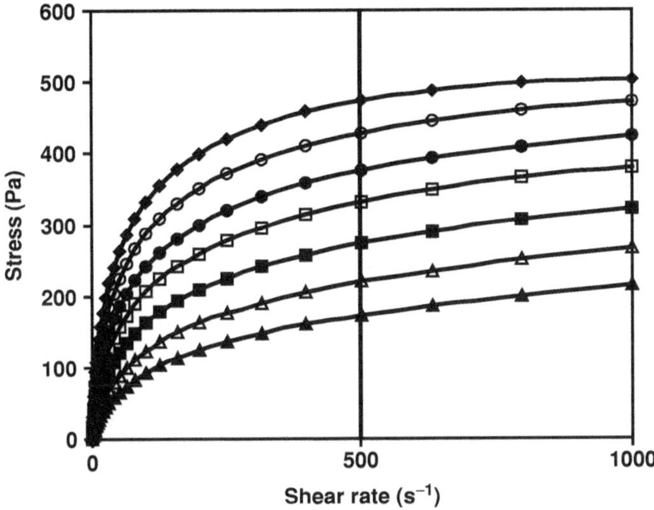

FIGURE 1.17. Rheograms of an emulsion at 2.5 wt% of dispersed phase and different alginate concentrations: (▲) 1.6 wt%, (△) 1.8 wt%, (■) 2.0 wt%, (□) 2.2 wt%, (●) 2.4 wt%, (○) 2.6 wt%, and (♦) 2.8 wt% (Adapted from [150].)

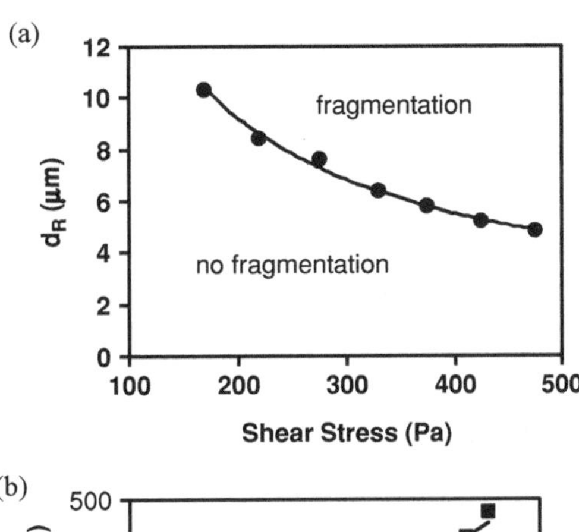

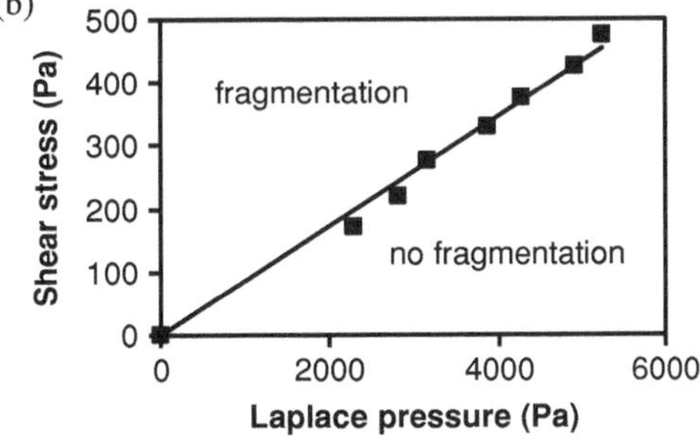

FIGURE 1.18. **(a)** Diameter resulting from the Rayleigh instability as a function of the shear stress. **(b)** Shear stress as a function of the Laplace pressure of the resulting drops. The linear fit gives a slope of 0.087. (Adapted from [149].)

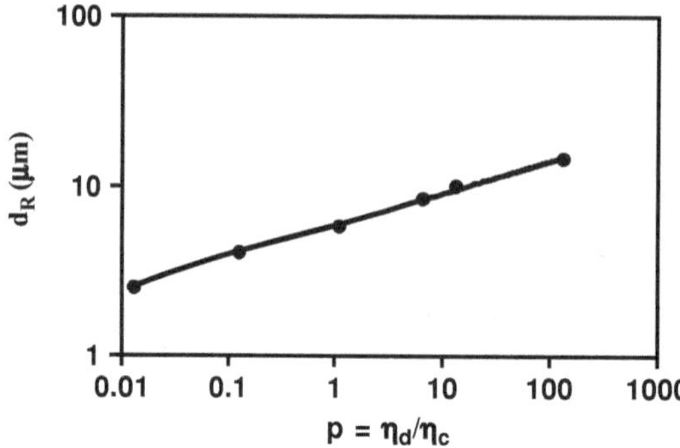

FIGURE 1.19. Effect of the viscosity ratio p on the diameter after breakup (Adapted from [149].)

place in the lower half plane. The experiments also provide an empirical law that can be used to predict the droplet size resulting from monodisperse fragmentation:

$$d_R = \alpha \frac{4\gamma_{\text{int}}}{\sigma} \tag{1.14}$$

where α is a dimensionless constant that should essentially depend on $p(\alpha \approx 0.1$ for $p = 1)$. For $p = 1$, the critical capillary number for droplet rupturing Ca_{cr}, determined in quasi-static conditions is of the order of 0.5 [139,140,158]. When the shear stress is applied suddenly, the stress and Laplace pressure are found to be proportional and from the slope of the experimental curve, it can be deduced that Ca_{cr} is of the order of 0.2 ($Ca_{\text{cr}} \approx 2\alpha = 0.2$), a value that is quite comparable to the one obtained in quasi-static conditions.

1.7.5.2. Viscosity Ratio

To explore the influence of p, the viscosity of the internal phase was varied over four decades, everything else being constant [149]. As can be seen in the log-log plot of Fig. 1.19, d_R (identically Ca_{cr}) scales with the viscosity ratio as $p^{0.2}$: the low value of the exponent indicates that d_R is only weakly dependent on p. In Fig. 1.20, the evolution of the polydispersity P as a function of p is plotted. As

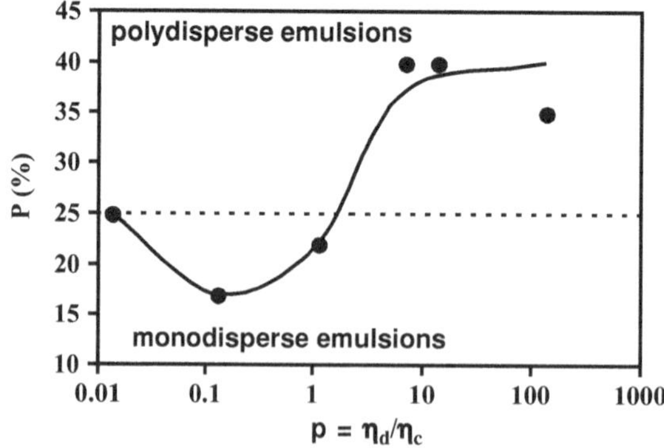

FIGURE 1.20. Effect of the viscosity ratio p on the emulsion polydispersity P. The dashed line represents the limit between polydisperse and monodisperse emulsions. (Adapted from [149]).

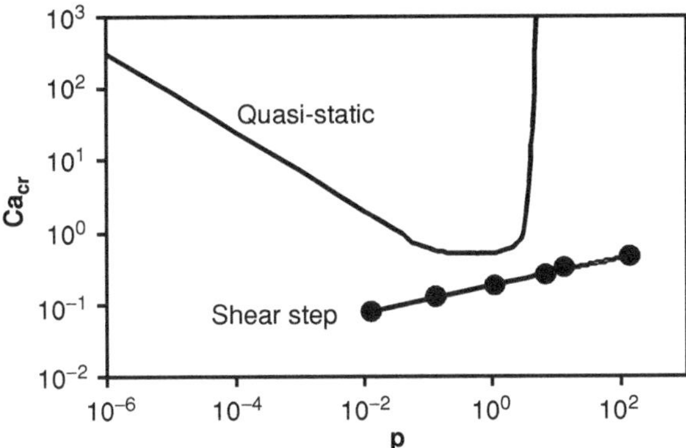

FIGURE 1.21. Comparison of the critical capillary number for simple shear, when the shear is applied in quasi-static conditions or suddenly.

long as p is in the range 0.01–2, the obtained emulsions are monodisperse. For $p > 0.1$, P increases, reflecting the progressive loss of wavelength selectivity during the development of the Rayleigh instability. In quasi-static conditions, Grace [139] demonstrated that Ca_{cr} diverges for $p > 3$, meaning that the fragmentation is no longer possible when the viscosity of the disperse phase is at least three times larger than that of the continuous phase. On sudden application of the shear, it becomes possible to break the drops even for very high p values ($p > 100$) (Fig. 1.21).

1.7.5.3. Rheological Behavior

To determine the influence of the rheological behavior of the emulsion, three dilute (droplet volume fraction $\phi = 2.5$ wt%) emulsions were sheared [150,156]. The three rheograms are shown in Fig. 1.22. One of them, denoted N, had a Newtonian behavior with a viscosity of 300 mPa·s. The interfacial tension between the continuous and dispersed phases was 4.3 mN/m. The two other emulsions, noted ST1 and ST2, were shear-thinning and were obtained by adding a nonadsorbing polymer in the continuous aqueous phase. For both emulsions, the interfacial tension was 6 mN/m. From Fig. 1.22, two intersecting points, A and B, can be defined, where the shearing conditions are identical (same σ and $\dot{\gamma}$) for N and ST1, and for N and ST2, respectively. The results are reported in Table 1.1. The sizes are in the same ratio as the interfacial tensions. At point A emulsions with a Newtonian or shear-thinning behavior (N and ST1) are both polydisperse, whereas at point B, the emulsions (N and ST2) are both monodisperse, thus proving that the fragmentation quality does not depend on the rheological behavior.

1.7.5.4. Shearing Protocol

In the previous experiments, the shear was applied as a step in less than 1 s. The shear can also be applied following a controlled ramp [150]. Four different protocols are schematically represented in Fig. 1.23. The shear is increased from

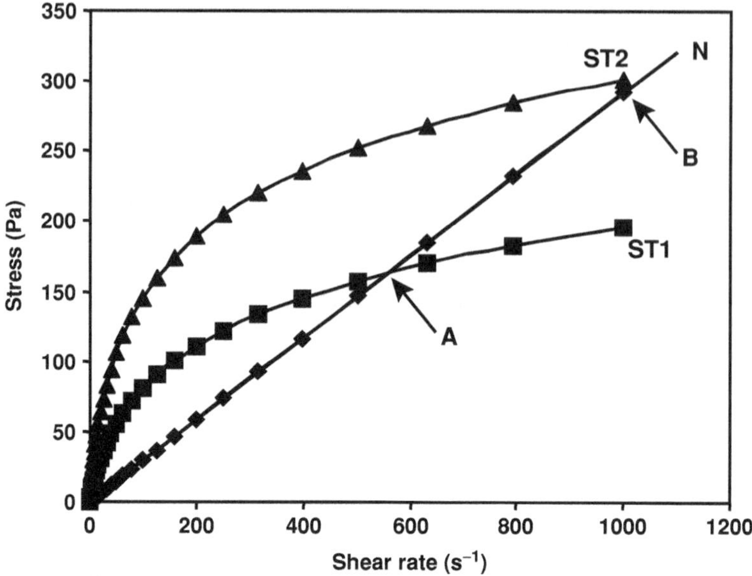

FIGURE 1.22. Rheograms of a newtonian (♦) N and two shear-thinning ST1 and ST2 emulsions obtained by adding 1.5 wt% (■) and 1.9 wt% (▲) of a nonadsorbing polymer in the continuous phase. (Adapted from [150].)

0 to 500 s^{-1} in less than 1 s (a), 10 s (b), and 100 s (c and d). The maximum shear of 500 s^{-1} is maintained for a duration τ and is suddenly interrupted for protocols a, b, and c. For protocol d, the shear is progressively decreased from 500 s^{-1} to zero in 100 s. To allow comparison, the resulting diameter is plotted (Fig. 1.24) versus total strain γ defined as:

$$\gamma(t) = \int_0^t \dot{\gamma}(t')\, dt' \tag{1.15}$$

Three emulsions with various values of p have been studied. For $p = 0.013$ (Fig. 1.24b) and $p = 1.1$ (Fig. 1.24a), all the emulsions were monodisperse. The

TABLE 1.1. Comparison of the mean sizes and distribution widths at the two intersecting points A and B (Fig. 1.22) of shear thinning and Newtonian rheogramms

		d (μm)	P (%)	Size ratio	Interfacial ratio
A	N	8.8	35	1.4	1.4
	ST1	12.2	35		
B	N	4.5	20	1.3	1.4
	ST2	5.9	22		

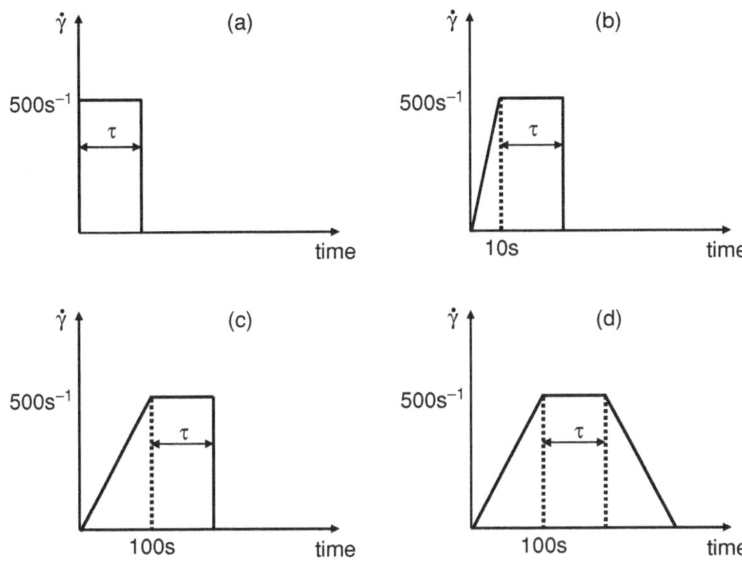

FIGURE 1.23. Four different ways of applying the shear. (Adapted from [150].)

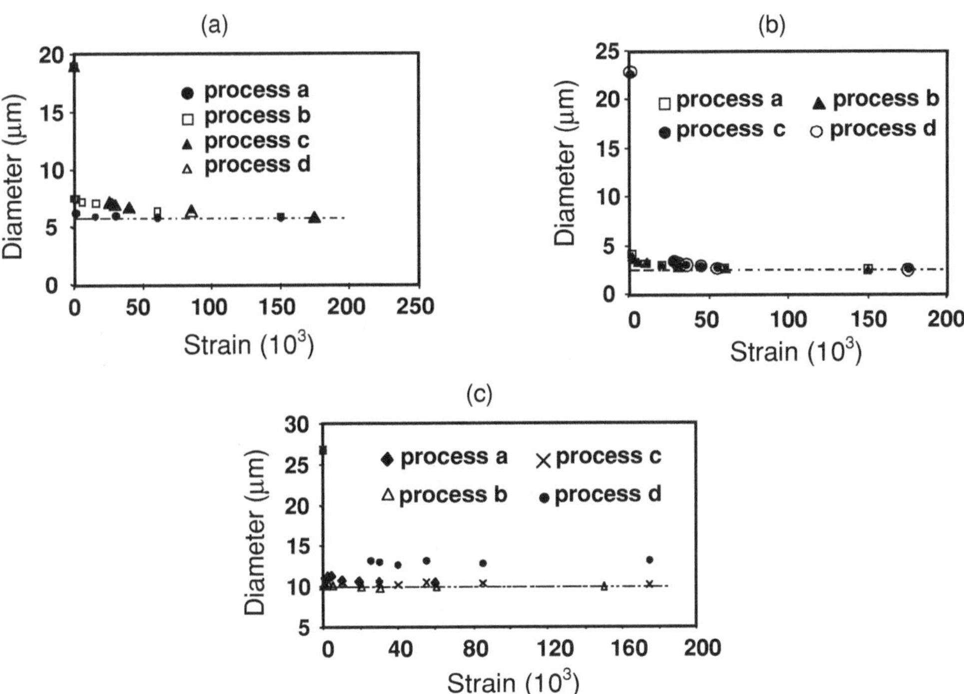

FIGURE 1.24. Comparison of the obtained diameter at same strain when **(a)** p = 1.1, **(b)** $p = 0.013$, and **(c)** $p = 13.4$. (Adapted from [150].)

resulting size was the same whatever the applied shearing protocol. For $p = 13.4$ (Fig. 1.24c), the initial polydisperse ($P = 46\%$) emulsion was fragmented into polydisperse emulsions (P close to 35%), in agreement with Section 1.7.5.2. The data corresponding to situations a, b, and c superimpose, showing that fragmentation does not depend on the way the shear is increased (in the non quasi-static limit). However, when the shear is abruptly stopped, the final size is 10 μm (a, b, and c) whereas for a progressive slowdown, the final size is 13 μm (d). The size ratio of 1.3 between the experiments corresponds roughly to $2^{1/3}$. It can be argued that when the shear is suddenly stopped, one drop breaks into two daughter droplets. In contrast, when the shear is stopped progressively (d), this last breakup does not occur.

It is worth noticing that the curves $Ca_{cr} = f(p)$ reported in Fig. 1.21 are two limit cases where the stress is applied either suddenly or quasi-statically. By varying progressively the rate of shear application, all the intermediary situations are observable [138,139].

1.7.6. Examples of Monodisperse Materials

All the previous experimental data provide a useful guideline for producing well-controlled materials. Monodisperse fragmentation is obtained if two experimental conditions are fulfilled:

1. The applied shear and stress are high enough to rapidly induce the Rayleigh instability.
2. The viscosity ratio lies in the range 0.01–2.

The most efficient way to produce monodisperse emulsions in large quantities is to use a mixer that applies a spatially homogeneous shear. That way, all the drops of the mother emulsion are directly submitted to the same shear rate and break into daughter drops (in less than 1 s). However, this is not a necessary condition for monodispersity. For example, Aronson [157] described the production of emulsions having a controlled droplet size and a fairly narrow distribution using a standard mixer that did not apply a spatially homogeneous shear. Concentrated emulsions were fragmented for a long period of time (30–60 min). This ensured that the whole volume was finally submitted to the same maximum stress. Thus, although the application of a spatially homogeneous shear is not a necessary condition, it allows a noticeable reduction of the shear time.

The monodisperse materials described hereafter were obtained with the Couette type cell designed by Bibette et al. [150,159]. It consists of two concentric cylinders (rotor and stator) separated by a very narrow gap (100 μm), allowing application of spatially homogeneous shear rates over a very wide range (from 0 to 14280 s^{-1}), with shearing durations of the order of 10 s.

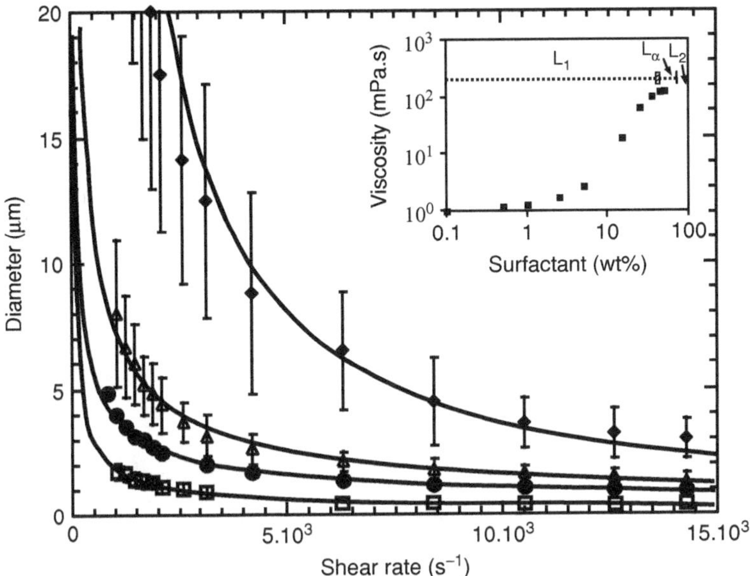

FIGURE 1.25. Influence of the emulsifier concentration C_{surf} on the size profiles: (♦) $C_{surf} =$ 15 wt%; (△) $C_{surf} = 25$ wt%; (●) $C_{surf} = 30$ wt%; and (□) $C_{surf} = 45$ wt% of surfactant. The dispersed phase is kept constant at 75 wt% of a 350 mPa·s silicone oil. The lines are guides to the eye. Inset: Zero-shear viscosity as a function of surfactant concentration. (Adapted from [137].)

1.7.6.1. Simple Emulsions

i. Effect of the Surfactant Amount

A crude polydisperse silicone oil-in-water emulsion stabilized by a nonionic surfactant was submitted to different shear rates [137]. The evolution of the mean diameter as a function of the applied shear rate is plotted in Fig. 1.25. The dispersed phase mass fraction was kept constant at 75%, while the emulsifier concentration in the continuous medium was varied from 15 to 45 wt%. Such large amounts of surfactant increase the continuous phase viscosity. The reported bars account for the distribution width deduced from light-scattering measurements. At a given shear rate, smaller droplets with lower polydispersity were produced as surfactant concentration increased. For example, at 45 wt% of surfactant, $P \leq 15\%$ whatever the applied shear rate, whereas $P \approx 25\%$ for 15 wt% of surfactant. Some microscopic pictures of the obtained emulsions are given in Fig. 1.26. The continuous phase viscosity increases with the surfactant concentration (see insert of Fig. 1.25). Hence for a given shear rate, the addition of surfactant has two main effects: (1) it increases the transmitted stress, thus promoting fragmentation and (2) it induces a variation of the viscosity ratio p (from

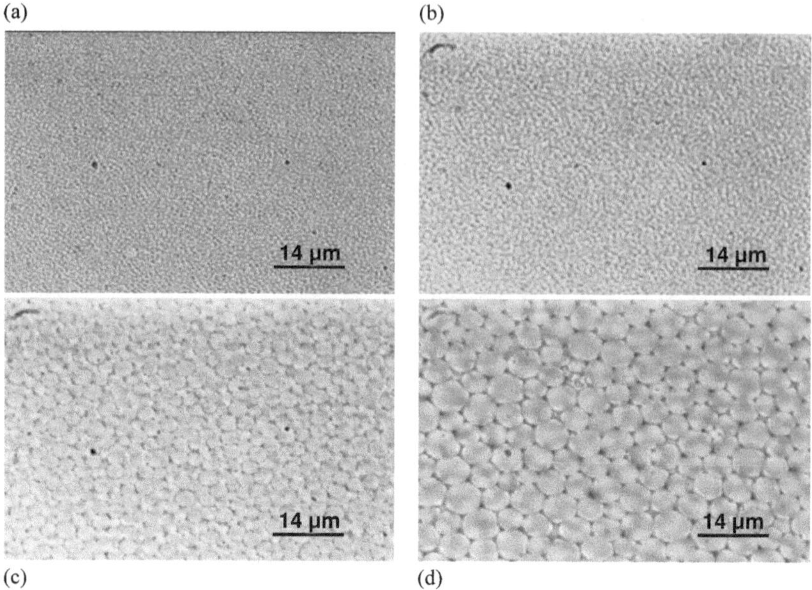

FIGURE 1.26. Images of monodisperse emulsions obtained at different surfactant concentrations, C_{surf}, and shear rates, $\dot{\gamma}$. **(a)** $C_{surf} = 55$ wt%, $\dot{\gamma} = 8400\ s^{-1}$. **(b)** $C_{surf} = 45$ wt%, $\dot{\gamma} = 2600 s^{-1}$. **(c)** $C_{surf} = 25$ wt%, $\dot{\gamma} = 4700\ s^{-1}$. **(d)** $C_{surf} = 15$ wt%, $\dot{\gamma} = 7350\ s^{-1}$. (Adapted from [137].)

$p = 0.06$ for 15 wt% to $p = 0.4$ for 45 wt% of surfactant), thus reducing the drop size distribution width.

ii. Addition of a Thickener

Adding surfactant to the continuous phase is a possible route to improve fragmentation. However, for economic and environmental reasons, it can be beneficial to reduce the amount of surfactant, while maintaining the viscosity of the continuous phase. This can be achieved by addition of a nonadsorbing polymer. In the following example, two aqueous phases with identical viscosity (1 Pa·s) were prepared: (1) 30 wt% of a nonionic surfactant in water and (2) 3 wt% of the same surfactant with 4 wt% of a nonadsorbing polymer in water [137]. Two mother emulsions were obtained by dispersing 30 wt% of the same silicone oil in the two aqueous phases. When sheared at $\dot{\gamma} = 14280\ s^{-1}$, the daughter emulsions had the same size distributions characterized by $d = 1$ μm and $P = 14\%$ (Fig. 1.27). This result proves the generality of the fragmentation concepts that are independent of the chemical nature of the compounds.

iii. Polymer-Stabilized Emulsions

Surfactants are commonly used to kinetically stabilize colloidal systems. An alternative way to achieve long-term metastability consists of adsorbing

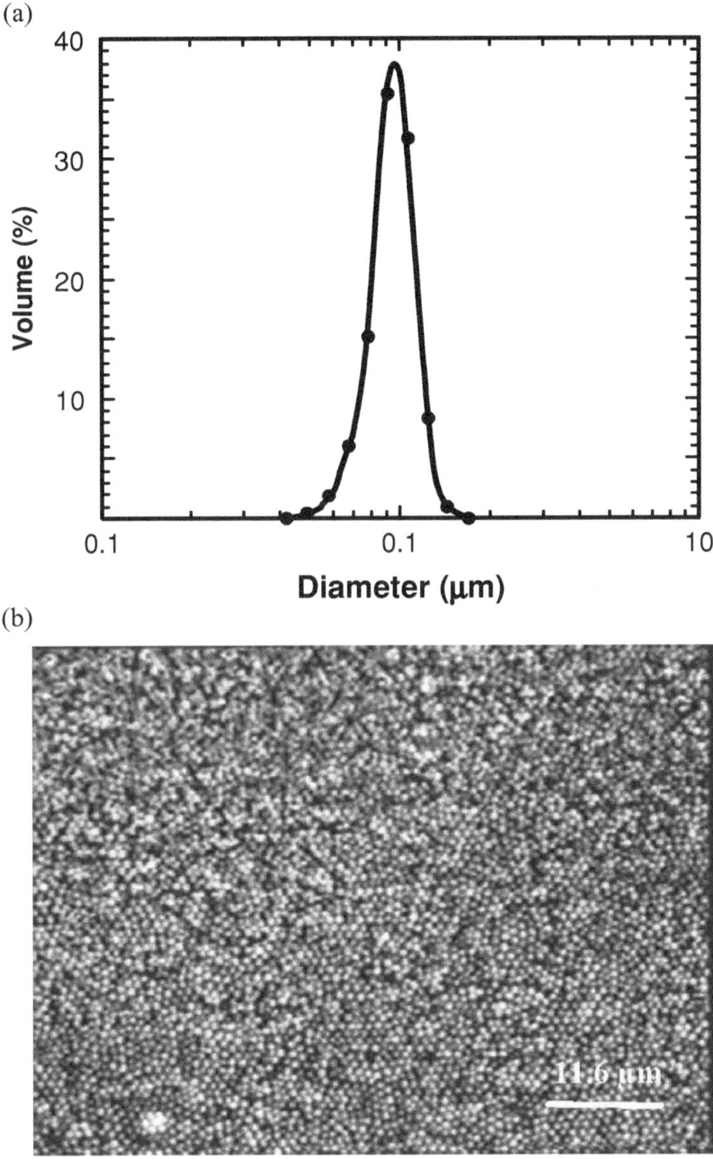

FIGURE 1.27. Emulsion obtained by shearing a premixed emulsion in presence of a nonadsorbing polymer. System composition: sodium alginate (nonadsorbing polymer) 4 wt%, nonionic surfactant (NP7) 3 wt%, oil fraction = 30 wt%. **(a)** Size distribution, **(b)** microscopic image. (Adapted from [137].)

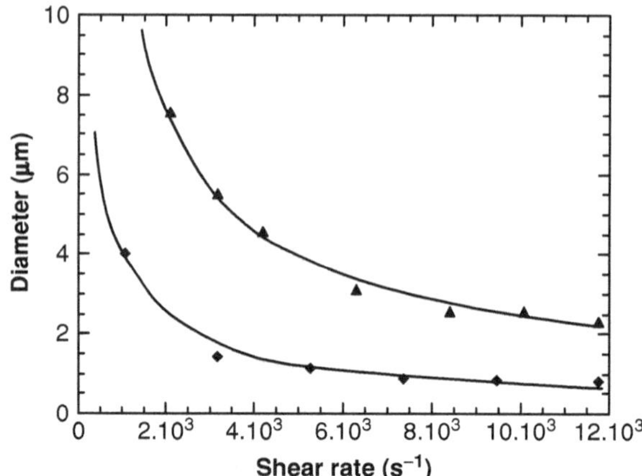

FIGURE 1.28. Role of interfacial tension on fragmentation. (♦) Surfactant concentration 5 wt% —5 mN/m; (▲) PVAAc 15 wt%—17 mN/m. The lines are visual guides. (Adapted from [137].)

macromolecules at the interface between the dispersed and the continuous phases. Polymer chains are densely adsorbed at the interfaces, where they form loops and tails extending in the continuous phase (see Chapter 2). There is no fundamental difference concerning the fragmentation of emulsions in the presence of small polymers [160,161]. The only feature that has to be pointed out is the importance of the interfacial tension [137,150]. Indeed, owing to their large molecular size, macromolecules reduce the interfacial tension between oil and water to a smaller extend than surfactants do. In Fig. 1.28, the fragmentation profiles of two premixed emulsions, one stabilized by a polymer, the other by a short surfactant, have been reported for comparison [137,150]. The compositions were adjusted so that the continuous phases possessed the same zero-shear viscosity (100 mPa·s). This was achieved with 15 wt% of a polyvinylic alcohol (PVA Ac) with a molecular weight of 15000 g/mol and 35 wt% of the nonionic tergitol NP7 (nonyl phenol ethoxy 7) surfactant in water. The sizes obtained for the polymer-stabilized emulsions are three times larger than those obtained with the surfactant, at the same shear rate. This can be explained considering that the interfacial tension between oil and the continuous phase is approximately three times larger with the polymer (17.2 mN/m) than with the surfactant (5 mN/m).

1.7.6.2. Suspensions

The monodisperse fragmentation process can be extended to produce monodisperse solid particles [156]. The general strategy consists of performing the emulsification in conditions such that the dispersed phase is in the liquid state, and to solidify the drops either by a temperature quench or through polymerization. The microscopic image in Fig. 1.29 illustrates this possibility. It corresponds to solid paraffin oil dispersed in water at room temperature. The emulsification was performed in the liquid state, at a temperature above the melting point of the

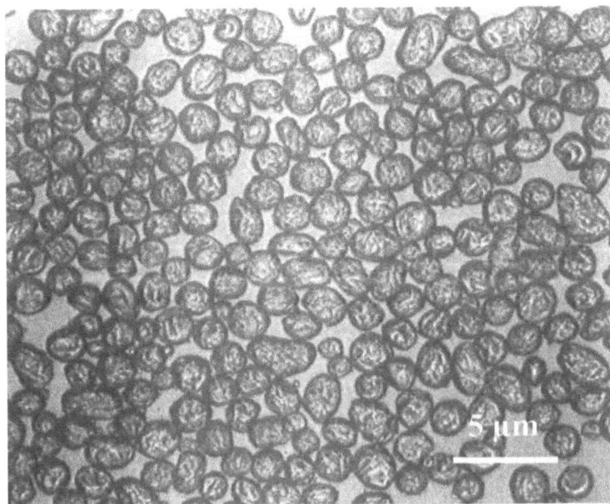

FIGURE 1.29. Microscopic image of a solid paraffin dispersed in water (room temperature). Adapted from [156].

paraffin oil. Once fragmented, the system was cooled at ambient temperature. On cooling, the spherical shape of the warm dispersed droplets, which is controlled by surface tension, evolved into a rough and rippled surface due to the formation of irregularly shaped/oriented crystals.

1.7.6.3. Double Emulsions

All the previous examples reveal that a large variety of emulsions may be produced by application of a controlled shear and that the drop size may be tuned from 0.3 to 10 μm. The same concepts can be applied with a dispersed phase which is an emulsion itself, as far as the characteristic length of the Rayleigh instability is large compared to the average size of the primary emulsion [162]. This allows one to fabricate the so-called double or multiple emulsions, which are materials suitable for the encapsulation and the sustained release of various substances (see Chapter 6). Because the rates of release in double emulsions are very sensitive to the droplet size [163,164], the most efficient control should be achieved in the presence of monodisperse emulsions [165]. W/O/W monodisperse emulsions are generally fabricated following a two-step procedure [166]. A monodisperse concentrated W/O emulsion stabilized by a lipophilic surfactant is first prepared. This emulsion is then emulsified in an aqueous phase containing a hydrophilic surfactant. The resulting W/O/W emulsions comprise oil globules containing smaller water droplets, both colloids having a well-defined and controlled diameter. We report here an example taken from [162].

1. **Inverted W/O emulsion.** The premixed emulsion was composed of aqueous droplets dispersed in an oil phase containing a lipophilic surfactant. The droplet mass fraction ϕ_i was set to 80%. This crude emulsion was sheared into the Couette-type cell [159] at constant shear rate $\dot{\gamma} = 10000\ \mathrm{s}^{-1}$ and a

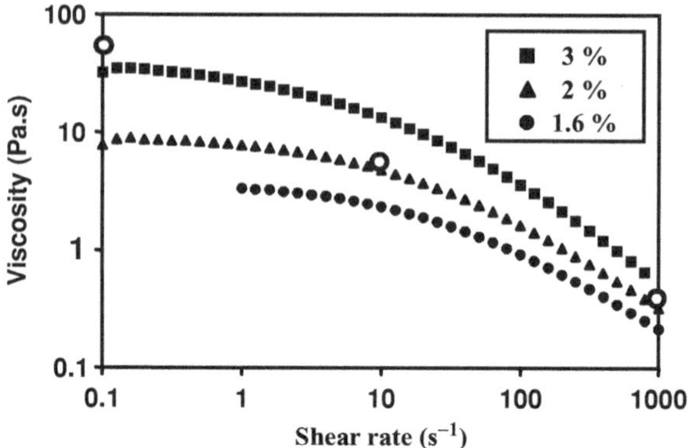

FIGURE 1.30. Evolution of the water phase viscosity η_c as a function of the shear rate and alginate concentration (filled symbols). For the sake of comparison, the viscosity η_d of the inverted emulsion at $\phi_i = 75$ wt% is reported in the same graph (open symbols). (Adapted from [162].)

monodisperse W/O emulsion with $d = 0.30$ μm and P $\approx 20\%$ was obtained. To vary the droplet volume fraction ϕ_i, the emulsion was diluted after the fragmentation process.

2. **Double emulsions.** A given amount of the primary inverted emulsion was dispersed in an aqueous phase containing the hydrophilic surfactant. Because the viscosity of the continuous phase was very low compared to that of the inverted emulsion, a thickening agent was dissolved in water. To predict the best conditions to obtain quasi-monodisperse fragmentation, the evolution of the aqueous phase viscosity η_c as a function of the shear rate and thickener concentration is plotted in Fig. 1.30. For the sake of comparison, the viscosity η_d of the inverted emulsion is reported in the same graph (up to 1000 s^{-1}). For a 2 wt% thickener solution, the viscosity ratio η_d/η_c was of the order of 1 at 1000 s^{-1}, so double globules with narrow size distribution were expected in this thickener composition range. Of course, the viscosity ratio became smaller than 1, in the presence of concentrated double globules, because the whole emulsion viscosity η_{eff} should be considered instead of η_c ($\eta_{eff} > \eta_c$). However, it was checked that the ratio remained in the adequate range, that is, between 0.01 and 2, over the whole set of experiments.

   The premixed double emulsion was sheared in the same Couette-type cell at different shear rates from 0 to 14200 s^{-1}. The obtained double emulsions had average diameters ranging from 7 to 2 μm and polydispersities between 15% and 30%. Figure 1.31 shows microscopic images of double emulsions fabricated at different shear rates. Large oil globules very uniform in size are visible, and the smaller inverted water droplets are also distinguishable. From these results, it can be concluded that the fragmentation method that has been developed for simple emulsions can be extended to produce quasi-monodisperse double emulsions. In other words, the capillary instability occurring in simple emulsions also takes place in the presence of materials such as concentrated emulsions,

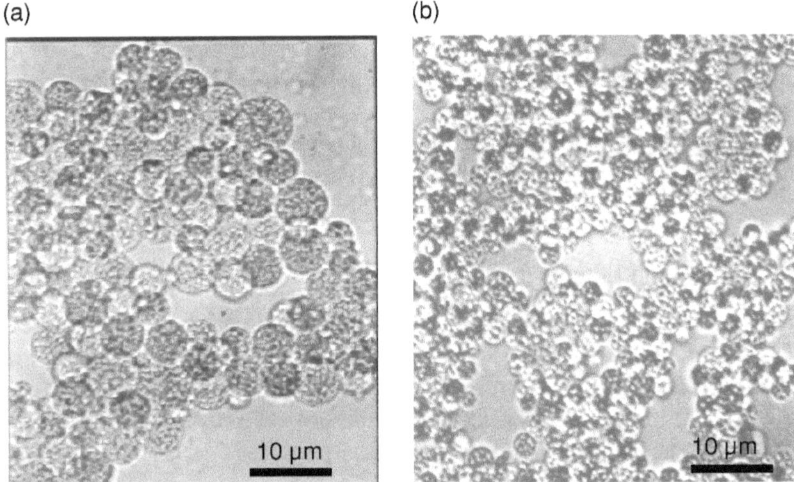

FIGURE 1.31. Images of W/O/ W emulsions obtained at different shear rates. The emulsions were diluted with water and glucose to facilitate observation. **(a)** $\dot{\gamma} = 1000\ \mathrm{s}^{-1}$ and **(b)** $\dot{\gamma} = 14{,}200\ \mathrm{s}^{-1}$. (Adapted from [162].)

within the same rheological and shearing conditions. It is worth noting that in the previous experiments, the final globule size was always significantly larger than the internal water droplet diameter (there is at least a factor of 10 between the two diameters). In such conditions, the W/O emulsion may be considered as an effective continuous medium obeying the same fragmentation properties as a simple fluid.

iv. Influence of the Internal Droplet Mass Fraction

The primary W/O emulsion was diluted to vary the mass fraction ϕ_i of the water droplets in the inverted emulsion. In Fig. 1.32 the evolution of the globule diameter

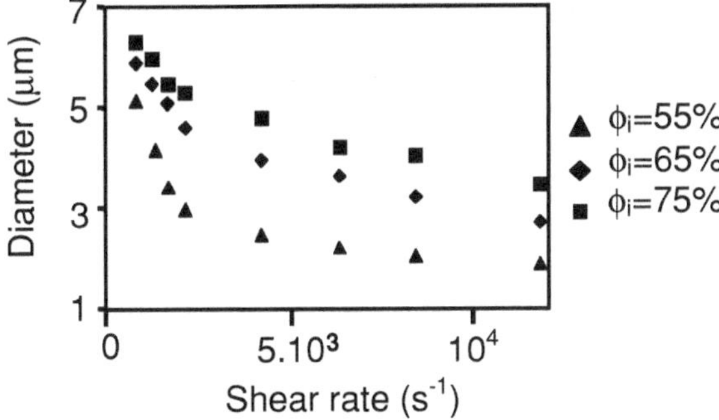

FIGURE 1.32. Dependence of the globule diameter in W/O/W emulsions on the steady shear rate $\dot{\gamma}$ for different ϕ_i values. (Adapted from [137].)

as a function of $\dot{\gamma}$, for three different mass fractions ϕ_i is reported. The mass fraction ϕ_g of the globules in the double emulsion was always equal to 70%. It can be observed that for a given shear rate, the globule size increases with the droplet mass fraction. By varying ϕ_i, the average viscosity of the inverted emulsion also changed. The results were qualitatively identical to the concepts discribed for the fragmentation of simple emulsions. By increasing the viscosity of the dispersed phase, they identically observed that the droplet diameter increases.

1.7.6.4. Ferrofluid Emulsions

Monodisperse ferrofluid emulsions comprise magnetic oil droplets dispersed in a water continuous phase. The specific magnetic and optical properties of these materials are currently exploited in research and/or technological fields. Leal-Calderon et al. [168] have developed a technique that allows one to directly determine the force distance law between tiny colloidal particles. Their technique exploits the fact that the anisotropy of the forces between dipoles causes the ferrofluid droplets to form chains. Because chains give rise to a strong Bragg diffraction of the visible light, the inter-droplet spacing is accurately measured. Moreover, because the attractive dipolar magnetic force can be varied trough the intensity of the external field, the balancing repulsive force can also be measured at various spacings. This technique has been used for the measurement of colloidal forces in the presence of many different surface active species (surfactants, polymers, proteins) and has provided interesting insights in the field of colloidal forces (see Chapter 2). Because this technique is based on a precise determination of the interdroplet spacing, it requires the use of highly monodisperse emulsions.

The preparation of a ferrofluid emulsions is quite similar to that described for double emulsions. The starting material is a ferrofluid oil made of small iron oxide grains (Fe_2O_3) of typical size equal to 10 nm, dispersed in oil in the presence of an oil-soluble surfactant. The preparation of ferrofluid oils was initially described in a US patent [169]. Once fabricated, the ferrofluid oil is emulsified in a water phase containing a hydrophilic surfactant. The viscosity ratio between the dispersed and continuous phases is adjusted to lie in the range in which monodisperse fragmentation occurs (0.01–2). The emulsification leads to direct emulsions with a typical diameter around 200 nm and a very narrow size distribution, as can be observed in Fig. 1.33.

1.8. Conclusion

The size characteristics of emulsions obtained through different techniques reviewed in this chapter are summarized in Table 1.2. The emulsification method should be selected depending on the required size for industrial use. For high-jet homogenization, the stabilizer concentration and the applied pressure should be properly controlled. The dispersed volume fraction is restricted to 30% to avoid

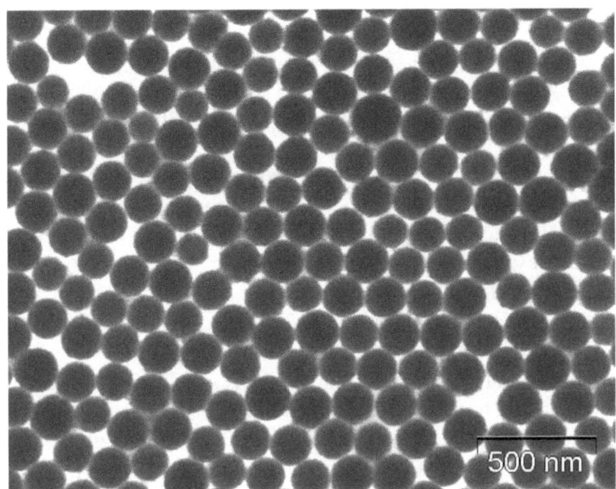

FIGURE 1.33. Scanning electron microscopy image of a ferrofluid emulsion (Courtesy of Ademtech Company.)

recombination. Higher volume fraction or viscous phases would also risk plugging the microchannels composing the apparatus.

Membrane emulsification allows a precise control of the droplet size and monodispersity but the scale up of this process is difficult. Microchannel emulsification is a promising technique but the low production rates restrict its use to highly monodisperse systems intended for high-technology applications.

Emulsification through inversion is often used at an industrial scale, especially in the cosmetics field. Its main restriction lies in the formulation dependence.

Mechanical stirring is widely used in industrial processes in which the flow is often generated by the displacement of a mobile (helix or rotor-stator) in a laminar or turbulent regime. The final size distribution is generally wide and results from a complex coupling between fragmentation and recombination (coalescence). In this chapter, we have described a simple regime of emulsification under mechanical stirring where the flow was spatially homogeneous and where recombination did not take place owing to the rapid adsorption of the surface active species. Under such idealized conditions, it is possible to produce monodisperse emulsions devoted to fundamental studies or to specific high-technology applications (Chapter 7).

TABLE 1.2. Summary of the accessible sizes

Emulsification method	Typical size
High-pressure homogenization ($\phi < 30\%$)	50 nm–5 μm
Membrane	0.2–100 μm
Microchannel	10–100 μm
Spontaneous	~100 nm
Inversion	≤1 μm
Mechanical stirring	1–15 μm

References

[1] S.Y. Soon, J. Harbidge, N.J. Titchener-Hooker, and P.A. Shamlou: "Prediction of Drop Breakage in an Ultra Velocity Jet Homogenizer." J. Chem. Eng. Jpn. **34**, 640 (2001).

[2] S. Mohan and G. Narsimham: "Coalescence of Protein-Stabilized Emulsions in a High-Pressure Homogenizer." J. Colloid Interface Sci. **192**, 1 (1997).

[3] P. Paquin: "Technological Properties of High-Pressure Homogenizers: The Effect of Fat Globules, Milk Proteins, and Polysachharides." Int. Dairy J. **9**, 329 (1999).

[4] S. Brösel and H. Schubert: "Investigation of the Role of Surfactants in Mechanical Emulsification Using a High-Pressure Homogenizer with an Orifice Valve." Chem. Eng. Process. **38**, 533 (1999).

[5] G. Narsimham and P. Goel: "Drop Coalescence During Emulsion Formation in a High-Pressure Homogenizer for Tetradecane-in-Water Emulsion Stabilized by Sodium Dodecyl Sulfate." J. Colloid Interface Sci. **238**, 420 (2001).

[6] L. Taisne, P. Walstra, and B. Cabane: "Transfer of Oil Between Emulsion Droplets." J. Colloid Interface Sci. **184**, 378 (1996).

[7] L. Lobo, A. Svereika, and M. Nair: "Coalescence During Emulsification. 1. Method Development." J. Colloid Interface Sci. **253**, 409 (2002).

[8] L. Lobo and A. Svereika: "Coalescence During Emulsification. 2. Role of Small Molecule Surfactants." J. Colloid Interface Sci. **261**, 498 (2003).

[9] P. Marie, J.M. Perrier-Cornet, and P. Gervais: "Influence of Major Parameters in Emulsification Mechanisms Using a High-Pressure Jet." J. Food Eng. **53**, 43 (2002).

[10] P. Walstra: "Formation of Emulsions." In P. Belcher (ed), *Encyclopedia of Emulsion Technology, Basic Theory*, Marcel Dekker, New York (1983).

[11] P. Walstra and P.E.A. Smulders: "Emulsion Formation." In B.P. Binks (ed), *Modern Aspects of Emulsion Science.*The Royal Society of Chemistry Cambridge, UK, 1998.

[12] J. Floury, A. Desrumaux, and J. Lardières: "Effect of High-Pressure Homogenization on Droplet Size Distributions and Rheological Properties of Model Oil-in-Water Emulsions." Innovat. Food Sci. Emergi. Technol. **1**, 127 (2000).

[13] E. Tornberg: "Functional Characterization of Protein Stabilized Emulsions: Emulsifying Behavior of Proteins in a Valve Homogenizer." J. Sci. Food Agric. **29**, 867 (1978).

[14] K. Kandori: "Application of Microporous Glass Membranes: Membrane Emulsification." In A.G. Gaonkar (ed), Food Processing: Recent Developments" Elsevier, Amsterdam (1995).

[15] R.A. Williams, S.J. Peng, D.A. Wheeler, N.C. Morley, D. Taylor, M. Whalley, and D.W. Houldsworth: "Controlled Production of Emulsions Using Crossflow Membrane Part II. Industrial Scale Manufacture." Chem. Eng. Res. Des. **76 A**, 902 (1998).

[16] S.J. Peng and R.A. Williams: "Controlled Production of Emulsions Using a Crossflow Membrane." Part. Part. Syst. Charact. **15**, 21 (1998).

[17] R. Katoh, Y. Asano, A. Furuya, K. Sotoyama, and M. Tomita: "Preparation of Food Emulsions Using Membrane Emulsification System." Proceedings of the 7th International Symposium on Synthetic Membranes in Science 407, Tübingen, Germany (1994).

[18] Y. Mine, M. Shimizu, and T. Nakashima: "Preparation and Stabilization of Simple and Multiple Emulsions Using Microporous Glass Membrane." Colloid Surfaces B Biointerfaces **6**, 261 (1996).

[19] V. Schröder, O. Behrend, and H. Schubert: "Effect of Dynamic Interfacial Tension on the Emulsification Process Using Microporous Ceramic Membranes." J. Colloid Interface Sci. **202**, 334 (1998).

[20] T. Nakashima, M. Shimizu, and M. Kukizaki: "Membrane Emulsification by Microporous Glass." Key Eng. Mater. **61/62**, 513 (1991).

[21] V. Schröder and H. Schubert: "Emulsification Using Microporous Ceramic Membranes." In Proceedings of the First European Congress on Chemical Engineering (ECCE 1) 2491, Florence Italy (1997).

[22] V. Schröder, Z. Wang, and H. Schubert: "Production of Oil-in-Water Emulsions by Microporous Membranes." In Proceedings of the Third International Symposium on Progress in Membrane Science and Technology, Euromembrane 1997 439, University of Twente (1997).

[23] G. Muschiolik and S. Dräger: "Emulsionbildung Mittels Mikroporösen Glas." Deutsche Milchwirtsch. **46**, 1041 (1995).

[24] R. Katoh, Y. Asano, A. Furuya, K. Sotoyama, and M. Tomita: "Preparation of Food Emulsions Using a Membrane Emulsification System." J. Membr. Sci. **113**, 131 (1996).

[25] R. Katoh, Y. Asano, A. Furuya, and M. Tomita: "Conditions for Preparation of O/W Food Emulsions Using a Membrane Emulsification System." Nippon Shokuhin Kagaku Kogaku Kaishi **42**, 548 (1995).

[26] S. Ban, M. Kitana, and A. Yamasaki: "Preparation of O/W Emulsions with Poly(Oxyethylene) Hydrogenated Castor Oil by Using Spg Membrane Emulsification." Nippon Kagaku Kogaku Kaishi **8**, 737 (1994).

[27] K. Kandori, K. Kishi, and T. Ishikawa: "Preparation of Monodispersed W/O Emulsions by Shirasu-Porous-Glass Filter Emulsification Technique." Colloid Surfaces **55**, 73 (1991).

[28] S.M. Joscelyne and G. Trägardh: "Membrane Emulsification—a Literature Review." J. Membr. Sci. **169**, 107 (2000).

[29] S. Van der Graaf, C.G.P.H. Schroën, and R.M. Boom: "Preparation of Double Emulsions by Membrane Emulsification—a Review." J. Membr. Sci. **251**, 7 (2005).

[30] T. Kawakatsu, Y. Kikuchi, and M. Nakajima: "Regular-Sized Cell Creation in Micro-Channel Emulsification by Visual Microprocessing Method." JAOCS **74**, 317 (1997).

[31] T. Kawakatsu, H. Komori, M. Nakajima, Y. Kikuchi, and T. Yonemoto: "Production of Monodispersed Oil-in-Water Emulsion Using Crossflow-Type Silicon Microchannel Plate." J. Chem. Eng. Jpn **32**, 241 (1999).

[32] S. Sugiura, M. Nakajima, S. Iwamoto, and M. Seki: "Interfacial Tension Driven Monodispersed Droplet Formation from Microfabricated Channel Array." Langmuir **17**, 5562 (2001).

[33] S. Sugiura, M. Nakajima, N. Kumazawa, S. Iwamoto, and M. Seki: "Characterization of Spontaneous Transformation-Based Droplet Formation During Microchannel Emulsification." J. Phys. Chem. B **106**, 9405 (2002).

[34] S. Sugiura, M. Nakajima, K. Yamamoto, S. Iwamoto, T. Oda, M. Satake, and M. Seki: "Prepartion Characteristics of Water-in-Oil-in-Water Multiple Emulsions Using Microchannel Emulsification." J. Colloid Interface Sci. **270**, 221 (2004).

[35] T. Kawakatsu, G. Trägardh, C. Trägardh, M. Nakajima, N. Oda, and T. Yonemoto: "The Effect of Hydrophobicity of Microchannels and Components in Water and Oil Phases on Droplet Formation in Microchannel Water-in-Oil Emulsification." Colloid and Surfaces A Physicochem. Eng. Aspects **179**, 29 (2001).

[36] S. Sugiura, M. Nakajima, and M. Seki: "Effect of Channel Structure on Microchannel Emulsification." Langmuir **18**, 5708 (2002).
[37] S. Sugiura, M. Nakajima, and M. Seki: "Prediction of Droplet Diameter for Microchannel Emulsification." Langmuir **18**, 3854 (2002).
[38] S. Sugiura, M. Nakajima, T. Oda, M. Satake, and M. Seki: "Effect of Interfacial Tension on the Dynamic Behavior of Droplet Formation During Microchannel Emulsification." J. Colloid Interface Sci. **269**, 178 (2004).
[39] I. Kobayashi, M. Nakajima, and S. Mukataka: "Preparation Characteristics of Oil-in-Water Emulsion Using Differently Charged Surfactants in Straight-Through Microchannel Emulsification." Colloid Surfaces A Physicochem. Eng. Aspects **229**, 33 (2003).
[40] S. Sugiura, M. Nakajima, and M. Seki: "Prediction of Droplet Diameter for Microchannel Emulsification: Prediction Model for Complicated Microchannel Geometries." Ind. Eng. Chem. Res. **43**, 8233 (2004).
[41] I. Kobayashi, and M. Nakajima: "Effect of Emulsifiers on the Preparation of Food-Grade Oil-in-Water Emulsions Using a Straight-through Extrusion Filter." Eur. J. Lipid Sci. Technol. **104**, 720 (2002).
[42] S. Sugiura, M. Nakajima, H. Itou, and M. Seki: "Synthesis of Polymeric Microspheres with Narrow Size Distributions Employing Microchannel Emulsification." Macromol. Rapid Commun. **22**, 773 (2001).
[43] S. Sugiura, M. Nakajima, and M. Seki: "Prepartion of Monodispersed Polymeric Microspheres over 50 μm Employing Microchannel Emulsification." Ind. Eng. Chem. Res. **41**, 4043 (2002).
[44] K. Nakagawa, S. Iwamoto, M. Nakajima, A. Shono, and K. Satoh: "Microchannel Emulsification Using Gelation and Surfactant-Free Coacervate Microencapsulation." J. Colloid Interface Sci. **278**, 198 (2004).
[45] S.L. Anna, N. Bontoux, and H.A. Stone: "Formation of Dispersions Using Flow Focusing in Microchannels." Appl. Phys. Lett. **82**, 364 (2003).
[46] Q. Xu and M. Nakajima: "The Generation of Highly Monodisperse Droplets through the Breakup of Hydrodynamically Focused Microthread in a Microfluidic Device." Appl. Phys. Lett. **85**, 3726 (2004).
[47] Y. Tan, V. Cristini, and A.P. Lee: "Monodispersed Microfluidic Droplet Generation by Shear Focusing Microfluidic Device." Sensors and Actuactors B: Chem. **144**, 350 (2006).
[48] J.D. Tice, A.D. Lyon, and R.F. Ismagilov: "Effects of Viscosity on Droplet Formation and Mixing in Microfluidic Channels." Anal. Chim. Acta **507**, 73 (2004).
[49] D.R. Link, S.L. Anna, D.A. Weitz, and H.A. Stone: "Geometrically Mediated Breakup of Drops in Microfluidic Devices." Phys. Rev. Lett **92**, 054503 (2004).
[50] S. Okushima, T. Nisisako, T. Torii, and T. Higuchi: "Controlled Production of Monodisperse Double Emulsions by Two-Step Droplet Breakup in Microfluidic Devices." Langmuir **20**, 9905 (2004).
[51] M. Pollack, R.B. Fair, and A.D. Shenderov: "Electrowetting-Based Actuation of Liquid Droplets for Microfluidic Applications." Appl. Phys. Lett. **77**(11), 1725 (2000).
[52] B. Zheng, L.S. Roach, and R.F. Ismagilov: "Screening of Protein Crystallization Conditions on a Microfluidic Chip Using Nanoliter-Size Droplets." J. Am. Chem. Soc. **125**, 11170 (2003).
[53] B. Zheng, J.D. Tice, and R.F. Ismagilov: "Formation of Droplets of Alternating Composition in Microfluidic Channels and Applications to Indexing of Concentrations in Droplets-Based Assays." Anal. Chem. **76**, 4977 (2004).

[54] B. Zheng, J.D. Tice, L.S. Roach, and R.F. Ismagilov: "A Droplet-Based, Composite PDMS/Glass Capillary Microfluidic System for Evaluating Protein Crystallization Conditions by Microbatch and Vapor-Diffusion Methods with on-Chip X-Ray Diffraction." Angew. Chem. Int. Ed. **43**, 2508 (2004).

[55] V. Srinivasan, V.K. Pamula, and R.B. Fair: "Droplet-Based Microfluidic Lab-on-a-Chip for Glucose Detection." Anal. Chim. Acta **507**, 145 (2004).

[56] H. Song, J.D. Tice, and R.F. Ismagilov: "A Mocrofluidic System for Controlling Reaction Networks in Time." Angew. Chem. Int. Ed. **42**, 767 (2003).

[57] H. Song, M.R. Bringer, J.D. Tice, C.J. Gerdts, and R.F. Ismagilov: "Experimental Test of Scaling of Mixing by Chaotic Advection in Droplets Moving through Microfluidics Channels." Appl. Phys. Lett. **83**, 4664 (2003).

[58] M.J. Groves: "Spontaneous Emulsification." Chem. Indust. **12**, 417 (1978).

[59] G. Quincke: "Ueber Emulsionbildung Und Den Einfluss Der Galle Bei Der Verdauung." Plüger Arch. Physiol. **19**, 129 (1879).

[60] G. Quincke: "Ueber Periodische Ausbreitung an Flüssigkeitsoberflächen Und Dadurch Hervorgerufene Bewegungserscheinungen." Wiedemanss Annalen der Physik und Chemie. Neue Folge **35**, 580 (1888).

[61] J.T. Davies and E.K. Rideal: "Diffusion Through Interfaces." In H. Willmer (ed), *Interfacial Phenomena*. Academic Press, New York (1961).

[62] J.T. Davies and D.A. Haydon: "Spontaneous Emulsification." In Proceedings of the International Congress of Surfactants Act. 2nd **1**, 417, (1957).

[63] J.W. McBain and T.M. Woo: "Spontaneous Emulsification and Reactions Overshooting Equilibrium." Proc. Roy. Soc. **A163**, 182 (1937).

[64] J. Schulman and E.G. Cockbain: "Molecular Interactions at Oil/Water Interfaces. Part I. Molecular Complex Formation and the Stability of Oil in Water Emulsions." Transact. Faraday Soc. **36**, 651 (1940).

[65] R.E.S. Gopal: "Principles of Emulsion Formation." In P. Sherman (ed), *Emulsion Science*, Academic Press, London (1968).

[66] C.A. Miller, R.-N. Hwan, W.J. Benton, and T.J. Fort: "Ultralow Interfacial Tensions and Their Relation to Phase Separation in Micellar Solutions." J. Colloid Interface Sci. **61**, 554 (1977).

[67] C.A. Miller: "Spontaneous Emulsification Produced by Diffusion—a Review." Colloid Surfaces **29**, 89 (1988).

[68] K.J. Ruschak and C.A. Miller: "Spontaneous Emulsification in Ternary Systems with Mass Transfer." Ind. Eng. Chem. Fundam. **11**, 534 (1972).

[69] M.J. Rang and C.A. Miller: "Emulsions and Microemulsions—Spontaneous Emulsification of Oil Drops Containing Surfactants and Medium-Chain Alcohols." Prog. Colloid Polym. Sci. **109**, 101 (1998).

[70] M.J. Rang and C.A. Miller: "Spontaneous Emulsification of Oils Containing Hydrocarbon, Nonionic Surfactant, and Oleyl Alcohol." J. Colloid Interface Sci. **209**, 179 (1999).

[71] J.T. Davies and E.K. Rideal: "Disperse Systems and Adhesion." In H. Willmer (ed), *Interfacial Phenomena*. Academic Press, New York (1961).

[72] N. Shahidzadeh, D. Bonn, and J. Meunier: "A New Mechanism of Spontaneous Emulsification: Relation to Surfactant Properties." Europhys. Lett **40**, 459 (1997).

[73] R.W. Greiner and D.F. Evans: "Spontaneous Formation of a Water-Continuous Emulsion from a W/O Microemulsion." Langmuir **6**, 1793 (1990).

[74] J.J. Rang, C.A. Miller, H.H. Hoffmann, and C. Thunig: "Behavior of Hydrocarbon/Alcohol Drops Injected into Dilute Solutions of an Amine Oxide Surfactant." Ind. Eng. Chem. Res. **35**, 3233 (1996).

[75] J.C. Lopez-Montilla, P.E. Herrera-Morales, and D.O. Shah: "New Method to Quantitatively Determine Spontaneity of Emulsification Process." Langmuir **18**, 4258 (2002).
[76] Y.A. Shchipunov and P. Schmiedel: "Phase Behavior of Lecithin at the Oil/Water Interface." Langmuir **12**, 6443 (1996).
[77] M.G. Wakerly, C.W. Pouton, B.J. Meakin, and F.S. Morton: "Self emulsification of vegetable oil non-ionic mixtures: a proposed mechanism of action." In *Phenomena in Mixed Surfactant Systems*. American Chemical Society, Washington, DC (1986).
[78] J.C. Lopez-Montilla, P.E. Herrera-Morales, S. Pandey, and D.O. Shah: "Spontaneous Emulsification: Mechanisms, Physicochemical Aspects, Modeling and Applications." J. Dispersion Sci. Technol. **23**, 219 (2002).
[79] T. Förster, W. Von Rybinski, H. Tesmann, and A. Wadle: "Calculation of Optimum Emulsifier Mixtures for Phase Inversion Emulsification." Int. J. Cosmet. Sci. **16**, 84 (1994).
[80] D.J. Miller, T. Henning, and W. Grübein: "Phase Inversion of W/O Emulsions by Adding Hydrophilic Surfactant—a Technique for Making Cosmetics Products." Colloids Surfaces **183–185**, 681 (2001).
[81] K. Shinoda and H. Arai: "The Correlation between Phase Inversion Temperature in Emulsion and Cloud Point in Solution of Nonionic Emulsifier." J. Phys. Chem. **68**, 3485 (1964).
[82] K. Shinoda and H. Saito: "The Stability of O/W Type Emulsions as a Function of Temperature and the HLB of Emulsifiers: The Emulsification by PIT Method." J. Colloid Interface Sci. **30**, 258 (1969).
[83] K. Shinoda and H. Kunieda: "Conditions to Produce So-Called Microemulsions: Factors to Increase the Mutual Solubility of Oil and Water by Solubilizer." J. Colloid Interface Sci. **42**, 381 (1973).
[84] W.D. Bancroft: "The Theory of Emulsification, V." J. Phys. Chem. **17**, 501 (1913).
[85] K. Shinoda and H. Saito: "The Effect of Temperature on the Phase Equilibrium and the Types of Dispersions of the Ternary System Composed of Water, Cyclohexane and Nonionic Surfactant." J. Colloid Interface Sci. **26**, 70 (1968).
[86] V.E. Wellman and H.V. Tartar: "The Factors Controlling Type of Water-Soap-Oil Emulsions." J. Phys. Chem. **34**, 379 (1930).
[87] H. Saito and K. Shinoda: "The Stability of W/O Type Emulsions as a Function of Temperature and of the Hydrophilic Chain Length of the Emulsifier." J. Colloid Interface Sci. **32**, 647 (1970).
[88] H. Kunieda and K. Shinoda: "Phase Behavior in Systems of Nonionic Surfactant/Water/Oil Around the Hydrophilic-Lypophilic-Balance-Temperature." J. Dispersion Sci. Technol. **3**, 233 (1982).
[89] K. Shinoda: "Solution Behavior of Surfactants: The Importance of Surfactant Phase and the Continuous Change in HLB of Surfactant." Prog. Colloid Polymer Sci. **68**, 1 (1983).
[90] R. Aveyard, B.P. Binks, T.A. Lawless, and J. Mead: "Interfacial Tension Minima in Oil + Water + Surfactant Systems. Effects of Salt and Temperature in Systems Containing Nonionic Surfactants." J. Chem. Soc. Faraday Trans. 1 **81**, 2155 (1985).
[91] R. Aveyard and T.A. Lawless: "Interfacial Tension Minima in Oil-Water-Surfactant Systems. Systems Containing Pure Nonionic Surfactants, Alkanes, and Inorganic Salts." J. Chem. Soc. Faraday Trans. 1 **82**, 2951 (1986).
[92] R. Aveyard, B.P. Binks, S. Clark, and J. Mead: "Interfacial Tension Minima in Oil-Water-Surfactant Systems. Behavior of Alkane-Aqueous Sodium Chloride Systems Containing Aerosol OT." J. Chem. Soc. Faraday Trans. I **82**, 125 (1986).

[93] R. Aveyard, B. Binks, T.A. Lawless, and J. Mead: "Nature of the Oil/Water Interface and Equilibrium Surfactant Aggregates in Systems Exhibiting Low Tensions." Can. J. Chem. **66**, 3031 (1988).

[94] P.D.I. Fletcher and D.I. Horsup: "Droplet Dynamics in Water-in-Oil Microemulsions and Macroemulsions Stabilized by Non-Ionic Surfactants." J. Chem. Soc. Faraday Trans. I **88**, 855 (1992).

[95] L.T. Lee, D. Langevin, J. Meunier, K. Wong, and B. Cabane: "Film Bending Elasticity in Microemulsions Made with Nonionic Surfactants." Prog. Colloid Polymer Sci. **81**, 209 (1990).

[96] D. Langevin: In S.-H. Chen, J. S. Huang, and P. Tartaglia (eds), "Low interfacial tensions in microemulsion systems." *Structure and Dynamics of Strongly Interacting Colloids and Supramolecular Aggregates in Solution*. 325. p. Kluwer, Dordrecht (1992).

[97] R. Strey: "Microemulsion, Microstructure and Interfacial Curvature." Colloid Polym. Sci. **272**, 1005 (1994).

[98] T. Sottmann and R. Strey: "Shape Similarities of Ultra-Low Interfacial Tension Curves in Ternary Microemulsion Systems of the Water-Alkane-CiEj Type." Ber. Bunsenges Phys. Chem. **100**, 237 (1996).

[99] T. Sottmann and R. Strey: "Ultralow Interfacial Tension in Water-N-Alkane-Surfactant Systems." J. Chem. Phys. **106**, 8606 (1997).

[100] M. Kahlweit, R. Strey, and G. Busse: "Weakly to Strongly Structured Mixtures." Phys. Rev. E **47**, 4197 (1993).

[101] K. Shinoda: "The Correlation between the Dissolution State of Nonionic Surfactant and the Type of Dispersion Stabilized with the Surfactant." J. Colloid Interface Sci. **24**, 4 (1967).

[102] F. Groeneweg, W.G.M. Agterof, P. Jaeger, J.J.M. Janssen, J.A. Wieringa, and J.K. Klahn: "On the Mechanism of the Inversion of Emulsions." Chem. Eng. Res. Des. **76**, 55 (1998).

[103] B.W. Brooks and H.N. Richmond: "Phase Inversion in Non-Ionic Surfactant-Oil-Water Systems, I. The Effect of Transitional Inversion on Emulsion Drop Size." Chem. Eng. Sci. **49**, 1053 (1994).

[104] J.L. Salager: "Phase Transformation and Emulsion Inversion on the Basis of Catastrophe Theory." In P. Becher (ed), *Encyclopedia of Emulsion Technology*. Vol. 3. *Basic Theory. Measurement. Applications*. Marcel Dekker, New York (1988).

[105] J.L. Salager, M. Perez-Sanchez, and Y. Garcia: "Physicochemical Parameters Influencing the Emulsion Drop Size." Colloid Polym. Sci. **274**, 81 (1996).

[106] P. Izquierdo, J. Esquena, T.F. Tadros, J.C. Dederen, M.J. Garcia, N. Azemar, and C. Solans: "Formation and Stability of Nano-Emulsions Prepared Using the Phase Inversion Method." Langmuir **18**, 26 (2002).

[107] T. Förster, F. Schambil, and W. Von Rybinski: "Production of Fine Dispersion and Long-Term Stable Oil-in-Water Emulsions by the Phase Inversion Temperature Method." J. Dispersion Sci. Technol. **13**, 183 (1992).

[108] A. Wadle, T. Förster, and W. Von Rybinski: "Influence of the Microemulsion Phase Structure on the Phase Inversion Temperature Emulsification of Polar Oils." Colloid and Surfaces A Physicochem. Eng. Aspects **76**, 51 (1993).

[109] L. Taisne and B. Cabane: "Emulsification and Ripening Following a Temperature Quench." Langmuir **14**, 4744 (1998).

[110] J.L. Salager: "Macro Emulsions Stabilized by an Ethoxylated Fatty Alcohol and an Alkyl Quat: Emulsion Type and Stability in View of the Phase Behavior." In Proceedings of the 3rd World Congress on Emulsions 1-F-107, Lyon, France (2001).

[111] J.L. Salager: "Evolution of Emulsion Properties Along a Transitional Inversion Produced by a Temperature Variation." In Proceedings of the 3rd Word Congress on Emulsions 1-F-094, Lyon, France (2001).

[112] P. Izquierdo, J. Esquena, T.F. Tadros, J.C. Dederen, J. Feng, J. Garcia-Celma, N. Azemar, and C. Solans: "Phase Behavior and Nano-Emulsion Formation by the Phase Inversion Temperature Method." Langmuir **20**, 6594 (2004).

[113] P. Fernandez, V. André, J. Rieger, and A. Kühnle: "Nano-Emulsion Formation by Emulsion Phase Inversion." Colloid and Surfaces A Physicochem. Eng. Aspects **251**, 53 (2004).

[114] J.K. Klahn, J.J.M. Janssen, G.E.J. Vaessen, R. de Swart, and W.G.M. Agterof: "On the Escape Process During Phase Inversion of an Emulsion." Colloid and Surfaces A: Physicochem. Eng. Aspects **210**, 167 (2002).

[115] J.L. Salager: "Properties of Emulsions at the Onset of Catastrophic Phase Inversion in the Normal to Abnormal Inversion." In Proceedings of the 3rd World Congress on Emulsions 1-F-185, Lyon, France (2001).

[116] N. Zambrano, E. Tyrode, I. Mira, L. Marquez, M.-P. Rodriguez, and J.L. Salager: "Emulsion Catastrophic Inversion from Abnormal to Normal Morphology. 1. Effect of the Water-to-Oil Ratio Rate of Change on the Dynamic Inversion Frontier." Ind. Eng. Chem. Res. **42**, 50 (2003).

[117] A.S. Kabalnov and H. Wennerström: "Macroemulsion Stability : The Oriented Wedge Theory Revisited." Langmuir **12**, 276 (1996).

[118] A.W. Nienow: "Breakup, Coalescence and Catastrophic Phase Inversion in Turbulent Contactors." Adv. Coll. Int. Sci. **108–109**, 95 (2004).

[119] A. Wadle, H. Tesmann, M. Leonard, and T. Förster: "Phase Inversion in Emulsions: Capico-Concept and Application." Surfactant Science Series **68**, 207 (1997).

[120] T. Iwanaga, M. Suzuki, and H. Kunieda "Effect of Added Salts or polyols on the Liquid Crystolline Structures of Polyesayethylene-Type Nonionic Surfactants" Langmuir **14**, 5775 (1998).

[121] J.L. Salager: "A 3rd Type of Emulsion Inversion Attained by Overlapping the Two Classical Methods: Combined Inversion." In Proceedings of the 3rd Word Congress on Emulsions 1-F-180, Lyon, France (2001).

[122] M. Perez, N. Zambrano, M. Ramirez, E. Tyrode, and J.L. Salager: "Surfactant-Oil-Water System near the Affinity Inversion. XII. Emulsion Drop Size Formulation and Composition." J. Dispersion Sci. Technol. **23**, 55 (2002).

[123] J.L. Salager, A. Forgiarini, L. Marquez, A. Pena, A. Pizzino, M.-P. Rodriguez, and M. Rondon-Gonzalez: "Using Emulsion Inversion in Inductrial Processes." Adv. Coll. Int. Sci. **108–109**, 259 (2004).

[124] J.L. Salager, M. Minana-Perez, M. Perez-Sanchez, M. Ramirez-Gouveia, and C.I. Rojas: "Surfactant-Oil-Water Systems near the Affinity Inversion. Part III: The Two Kinds of Emulsion Inversion." J. Dispersion Sci. Technol. **4**, 313 (1983).

[125] I. Mira, N. Zambrano, E. Tyrode, L. Marquez, A.A. Pena, A. Pizzino, and J.L. Salager: "Emulsion Catastrophic Inversion from Abnormal to Normal Morphology. 2. Effect of the Stirring Intensity on the Dynamic Inversion Frontier." Ind. Eng. Chem. Res. **42**, 57 (2003).

[126] S. Sajjadi, F. Jahanzad, and M. Yianneskis: "Catastrophic Phase Inversion of Abnormal Emulsions in the Vicinity of the Locus of Transitional Inversion." Colloid and Surfaces A Physicochem. Eng. Aspects **240**, 149 (2004).

[127] N. Uson, M.J. Garcia, and C. Solans: "Formation of Water-in-Oil (W/O) Nano-Emulsions in a Water/Mixed Non-Ionic Surfactant/Oil Systems Prepared by a Low-Energy Emulsification Method." Colloid and Surfaces A Physicochem. Eng. Aspects **250**, 415 (2004).

[128] J. Allouche, E. Tyrode, V. Sadtler, L. Choplin, and J.L. Salager: "Simultaneous Conductivity and Viscosity Measurements as a Technique to Track Emulsion Inversion by the Phase-Inversion-Temperature Method." Langmuir **20**, 2134 (2004).

[129] L. Liu, O.K. Matar, E.S. Perez de Ortiz, and G.F. Hewitt: "Experimental Investigation of Phase Inversion in a Strirred Vessel Using Lif." Chem. Eng. Sci. **60**, 85 (2005).

[130] R. Pons, I. Carrera, P. Erra, H. Kunieda, and C. Solans: "Novel Preparation Methods for Highly Concentrated Water-in-Oil Emulsions." Colloids Surf., A Physicochem. Eng. Asp. **91**, 259 (1994).

[131] H. Kunieda, Y. Fukui, H. Uchiyama, and C. Solans: "Spontaneous Formation of Highly Concentrated Water-in-Oil Emulsions (Gel-Emulsions)." Langmuir **12**, 2136 (1996).

[132] K. Ozawa, C. Solans, and H. Kunieda: "Spontaneous Formation of Highly Concentrated Oil-in-Water Emulsions." J. Colloid Interface Sci. **188**, 275 (1997).

[133] C. Solans, R. Pons, and H. Kunieda: "Gel Emulsions - Relationship between Phase Behaviour and Formation." In B.P. Binks (ed), *Modern Aspects of Emulsion Science*. The Royal Society of Chemistry, Cambridge (1998).

[134] F. Leal Calderon, J. Bibette, and F. Guimberteau: "Method for Preparing Concentrated and Calibrated Emulsions in a Highly Viscous Phase, in Particular Bitumen Emulsions." US Patent 6602917 (1998).

[135] T.G. Mason and J. Bibette: "Emulsification in Viscoelastic Media." Phys. Rev. Let **77**, 3481 (1996).

[136] J. Bibette and T.G. Mason: "Procédé de Préparation d'une Émulsion." French Patent 96 04736 PCT SR97/00690 (1996).

[137] C. Mabille, V. Schmitt, P. Gorria, F. Leal Calderon, V. Faye, and B. Deminière: "Rheological and Shearing Conditions for the Preparation of Monodisperse Emulsions." Langmuir **16**, 422 (2000).

[138] G.I. Taylor: "The Formation of Emulsions in Definable Fields of Flow." Proc. R. Soc. **A146**, 501 (1934).

[139] H.P. Grace: "Dispersion Phenomena in High Viscosity Immiscible Fluid Systems and Application of Static Mixers as Dispersion Devices in Such Systems." Chem. Eng. Commun **14**, 225 (1982).

[140] R.A. De Bruijn: "Deformation and Breakup of Drops in Simple Shear Flows." Ph. D. Thesis Eindhoven University of Technology (1989).

[141] B.J. Bentley and L.G. Leal: "An Experimental Investigation of Drop Deformation and Breakup in Steady Two-Dimensional Linear Flows." J. Fluid. Mech. **167**, 241 (1986).

[142] J.W.S. Rayleigh: "On the Instability of Jets." Proc. London Math. Soc. **10**, 4 (1878).

[143] J.W.S. Rayleigh: "On the Capillary Phenomena of Jets." Proc. Roy. Soc. **29**, 71 (1879).

[144] J.W.S. Rayleigh: "On the Instability of a Cylinder of Viscous Liquid Under Capillary Force." Philos. Mag. **34**, 145 (1892).

[145] F.D. Rumscheidt and S.G. Mason: "Breakup of Stationary Liquid Threads." J. Colloid Sci. **17**, 260 (1962).

[146] R.A. De Bruijn: "Tip Streaming of Drops in Simple Shear Flow." Chem. Eng. Sci. **48**, 277 (1993).

[147] H.A. Stone, B.J. Bentley, and L.G. Leal: "An Experimental Study of Transient Effects in the Breakup of Viscous Drops." J. Fluid Mech. **173**, 131 (1986).

[148] E.J. Hinch and A. Acrivos: "Long Slender Drops in a Simple Shear Flow." J. Fluid. Mech. **98**, 305 (1980).

[149] C. Mabille, F. Leal-Calderon, J. Bibette, and V. Schmitt: "Monodisperse Fragmentation in Emulsions: Mechanisms and Kinetics." Europhys. Lett **61**, 708 (2003).

[150] C. Mabille: "Fragmentation in Emulsions Submitted to a Simple Shear." Ph.D Thesis, Bordeaux I University (2000).

[151] T. Tomotika: "On the Instability of a Cylindrical Thread of a Viscous Fluid." Proc. Roy. Soc. (Lond.) **A150**, 322 (1935).

[152] D.C. Chappelear: "Interfacial Tension between Molten Polymers." Polym. Prep. **5**, 363 (1964).

[153] P.H.M. Elemans, J.M.H. Janssen, and H.E.H. Meier: "The Measurement of Interfacial Tension in Polymer/Polymer Systems: The Breaking Thread Method." J. Rheol. **34**, 1311 (1990).

[154] D.V. Khakhar and J.M. Otino: "Deformation and Breakup of Slender Drops in Linear Flows." J. Fluid Mech. **166**, 265 (1986).

[155] D. Rusu: "Etude in-Situ, Par Diffusion De La Lumière, De La Morphologie De Mélanges De Polymères Immiscibles Durant Un Cisaillement." Thèse de l'ecole des Mines de Paris, Sophia Antipolis (1997).

[156] V. Schmitt, F. Leal-Calderon, and J. Bibette: "Preparation of Monodisperse Particles and Emulsions by Controlled Shear." In M. Antonietti (ed), *Colloid Chemistry II*. Springer-Verlag, Berlin (2003).

[157] M.P. Aronson: "The Role of Free Surfactant in Destabilizing Oil-in-Water Emulsions." Langmuir **5**, 494 (1989).

[158] G.I. Taylor: "The Viscosity of a Fluid Containing Small Drops of Another Fluid." Proc. R. Soc. **A138**, 41 (1932).

[159] J. Bibette and T.G. Mason: "Emulsion Manufacturing Process." US Patent 5,938,581 (1999).

[160] P. Perrin: "Amphiphilic Copolymers: A New Route to Prepare Ordered Monodisperse Emulsions." Langmuir **14**, 5977 (1998).

[161] P. Perrin and F. Lafuma: "Low Hydrophobically Modified Poly (Acrylic Acid) Stabilizing Macroemulsions: Relationship Between Copolymer Structure and Emulsions Properties." J. Colloid Interface Sci. **197**, 317 (1998).

[162] C. Goubault, K. Pays, D. Olea, J. Bibette, V. Schmitt, and F. Leal-Calderon: "Shear Rupturing of Complex Fluids: Application to the Preparation of Quasi-Monodisperse W/O/W Double Emulsions." Langmuir **17**, 5184 (2001).

[163] K. Pays, J. Giermanska-Kahn, P. Pouligny, J. Bibette, and F. Leal-Calderon: "Double Emulsions: A Tool for Probing Thin Film Metastability." Phys. Rev. Lett. **87**, 178304 (2001).

[164] K. Pays, J. Kahn, B. Pouligny, J. Bibette, and F. Leal-Calderon: "Coalescence in Surfactant-Stabilized Double Emulsions." Langmuir **17**, 7758 (2001).

[165] J. Bibette, F. Leal-Calderon, and P. Gorria: "Polydisperse Double Emulsion, Corresponding Monodisperse Double Emulsion and Process to Fabricate the Monodisperse Emulsion." Patent WO0121297 (1999).

[166] K. Pays, F. Leal Calderon, and J. Bibette: "Method to Produce a Monodisperse Double Emulsion." French Patent 00 05880 (2000).

[167] T.G. Mason and J. Bibette: "Shear Rupturing of Droplets in Complex Fluids." Langmuir **13**, 4600 (1997).

[168] F. Leal-Calderon, T. Stora, O. Mondain Monval, P. Poulin, and J. Bibette: "Direct Measurement of Colloidal Forces." Phys. Rev. Lett. **72**, 2959 (1994).

[169] G. Reimers and S. Khalafalla: "Production of Magnetic Fluids by Peptization Techniques." US Patent 38,435,40 (1974).

2 Force Measurements

2.1. Introduction

Knowledge of the forces acting between colloidal particles is of primary importance in understanding the behavior of dispersed systems. Emulsions, foams, and colloidal dispersions require repulsive surface forces to become metastable. In the case of liquid dispersions, coalescence may occur when the thin liquid films separating the interfaces break. If the net force acting at short distances is highly repulsive, coalescence is inhibited and the system can remain stable for a long period of time. A net attractive force may lead to aggregation of droplets. Whether droplets are aggregated or not determines the phase behavior and the rheological properties of emulsions. This is why it is important to know the variation of the net force with distance, and many different techniques have been devised for that purpose.

This chapter comprises two sections. The first describes the most usual techniques to directly measure force versus distance profiles between solid or liquid surfaces. We then describe different long-range forces (range >5 nm) accessible to evaluation via these techniques for different types of surface active species. The second section is devoted to attractive interactions whose strong amplitude and short range are difficult to determine. In the presence of such interactions, emulsion droplets exhibit flat facets at each contact. The free energy of interaction can be evaluated from droplet deformation and reveals interesting issues.

2.2. Long-Range Forces

The development of various techniques has led to important advances. The possibility to measure intermolecular and intercolloidal forces directly represents a qualitative change from the indirect way such forces had been inferred in the past from aggregation kinetics or from bulk properties such as the compressibility (deduced from small angle scattering) or phase behavior. Both static (i.e., equilibrium) and dynamic (e.g., viscous) forces can now be directly measured, providing information not only on the fundamental interactions in liquids but also on the structure of liquids adjacent to surfaces and other interfacial phenomena. Two techniques are available for measuring forces between macroscopic solid surfaces as a function of distance: the surface force apparatus (SFA) [1–3] and the measurement and analysis of surface interaction forces (MASIF) [4,5]. Forces between a macroscopic surface and a particle can be measured by means of the atomic force microscope (AFM) using a colloidal probe [6], or by employing total internal reflection microscopy (TIRM) to monitor the position of a particle trapped by a laser beam [7]. Measurement of forces between two liquid interfaces may be performed via the thin film balance technique (TFB) for macroscopic single films [8] or via the liquid surface force apparatus (LSFA) for very small films formed between a droplet and a macroscopic liquid surface [9]. The magnetic chaining technique (MCT) used for direct measurement of force–distance profiles between liquid particles of colloidal size involves the application of magnetic fields to ferrofluid emulsions [10]. Osmotic stress techniques combined with X-ray scattering are commonly used for studying interactions in liquid crystalline surfactant phases or in concentrated dispersions [11]. All these techniques (and others not mentioned here) have provided very detailed and useful information on surface forces, especially in the presence of species adsorbed at the interfaces (ions, surfactants, polymers, phospholipids, proteins, etc.). They have been used under a sufficiently large variety of conditions to allow comparisons and to infer some general rules that govern the behavior of thin liquid films. In this section, we first briefly describe three of them, which have been chosen because they are representative of the techniques that can be used for measuring forces between solid (SFA) and "soft" interfaces (TFB and MCT). We then report some of the most recent advances in the field of force measurements between liquid interfaces, emphasizing the influence of adsorbed and nonadsorbed surfactant and polymer molecules.

2.2.1. Techniques for Surface Force Measurements

2.2.1.1. Surface Force Apparatus (SFA) (Fig. 2.1)

In this technique, two molecularly smooth mica surfaces are mounted in a crossed cylinder configuration with a typical radius, R, of 2 cm. They are silvered on their backside and glued onto cylindrical silica disks. Collimated white light oriented normal to the surfaces passes through one surface and is multiply reflected between the silver layers. Because of constructive interferences, multiple-beam interference fringes are transmitted through the second surface. The surface separation h can be determined by comparing the interference fringes when the surfaces are in contact and apart. The distance resolution is of the order of 0.1 nm. The distance between the two interacting surfaces is generally changed by applying a variable voltage to a piezoelectric crystal. The force between them is deduced from the deflection of a cantilever spring of constant k on which one of the surfaces is attached. The force F is measured by expanding or contracting the piezoelectric crystal to a known amount Δh and by measuring by interferometry the actual displacement Δh_0 of the two surfaces relative to one another. The difference of the two distances

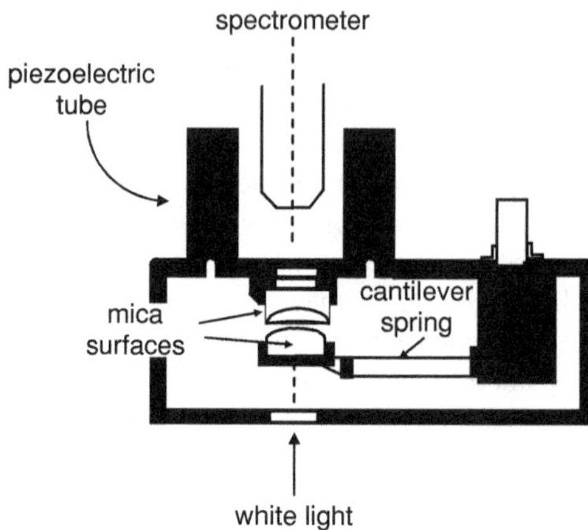

FIGURE 2.1. Scheme of the surface force apparatus (SFA).

gives, using Hookes's law, the difference in the forces applied before and after displacement:

$$\Delta F = k(\Delta h - \Delta h_0) \tag{2.1}$$

The force resolution is generally of the order of 10^{-7} N. One of the main advantages of the SFA technique compared to the two others presented here is that it allows the measurement of both attractive and repulsive forces.

2.2.1.2. Thin Film Balance (TFB) (Fig. 2.2)

Initially devised to measure interactions in single soap films (air/water/air) [8], the TFB technique has been progressively improved and its application has been broadened to emulsion films (oil/water/oil) [12] and asymmetric films (air/water/oil or air/water/solid) [13,14]. In a classical setup, a thin porous glass disk is fused on the side to a capillary tube and a small hole is drilled in the center of the disk. The liquid solution fills the disk, part of the capillary, and a thin horizontal film is formed across the hole. The disk is enclosed in a hermetically sealed box, with the capillary tube exposed to a constant reference pressure P_r. Under the effect of the pressure difference ΔP between the box and the reference, the

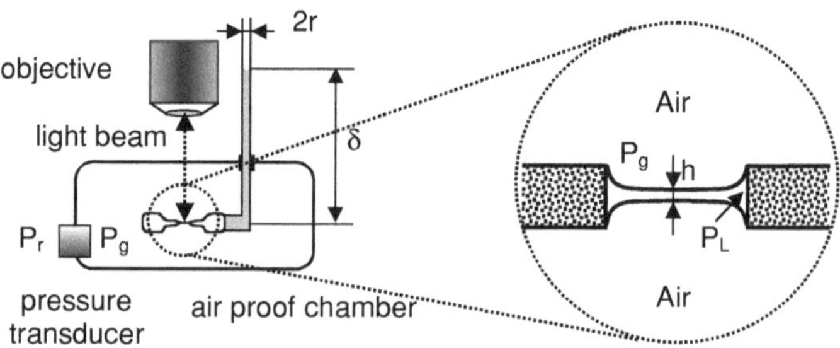

FIGURE 2.2. Scheme of the thin film balance technique (TFB).

film thins and finally stabilizes at a thickness h if the surface force per unit area Π balances the applied pressure. Π is a function of h and is called disjoining pressure after Derjaguin [15]. The pressure difference is typically controlled within a few Pascals through changes in the box pressure via a precise syringe pump. As in the SFA technique, the thickness h is measured by monitoring the light reflectivity at normal incidence, and by using classical interferometric formulas. Typically, the optical system consists of a reflected light microscope supplied with a heat-filtered light source. At equilibrium and in the flat portion of the film, the disjoining pressure equals the capillary pressure, that is, the pressure difference between the box and the liquid in the Plateau-border region:

$$\Pi = P_g - P_L = P_g - P_r + \frac{2\gamma_{surf}}{r} - \rho g \delta \tag{2.2}$$

where γ_{surf} is the air–solution surface tension, r is the capillary radius, ρ is the density of the solution, δ is the height of the solution in the capillary tube, and g is the gravitational constant. The first term $P_g - P_r$ can be measured with a differential pressure transducer and the others terms are easily obtained from standard measurement methods. It should be noted that only globally repulsive forces can be measured via this technique, in contrast to the SFA.

2.2.1.3. Magnetic Chaining Technique (MCT) (Fig. 2.3)

In TFB technique, the thin film radius is typically of the order of 100 μm, far larger than the contact film radius likely to be formed when two micron-sized droplets approach. The magnetic chaining technique overcomes this limitation, allowing the direct measurement of force–distance profiles between liquid colloidal droplets. This technique exploits the properties of monodisperse ferrofluid

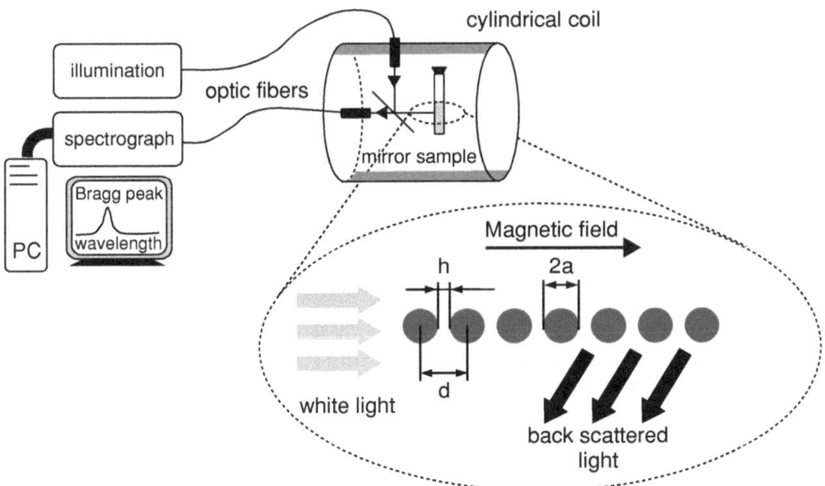

FIGURE 2.3. Scheme of the magnetic chaining technique (MCT).

emulsions. The dispersed phase of a ferrofluid emulsion is an octane ferrofluid, that is, octane containing small ferrimagnetic oxide grains (Fe_2O_3, 100 Å in size). Highly Monodisperse oil-in-water emulsions are obtained using the fragmentation method described in Chapter 1 (Section 1.7.6.4) followed by the fractionated crystallization technique described in Chapter 3 (Section 3.2.1.2).

At very low droplet volume fractions ($\phi < 0.1\%$) and on application of an external field, the droplets form chains that are only one droplet thick and that remain well separated. If the sample is illuminated by a white light source, the emulsion appears beautifully colored in the back-scattering direction. These colors originate from Bragg diffraction and provide a straightforward measure of the spacing between droplets within the chains. For perfectly aligned particles with a separation d, illuminated by incident white light parallel to the chains, the first-order Bragg condition reduces to:

$$2d = \lambda_0/n \tag{2.3}$$

where n is the refractive index of the suspending medium ($n = 1.33$ for water), and λ_0 is the wavelength of the light Bragg-scattered at an angle of 180° with respect to the incident beam direction. The wavelength of the Bragg peak provides a direct measure of the spacing between the drops, through Eq. (2.3). Because the drops are monodisperse and negligibly deformable owing to their large capillary pressure ($\sim$1 atm), it is possible to determine the interfacial separation, $h = d - 2a$, where a is the droplet radius. The average interfacial separation is resolved with a precision of about 1.5 nm.

The repulsive force, F_r, between the droplets must exactly balance the attractive force, F_m, between the dipoles induced by the applied magnetic field. Since the dipoles are aligned parallel to the field, this force can be calculated exactly and is given by [16]:

$$F_r(d) = F_m(d) = -\frac{1.202}{2\pi\,\mu_0} * \frac{3\,m^2}{d^4} \tag{2.4}$$

where μ_0 is the magnetic permeability of free space and m is the induced magnetic moment of each drop. The induced magnetic moment must be determined self-consistently from the susceptibility of the ferrofluid, and the presence of the neighboring droplets. Thus,

$$m = \mu_0 \frac{4}{3}\pi\, a^3\, \chi_s\, H_T \tag{2.5}$$

where H_T is the total magnetic field acting on each drop, and χ_s is the susceptibility of a spherical droplet. The total applied field, H_T, is given by the sum of the external applied field and the field from the induced magnetic moments in all the neighboring drops in the chain. This can easily be calculated for an infinite chain, assuming point dipoles, giving:

$$H_T = H_{ext} + 2^{*}1.202^{*}\frac{2m}{4\pi\,\mu_0\, d^3} \tag{2.6}$$

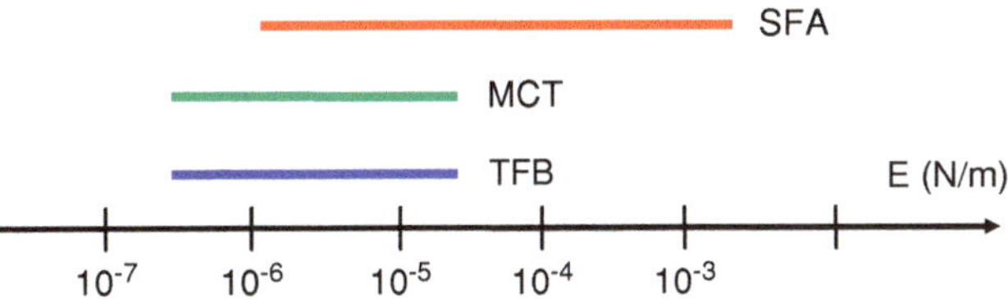

FIGURE 2.4. Comparison of the free energy levels that are accessible using SFA, TFB, and MCT techniques.

This technique allows measuring interparticle forces as small as 10^{-13} N, corresponding to the minimum force required for forming chains. One advantage of this particular experiment is the "built-in" averaging over extremely large number of emulsion films.

2.2.1.4. Comparison of the Different Techniques—Free Energy per Unit Area

It is instructive to compare the data emanating from different force measurement techniques. This requires a conversion of the disjoining pressure in energy per unit area. By integration over the thickness of the disjoining pressure, one obtains the corresponding energy per unit area, $E(h)$, between two infinite planes:

$$E(h) = \int_h^\infty \Pi(h)\,dh \tag{2.7}$$

The so-called Derjaguin equation relates in a general way the force $F(h)$ between curved surfaces to the interaction energy per unit area $E(h)$, provided the radius of curvature R is larger than the range of the interactions [17]. Adopting the Derjaguin approximation, one obtains:

$$E(h) = F(h)/\pi a \tag{2.8}$$

In Eq. (2.8), a is the radius of the spheres (MCT). For crossed cylinders with radius of curvature R (SFA), one obtains:

$$E(h) = F(h)/2\pi R \tag{2.9}$$

In Fig. 2.4, we compare the free energy levels that are accessible using SFA, TFB, and MCT techniques. For TFB and MCT, the accessible range is narrower but these two techniques offer the advantage to reach lower levels of free energy per unit area compared to SFA.

2.2.2. *Recent Advances*

2.2.2.1. Surfactant-Covered Interfaces

By using MCT, Leal-Calderon et al. [10] measured the total repulsive force between tiny colloidal droplets stabilized with sodium dodecyl sulfate (SDS) (Fig. 2.5). The measurements were performed for emulsions with three different concentrations

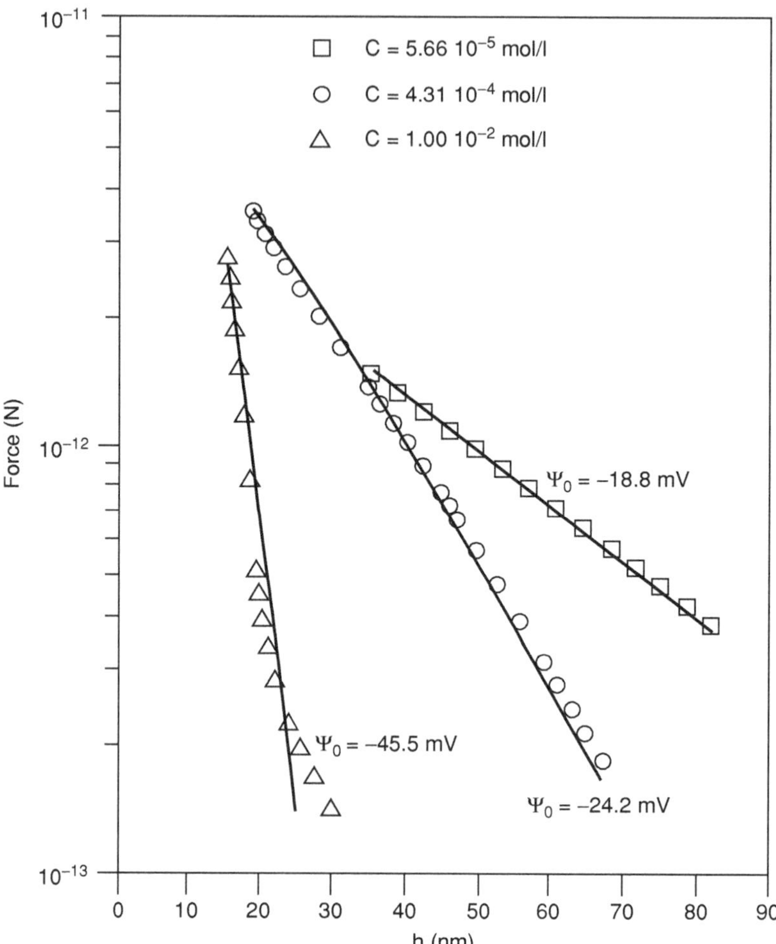

FIGURE 2.5. Evolution of the force with the spacing h for three surfactant (SDS) concentrations. Points correspond to experimental values and solid lines to theoretical predictions. Droplet radius = 94 nm. (Adapted from [10].)

of SDS in the continuous phase, ranging from 6 10^{-5} mol/l to10^{-2} mol/l. As can be seen, the range of the repulsive interaction clearly increases as the SDS concentration decreases. An electrostatic repulsion arises from the presence of anionic SDS molecules adsorbed on the droplets. However, it is not a simple Coulomb repulsion, since, to conserve electroneutrality, the particle charge is surrounded by a diffuse ion atmosphere forming an electric double-layer with the surface charge. For the case of particles having small charge densities, the force, F_e, between two particles is best represented by [18]:

$$F_e(h) = 4\pi \varepsilon \, \psi_0^2 \, a^2 \left[\frac{\kappa}{h + 2a} + \frac{1}{(h + 2a)^2} \right] \exp(-\kappa h) \tag{2.10}$$

In this expression, ε is the dielectric permittivity of the suspending medium, ψ_0 is the electric surface potential, and κ is the inverse Debye length [17], defined as:

$$\kappa^{-1} = (8\pi L_b I)^{-1/2} \tag{2.11}$$

where

$$I = 1/2 \sum_i Z_i^2 C_i \tag{2.12}$$

is the ionic strength of the solution and L_b is the Bjerrum length ($L_b = 0.708$ nm at $T = 25°$C in water). In Eq. (2.12), the summation involves all ionic species with charge number Z_i, at bulk concentration C_i. For a 1:1 electrolyte, Eq. (2.11) reduces to:

$$\kappa^{-1} = (8\pi L_b C_S)^{-1/2} \tag{2.13}$$

where C_s is the electrolyte concentration in the continuous phase which in this case is equal to the surfactant concentration. Equation (2.10) is valid for extended double layers where $\kappa a < 5$. If the particles have a thin double-layer, $\kappa a > 5$, the alternate approximation [19] provides the appropriate interaction force with the assumption that ψ_0 remains constant and independent of h:

$$F_e(h) = 2\pi\varepsilon\, \psi_0^2\, a\, \kappa \frac{\exp(-\kappa h)}{1 + \exp(-\kappa h)} \tag{2.14}$$

Because the inverse Debye length is calculated from the ionic surfactant concentration of the continuous phase, the only unknown parameter is the surface potential ψ_0; this can be obtained from a fit of these expressions to the experimental data. The theoretical values of $F_e(h)$ are shown by the continuous curves in Fig. 2.5, for the three surfactant concentrations. The agreement between theory and experiment is spectacular, and as expected, the surface potential increases with the bulk surfactant concentration as a result of the adsorption equilibrium. Consequently, a higher surfactant concentration induces a larger repulsion, but is also characterized by a shorter range due to the decrease of the Debye screening length.

The repulsive electrostatic force may coexist with other types of interactions and specially the so-called depletion force. This attractive force arises when small particles are present within the continuous phase of the emulsions, for example, surfactant aggregates. Indeed, surfactant form small (nanometer sized) and approximately spherical aggregates called micelles when present at a concentration above the critical micellar concentration (CMC). Emulsion droplets mixed with polymer coils also exhibit this kind of attraction. The depletion attraction has an entropic origin: if two large oil droplets approach one another, micelles or polymer coils are excluded from the region in between, leading to an uncompensated osmotic pressure within the depleted region. Therefore, the so-called depletion interaction scales with the osmotic pressure P_{osm} of the micelles or polymer coils and also with the depleted volume in between the two large oil droplets (Fig. 2.6). The simplest description of the depletion interaction consists in ascribing a characteristic separation at which the small particles are excluded [20]. This simple model is in agreement with direct force measurements performed with the surface

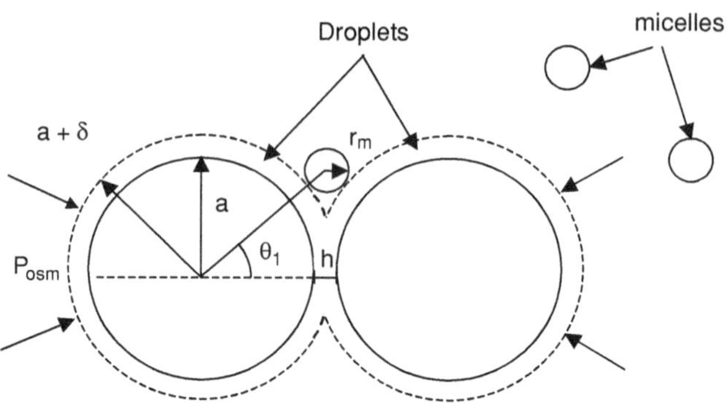

FIGURE 2.6. Schematic representation of the depletion mechanism. Because micelles are excluded from the gap, a depletion force takes place and is calculated by integrating the uncompensated osmotic pressure over the accessible surface.

force apparatus (SFA), when depletion interactions are induced by charged surfactant micelles [21]. Indeed, small surfactant micelles are good candidates to test this simple limit: they may be considered as non deformable objects and the repulsion that eventually arises from their charge may be simply accounted for by considering an effective exclusion diameter.

In the case of charged droplets mixed with identically charged micelles, the net interaction between the emulsion droplets is certainly a combination of both double-layer repulsion and some depletion mechanism. By using MCT, Mondain-Monval et al. have measured the repulsive force–distance profiles between emulsion droplets stabilized by a cationic surfactant (cethyltrimethylammonium bromide [CTAB]) in the presence of the same surfactant micelles [22]. Figure 2.7 shows the force–distance profiles for four different surfactant concentrations $C_s = 1, 5$, 10, and 20 CMC (CMC $= 9\ 10^{-4}$ mol/l). For the lowest surfactant concentration (C_s = CMC), for which a normal electrostatic repulsion is expected, the force profile can be accounted for by Eq. (2.14). The profile is linear in a semilog plot and the slope obtained is in perfect agreement with the experimentally controlled Debye length given by Eq. (2.13). At C_s larger than CMC, the behavior is no longer linear; instead, it exhibits a larger slope at larger separation, which suggests that micelles contribute to an attractive force that becomes comparatively more pronounced at large distance. Using the surface force apparatus, Richetti et al. have measured a highly repulsive regime at short separation which was shown to follow the classical double-layer theory [21]. However, the screening length was larger than that expected from the total amount of charged surfactant. The Debye length was empirically deducible from the amount of free ions only and did not include the presence of charged micelles:

$$\kappa^{-1} = \{4\pi L_b [2CMC + (C_S - CMC)Q)]\}^{-1/2} \tag{2.15}$$

where Q is the fraction of dissociated CTAB molecules in a micelle ($Q \approx 25\%$). The deviation at large separation was attributed to the depletion force, which was superimposed to the electrostatic repulsion. However, in the net repulsive regime,

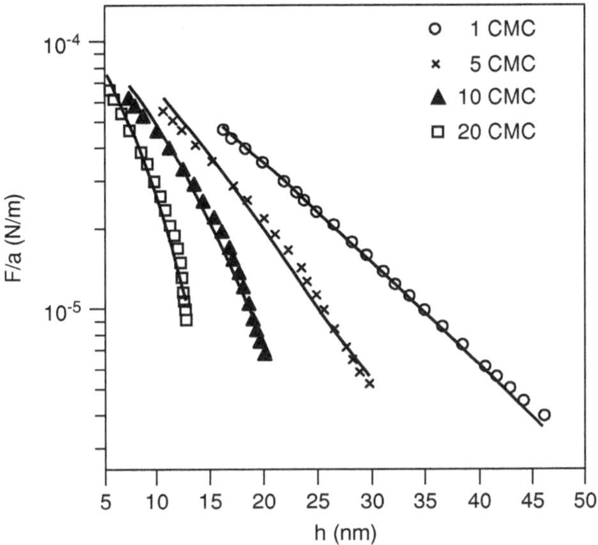

FIGURE 2.7. Force–distance profiles at different CTAB surfactant concentrations. Droplet radius = 98 nm. The continuous lines are the best fits obtained with Eqs. (2.14), (2.15) (for double-layer repulsion), and (2.17) (for depletion attraction). (Adapted from [22].)

the double-layer repulsion was always largely predominant with respect to the depletion force, which precluded a reliable observation of their interplay.

To account for their data (Fig. 2.7), Mondain-Monval et al. hypothesized that these two forces simply add and that the repulsion between micelles and droplets increases the effective diameter of the droplets (or micelles) [22]. This force is derived by integrating the osmotic pressure P_{osm} over the accessible zone for micelles of diameter $2r_m$ ($r_m = 2.35$ nm) from $\theta = \pi$ to $\theta = \pi - \theta_1$, with θ_1 defined in Fig. 2.6. The distance at which the small micelles are excluded from the gap between the droplets is evidently influenced by the electrostatic micelle–droplet repulsion. To account for this repulsion, droplets (or micelles) may be considered as particles of effective radius $(a + \delta)$ [or micelles of radius $(r_m + \delta)$]. From

$$F_d(h) = -2P_{osm}\pi(a + r_m + \delta)^2 \int_{\pi}^{\pi-\theta_1} \sin\theta \cos\theta d\theta \tag{2.16}$$

one gets:

$$\begin{aligned} F_d(h) &= -P_{osm}\pi(a + r_m + \delta)^2 \left[1 - \left(\frac{a + h/2}{a + r_m + \delta}\right)^2\right] && \text{for } 0 \le h \le 2(r_m + \delta) \\ F_d(h) &= 0 && \text{for } h > 2(r_m + \delta) \end{aligned} \tag{2.17}$$

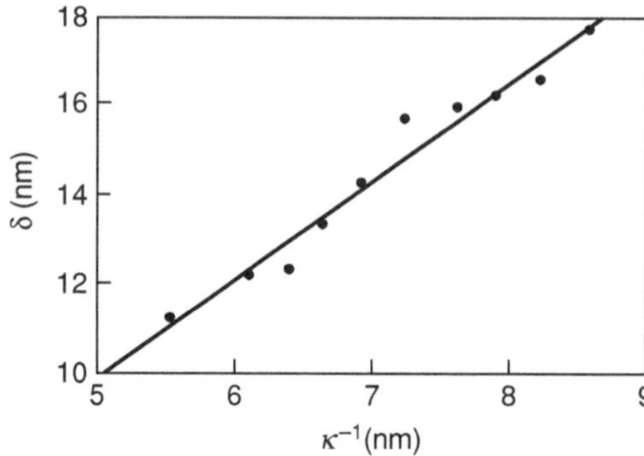

FIGURE 2.8. Evolution of the parameter δ deduced from the best fits of the data as a function of κ^{-1} deduced from Eq. (2.15). (Adapted from [22].)

The osmotic pressure is given by $P_{osm} = n_m k_B T$ (in the perfect gas approximation), $k_B T$ being the thermal energy, and n_m being the micelle concentration related to $C_s[n_m = (C_s - \text{CMC})N_a/N$, where N is the aggregation number equal to 90, and N_a the Avogadro number]. Figure 2.7 provides a comparison of the data and the theoretical curve (continuous lines) for the three surfactant concentrations above CMC. Clearly, the net repulsive regime is properly described by the sum of a screened electrostatic force where κ^{-1} is set by free ions only (Eq. (2.15)) and a depletion force that includes the role of the micelles' free zone around droplets (extra thickness δ).

On the basis of this description, a relationship between the two lengths δ and κ^{-1} can be established. Different δ values are obtained by gradually increasing the amount of micelles and fitting the force profiles. The evolution of δ as a function of the calculated Debye length κ^{-1} is plotted in Fig. 2.8. The thickness δ increases linearly with κ^{-1}. The inherent coupling between depletion and double-layer forces is reflected by this empirical linear relationship which is a consequence of the electrostatic repulsion between droplets and micelles. The thickness δ may be conceptually defined as a distance of closer approach between droplets and micelles and thus may be empirically obtained by writing:

$$B = A \exp(-\kappa\delta) \tag{2.18}$$

where B is a threshold energy of the order of the thermal energy, $k_B T$, and A is a constant depending on surface potentials of droplets and micelles. From this assumption, it can be deduced that $\kappa\delta$ scales as $\ln[B/A]$ which may be considered as a constant. Hence, κ^{-1} and δ should be linearly related as observed experimentally. The slope is a nonuniversal quantity that depends on the surface potentials and respective diameters.

In the presence of large amount of micelles, the total force between surfaces may oscillate due to the occurrence of oscillatory structural forces. Structural forces are a consequence of variations in the density of packing of small particles around a surface on the approach of a second one (stratification). Stratification of particles such as micelles in thin liquid films explains, for example, the stepwise

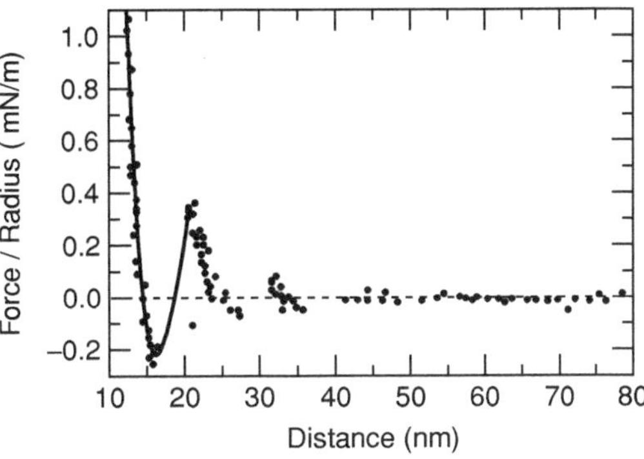

FIGURE 2.9. Measured force F (normalized by the mean radius of curvature R of the surfaces) as a function of the surface separation between crossed mica cylinders coated with an adsorbed bilayer of CTAB and immersed in a micellar solution of CTAB (volume fraction of 0.073). In addition to the depletion attractive minimum, two oscillations due to structural forces turn up. (Reproduced from [21], with permission.)

thinning occurring in large soap and emulsion films during water drainage [23]. Theoretical models [24] predict that at high particle concentration, the structural forces have an oscillatory profile: the force varies between attraction and repulsion with a periodicity close to the mean diameter of the small particles. At low particle concentration, the force between the surfaces becomes monotically attractive and structural forces transform into the depletion attraction. As the particle volume fraction increases, the depletion attraction is still present at short separations and the force begins to oscillate at larger separations, the amplitude of the oscillations increasing with the particle volume fraction. Experimentally, oscillatory structural forces between mica surfaces in the presence of CTAB micelles have been measured by Richetti and Kékicheff [21] using the SFA technique. As predicted, at high enough micellar concentration, they observe the addition of an oscillatory potential to a depletion minimum (Fig. 2.9). Structural forces have also been measured on foam [25] and emulsion films [12] using the TFB technique. One important difference between the SFA and the TFB measurements is the magnitude of the oscillatory forces: in foam and emulsion films, the magnitudes of the oscillations are significantly lower than in SFA measurements [25]. This difference probably originates from the physical nature of the interfaces that confine micelles. In SFA measurements, micelles are confined between two perfectly smooth solid surfaces while fluid interface are deformable and may experience thermal surface fluctuations that reduce the ordering responsible for the oscillatory structural forces.

2.2.2.2. Polymer-Covered Interfaces

Double-layer forces are commonly used to induce repulsive interactions in colloidal systems. However, the range of electrostatic forces is strongly reduced by increasing the ionic strength of the continuous phase. Also, electrostatic effects are strong only in polar solvents, which is a severe restriction. An alternative way to create long-range repulsion is to adsorb macromolecules at the interface between the dispersed and the continuous phase. Polymer chains may be densely adsorbed on surfaces where they form loops and tails with a very broad distribution of sizes and extending in the continuous phase [26]. The repulsive forces between adsorbed polymer layers are essentially due to steric effects between the two layers when they overlap.

i. Dilute Regime

The structure, the thickness, and the interactions generated by the presence of the adsorbed polymer layers have been extensively studied [17]. In particular, the force between two polymer-covered mica sheets in various solvency conditions has been probed via the SFA technique [27,28]. The force is purely repulsive in a good solvent and becomes attractive as the solvent gets poorer. However, these studies concern only a regime of large interaction compared to the thermal energy and are restricted to interactions between solid surfaces.

Following the pioneering experiments of Lyklema and van Vliet [29], Mondain-Monval et al. [30,31] have measured the repulsive forces between polymer-covered liquid interfaces in the low interaction regime (force/radius $\leq 10^{-4}$ N/m). They used two different force measurement techniques, i.e., MCT and TFB techniques. All the experiments were performed in the dilute regime ($\phi_b \ll \phi_b^*$, where ϕ_b^* is the dilute to semidilute bulk polymer volume fraction). The force–distance profiles between the ferrofluid droplets and the air–water films are displayed for three molecular weights for PVA–Vac polymer (statistical copolymer of vinyl alcohol [88%] and vinyl acetate [12%]) on Fig. 2.10a. The disjoining pressure Π measured via the TFB technique was transformed into a force through the Derjaguin approximation [17] (Eqs. (2.7) and (2.8)). The two profiles are qualitatively similar and show a linear decay with the same slope on a semilogarithmic scale. However, the distances corresponding to equal forces are very different in the two experiments (they are larger in the air–liquid films). The force as a function of the distance may be written as:

$$F_s(h) = \alpha\, h \exp(-h/\lambda) \tag{2.19}$$

where λ is a characteristic decay length and α is a constant prefactor. These exponentially decaying profiles are insensitive to the presence of electrolytes and thus cannot be attributed to an electrostatic repulsion that could exist if, for example, some parasitic charges were present at the interface. In both techniques, the thickness λ is of the order of the radius of gyration of the chains in solution despite the fact that the interaction that drives the adsorption can be very different from one interface to the other. Using a different polymer, it can be shown that this exponential behavior is not specific to PVA–Vac. A second set of experiments has been performed with a weak polyelectrolyte (polyacrylic acid [PAA]) in solutions of high ionic strength and low pH (Fig. 2.10b). The presence of salt ensures that any long-range repulsion is not due to electrostatics. Here also, exponentially decaying profiles were measured with characteristic distances varying with molecular weight. The polymer radius of gyration R_g may be varied by increasing the temperature from 20°C to 80°C, i.e., in a regime in which the

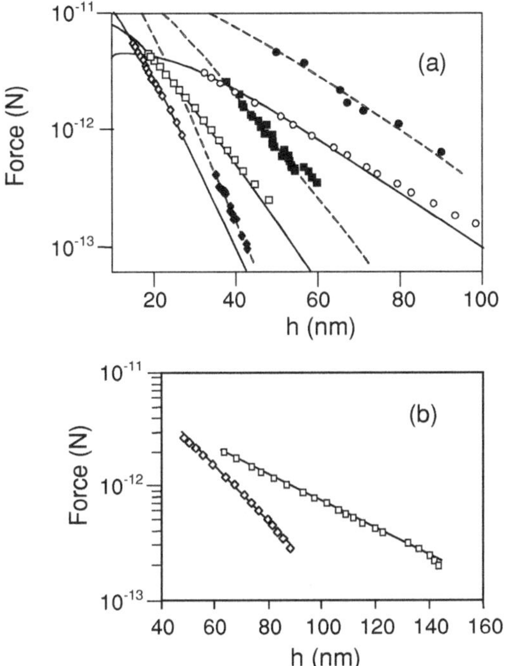

FIGURE 2.10. **(a)** Influence of the polymer molecular weight on the force-distance profiles between air-water (filled symbols) and oil-water interfaces (empty symbols). Polymer concentration = 0.5 wt%. Average molecular masses of the polymers: oil–water interface: (◇) M_w = 10000 g/mol; (□) M_w = 55000 g/mol; (○) M_w = 155000 g/mol. air–water interface: (◆) M_w = 10000 g/mol; (■) M_w = 55000 g/mol; (●) M_w = 155000 g/mol. **(b)** Forces in PAA solutions at the oil-water interface. Polymer concentration = 0.1% wt%. Experiments were performed in the presence of NaCl (0.2 mol/l) at pH = 3. Average molecular masses of the polymers: (◇) M_w = 100000 g/mol; (□) M_w = 320000 g/mol. In both figures, the lines are the best fits to the data using Eq. (2.19) in the text. (Adapted from [31].)

radius of gyration is significantly reduced but still in good solvent conditions. The profiles always remain exponential but the range decreases significantly. Figure 2.11 shows the experimental evolution of the adsorbed layer thickness λ as a function of the radius of gyration: the variation is reasonably linear.

Using both a mean field and a scaling approach, Semenov et al. [32] compared these experimental results with the theoretical predictions. The theory distinguishes the loop and tail sections of the adsorbed chains and involves three length scales: the adsorbed layer thickness λ, an adsorption length z^* that separates the regions where the monomer concentration is dominated by loops and by tails, and a microscopic length b inversely proportional to the adsorption strength. Two regimes must be distinguished depending on the strength of the adsorption measured by the ratio λ/b.

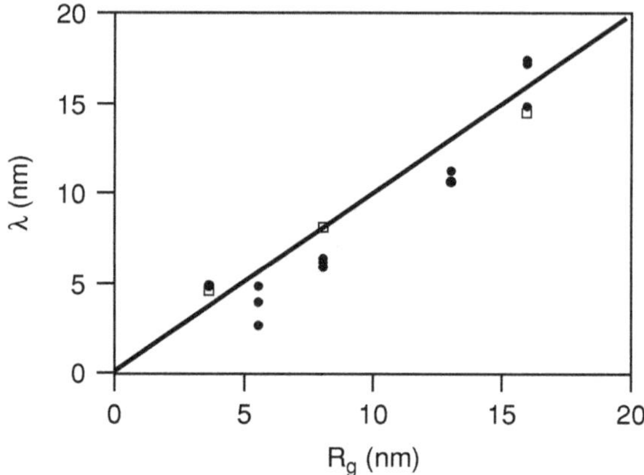

FIGURE 2.11. Evolution of the experimental characteristic distance λ (deduced from the best fit to the data using Eq. (2.19)) as a function of the polymer coil hydrodynamic radius R_g (deduced from viscosimetric measurements) for the PVA–Vac polymer. (□) Air–water interface; (●) Oil–water interface. (Adapted from [31].)

In the strong adsorption limit ($\lambda/b \gg 1$) the expression of the adsorbed layer thickness λ, corresponding to the size of the largest loops or tails in the layer, reads in the mean field theory:

$$\lambda = R_g/(\ln(1/\phi_b\, vb^2))^{1/2} \tag{2.20}$$

where v is the Flory excluded volume parameter and ϕ_b is the bulk polymer volume fraction. The adsorbed layer thickness is thus proportional to the chain radius of gyration and varies only weakly with the polymer concentration ϕ_b and the adsorption strength $1/b$. The scaling theory in a good solvent leads to similar conclusions.

If the distance between the two surfaces is smaller than λ, the polymer-mediated interaction decays as a power law (h^{-4} in the mean field theory and h^{-3} in the scaling theory). The sign of the force depends, however, on the reversibility of the adsorption and the force is repulsive at short distances only if the adsorption is irreversible. At distances larger than λ, the concentration is dominated by the tails and the force is always repulsive and decays exponentially with the distance. Using the Derjaguin approximation, one gets, in the scaling theory which is more appropriate to describe polymers in good solvents, the force between spherical droplets of radius a:

$$F_s(h) = \left(k_B T \pi a/\lambda^3\right) h \exp\left(-h/\lambda\right) \tag{2.21}$$

This expression is valid only if λ is large enough (essentially larger than z^*), i.e., for a polymer adsorbed amount close to the saturation value.

The weak adsorption limit ($\lambda/b \ll 1$) has been studied less extensively. The only relevant length scale in this limit is the chain radius of gyration and the adsorbed

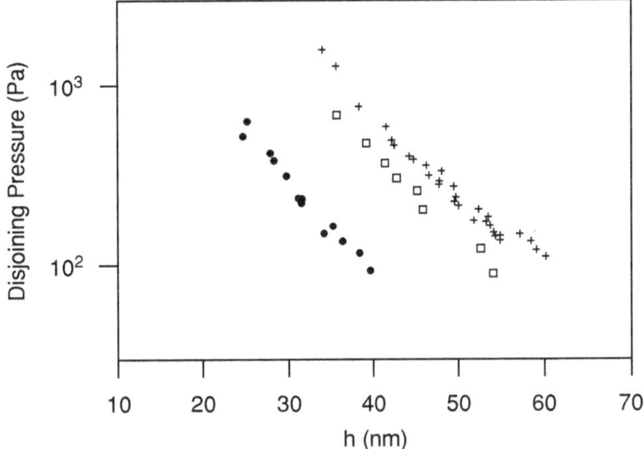

FIGURE 2.12. Influence of a nonionic surfactant concentration (NP10 of CMC = 7 10^{-5} mol/l) on the pressure isotherms (air–water interface). PVA–Vac concentration = 0.5 wt%. (+) [NP10] = 0; (□) [NP10] = CMC/10; (●) [NP10] = CMC. (Adapted from [30].)

layer thickness is proportional to the radius of gyration. The force is proportional to the number of chains adsorbed at the interface, or to the polymer-adsorbed amount Γ. One expects that in the crossover range between the weak and strong adsorption regimes the force increases (in an unknown way) with the polymer adsorbed amount.

The model in the strong adsorption limit is in qualitative agreement with the data of Mondain-Monval et al. For the PVA–Vac polymer, an estimation of the theoretical prefactor $(k_B T \pi a/\lambda^3)$ using λ deduced from the experimental slopes (semilogarithmic plots) leads to a value between 10^{-11} and 4.10^{-13} N/m. From the fit to the data, the value typically obtained for the prefactor α in Eq. (2.19) lies in the same range. The dependence of α on λ could not be observed owing to a lack of precision. To test the variation of the force with Γ, increasing quantities of nonyl phenol oxyethylene (NP10) were added to the polymer solution. This nonionic surfactant is known to adsorb preferentially at the interface and to displace the polymer [33]. The evolution of the disjoining pressure–distance profiles with increasing NP10 concentrations are plotted in Fig. 2.12. The characteristic distance λ remains unchanged while the prefactor decreases as the adsorbed polymer amount is reduced. A similar behavior is observed at the oil–water interface. This variation of the force–distance profile with the adsorbed amount provides an explanation for the different equilibrium distances at identical repulsive forces observed in the two experiments. These observations are in disagreement with the predictions of the theory (for $\lambda/b \gg 1$) in which both λ and the preexponential factor vary only weakly with the adsorption strength that controls the adsorbed amount. A reasonable reason could be that the polymer desorption that is driven by NP10 concentration causes a change of regime from the large (for the lowest NP10 concentrations) to the weak adsorption limit (at high NP10 concentrations).

Omarjee et al. [34] have measured and compared the forces between different polymers adsorbed at different surfaces, as determined by three different methods involving liquid–liquid (MCT), liquid–air (TFB), and solid–solid (SFA) interfaces. In the distal regime (weak interactions at the onset of overlap), the forces are exponentially decaying with characteristic distances depending linearly on the polymer chain gyration radius. Such qualitative behavior is very general and is not expected to depend on the type of interfaces or details of the polymer–good solvent system. In contrast, the preexponential factors or absolute magnitudes are strongly system-dependent. The differences in the prefactor are attributable to the different adsorption abilities of the polymers at the different interfaces. A higher absorbance leads to higher segment concentration in the overlapping regions and therefore stronger repulsion. The mobility of the adsorbed polymer on the highly curved fluid emulsion surfaces may also result in some reduction of the absolute repulsion when the layers are compressed (as chains can respond by moving sideways). All these effects would result in a lower absolute repulsion between the fluid-adsorbed compared to the solid-adsorbed polymers.

If the surfaces are not fully covered by polymers, an attraction can occur even in good solvent. In this case, polymer coils can extend from one surface to the other and make molecular contacts on both sides. One surface then provides the missing polymer to the other, resulting in an attractive interaction usually called bridging. At equilibrium adsorption, polymers are used to stabilize colloids, since the steric repulsions exceed the bridging attractions at all separations. Reducing the adsorption to less than its equilibrium value can therefore change the net repulsion to a net bridging attraction [35,36]. To study the kinetics of polymer bridging, Cohen-Tannoudji et al. [37] have developed a new tool based on the self-assembling properties of superparamagnetic beads under a field. It consists of following the kinetics of formation of permanents links between self-organized magnetic beads with optical microscopy. It has been shown that polymer bridging is activated by temperature but it cannot be understood simply in terms of an Arrhenius model. Indeed, a more precise approach is required that takes into account the necessary removal of adsorbed polymer from the near-contact region and the slowdown of the dynamics in the near-surface layer, owing to the proximity of a surface glassy state. This difficulty to induce bridging at equilibrium adsorbance explains the efficiency of adsorbing polymers at the surface of colloids to improve their kinetic stability.

The magnetic beads were spherical, monodisperse, strongly magnetic, and Brownian. They consisted of calibrated emulsion droplets of diameter 800 nm of an organic ferrofluid in water. The method to follow the kinetics of bridging is based on the formation of magnetic chains that persist after removal of the field. A colloidal sample was introduced by capillarity into a square tube of 50 μm (Fig. 2.13). The tube was submitted to a given field, at a controlled temperature, for a given incubation time. After the field was removed, the sample was observed via optical microscopy and photographs of different parts of the tube were taken.

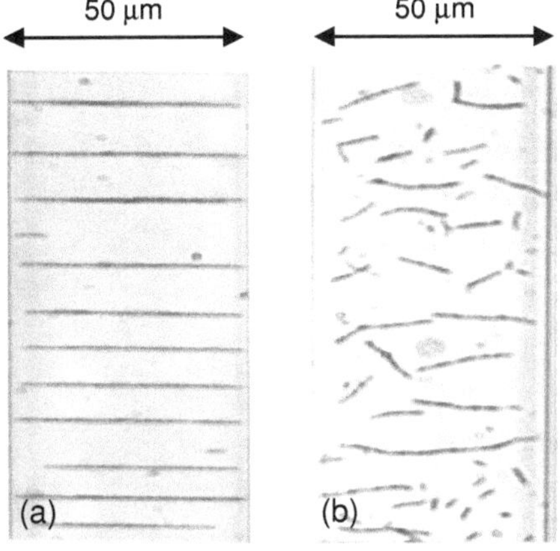

FIGURE 2.13. Superparamagnetic beads observed under an optical microscope (×40 objective): **(a)** under magnetic field; **(b)** after field removal. (Reproduced from [37], with permission.)

For each photograph, the sum of the lengths of all the chains was determined numerically. The ratio of this total length by the diameter of a bead provided the number N of adhesive links and was averaged over the photographs. This was repeated for different incubation times, which gave the kinetics of the bridging process. The polymer used was PAA of average molecular weight 250000 g/mol. It had a pK_a of 5.8, so that in the conditions of the experiments, at pH = 3.5, it was essentially neutral. To ensure that equilibrium adsorption was attained, the beads were incubated for 20 h in the PAA solution prior to application of any field [30]. When subjected to a magnetic field for a sufficient time, the beads covered by PAA formed permanent chains. Chain formation was attributed to a net attraction between the beads arising from bridging by the polymers. This net attraction can arise only through reduction of the adsorption to subequilibrium levels by removal of polymer from the region of closest approach. Polymers may be squeezed out of the gap by the lateral force $F_{sq}(r)$ on each polymer chain arising from the lateral pressure gradient $(\partial P/\partial r)$ acting between the curved surfaces, which results from the decrease in osmotic pressure $P(r)$ at a distance r away from the point of closest approach h. $P(r)$ can be written as [26]:

$$P(r) \approx \left(k_B T/l^3\right) \phi_s^2, \tag{2.22}$$

where l is a monomer size and ϕ_s is the monomer volume fraction:

$$\phi_s = 2\Gamma \bigg/ \left(h + \frac{r^2}{a}\right), \tag{2.23}$$

a being the bead radius and Γ the polymer adsorbance. Thus, $F_{sq}(r)$ may be roughly estimated

$$F_{sq}(r) \approx (\text{volume of polymer chain})\,(\partial P/\partial r) \approx \left(n_p^{3/2} k_B T \Gamma^2 r/ah^3\right) \tag{2.24}$$

where n_p is the degree of polymerization ($\approx$ 3500 for the PAA used). For typical values of the parameters $F_{sq}(r) \approx 10^{-11}$ N. The typical net monomer adsorption energy is $\varepsilon \approx 0.02 - 0.1\, k_B T$ [36], requiring a tension of order

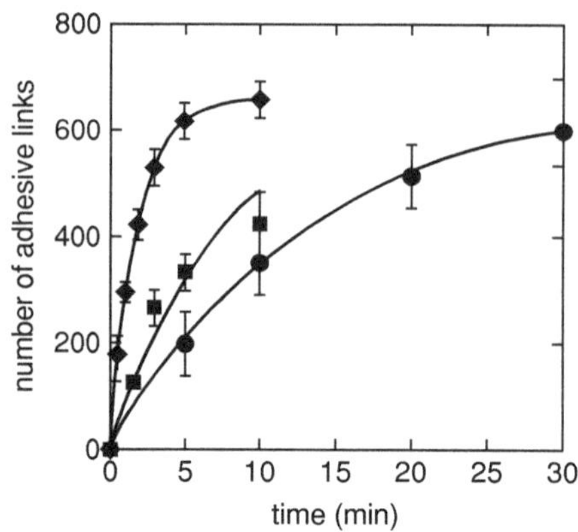

FIGURE 2.14. $N(t)$ under a magnetic field of 12 mT at 35°C (●), 40°C (■), and 45°C (◆); curves are fits to Eq. (2.25). (Reproduced from [37], with permission.)

$(\varepsilon/1\ \text{Å}) \approx (1{-}5)10^{-12}$ N to detach. The outward lateral force on each chain is thus comparable with that required to detach the adsorbed monomers, enabling the creep of chains along the surface and their squeeze-out to take place, and therefore bridging to become dominant.

Figure 2.14 shows the number N of adhesive links as a function of the incubation time t under a magnetic field of 12 mT, for three different temperatures. $N(t)$ exhibits a monoexponential growth, well modeled by a first-order kinetics:

$$N(t) = N_0\left[1 - \exp(-t/\tau)\right], \tag{2.25}$$

where N_0 ($\approx$660) is the averaged total number of particles in the volume captured in a photograph and τ is the characteristic time of the kinetics. The only fitting parameter in Eq. (2.25) is thus the characteristic time τ. The process of polymer creep and relaxation along the surfaces is complex. Because it occurs by the detachment of monomers from the surface, it is likely to be an activated process but will depend also on the dynamics of the near-surface layer. Using an Arrhenius law to fit the evolution of τ with T,

$$\frac{1}{\tau} = \frac{1}{\tau_0}\exp\left(-\frac{E_a}{k_B T}\right), \tag{2.26}$$

leads to an estimated activation energy E_a without field of about 150 $k_B T_r$,T_r being the room temperature (20°C) (fits not shown). The rate at which bridging is attained will depend on the magnetic field through its effect on h, which affects the lateral squeezing force $F_{sq}(\propto 1/h^3)$. The value of 150 $k_B T_r$ may reflect the need for cooperativity in detachment of polymer segments. Surprisingly, attempt frequencies $1/\tau_0$ of order 10^{30} Hz were found, which is much too high for a molecular phenomenon and cannot have any physical meaning. Evidently, the Arrhenius law is thus unable to model the activation of polymer bridging, and another approach has to be considered that takes into account the existence of a glass transition in the system [38–42]. The relevance of this relies on the approach to a glassy state of the near-surface layers [43]. The so-called

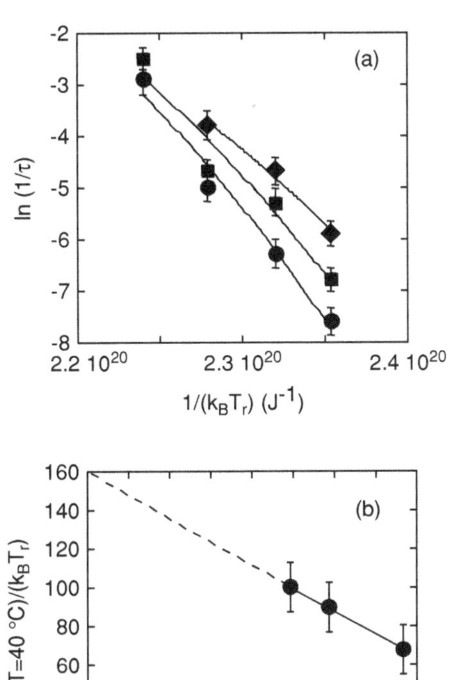

FIGURE 2.15. **(a)** $\ln(1/\tau)$ as a function of $1/(k_B T_r)$ under a magnetic field of 10 mT (♦), 12 mT (■), and 15 mT (●); lines are fits to Eq. (2.27) with $T_0 = -30°$C; (b) $E_a(T = 40°\ C)/(k_B T_r)$ as a function of the magnetic field B; the dotted line is a linear extrapolation to $B = 0$. (Reproduced from [37], with permission.)

Vogel–Tamman–Fulcher (VTF) model reads:

$$\frac{1}{\tau} = \frac{1}{\tau_0} \exp\left[-\frac{A}{k_B(T - T_0)}\right], \tag{2.27}$$

where $1/\tau_0$ is still an attempt frequency, A is an energetic parameter, T_0 is the empirical Vogel temperature. T_0 is the temperature at which all the relaxation times of the system diverge [42] and is a few tens of degrees below the glass transition temperature [41]. It is still possible to define an apparent activation energy, which depends on temperature [42], and which is simply the slope of the $\ln(\tau)$ vs. $1/k_B T$ curve:

$$E_a(T) = k_B \frac{\partial\,[Ln(\tau)]}{\partial\,(1/T)} \tag{2.28}$$

Figure 2.15a shows a good fit of the experimental points to the VTF model for three field intensities. T_0 is an increasing function of the polymer concentration [41], and it is known that the local concentration of polymer near the surface on which it is adsorbed is much higher than the bulk concentration [44]. As the polymer concentration near the surface was unknown, it was impossible to have any precision on the value of T_0.$T_0 \approx -120°$C for dilute solutions of PAA in water (less than 1 wt%) [45] and $T_0 \approx 0°$C for 75% wt% PAA in water [46]. The adsorbed polymers involved in the bridging should thus have a Vogel temperature ranging roughly between $-70°$C and 0°C. Since the choice of T_0 had almost no influence on the apparent activation energy deduced from the fit, T_0 was arbitrarily fixed at $-30°$C in Fig. 2.15. A value of 160 $k_B T_r$ for $E_a(40°\text{C})$ was obtained at zero field (Fig. 2.15b), which compares very well with the value of 150 $k_B T_r$ found with the Arrhenius law. In contrast, the frequency $1/\tau_0$ determined by the VTF fit is very sensitive to T_0. It ranges from 100 Hz for $T_0 = 0°$C to 10^{10} Hz for $T_0 = -70°$C, which is physically reasonable. This proves that the VTF model for the thermal activation of polymer bridging is a much better description than the Arrhenius law.

As mentioned earlier, the squeezing away of the polymer is due to the osmotic outward lateral force $F_{sq}(r)$, which is an increasing function of the degree n_p of polymerization of the considered polymer. Experimentally, this implies that the bridging phenomenon should be faster when the adsorbed polymer is longer. Some experiments were carried out at 45°C under a magnetic field of 11 mT with three PAA of different average n_p. For n_p of about 28, 1400, and 3500, the characteristic times τ were equal to 1800, 74, and 54 s, respectively. This confirms the role of the squeezing out of polymer in the bridging process. The interpretation of polymer bridging was also confirmed by performing measurements at 25°C under a magnetic field of 12 mT after different equilibration times of the beads with the polymer. Whereas a characteristic time $\tau \approx 12$ min was found if the adsorption equilibrium was reached (after 20 h), the kinetics is much more rapid ($\tau \approx 0.9$ min) if the PAA had only 10 min to adsorb prior to application of the field. When the time was insufficient for the polymer to be completely adsorbed, bridging dominance required less polymer to be squeezed away.

Magnetically induced bridging by adsorbed polymers has been exploited by Goubault et al. [47] to produce long filaments with an extremely uniform diameter of one particle and lengths of several hundreds of micrometers as can be observed in Fig. 2.16. Such filaments can find numerous applications in chemistry and biology. Self-organized in a microchip, they provide a new promising sewing matrix that can be exploited for rare cells sorting.

ii. Semi-dilute Regime

The stability of films in the presence of polymers (adsorbed or nonadsorbed) in a semidilute regime ($\phi_b > \phi_b^*$) has also been investigated. In addition to the eventual steric interaction (in the case of adsorbing polymers), depletion forces are expected due to the exclusion of the polymer chains. However, polymer molecules in the semidilute regime can no longer be treated as hard spheres and the simple picture given for micelles (total exclusion when surface separation is lower than the micellar effective diameter) is no longer valid. Instead, in the semidilute regime the polymer solution may be regarded as an entangled network of flexible chains with average mesh size ξ [26]. Owing to their flexibility, the polymer molecules confined between two walls will not totally vacate the confined region and there

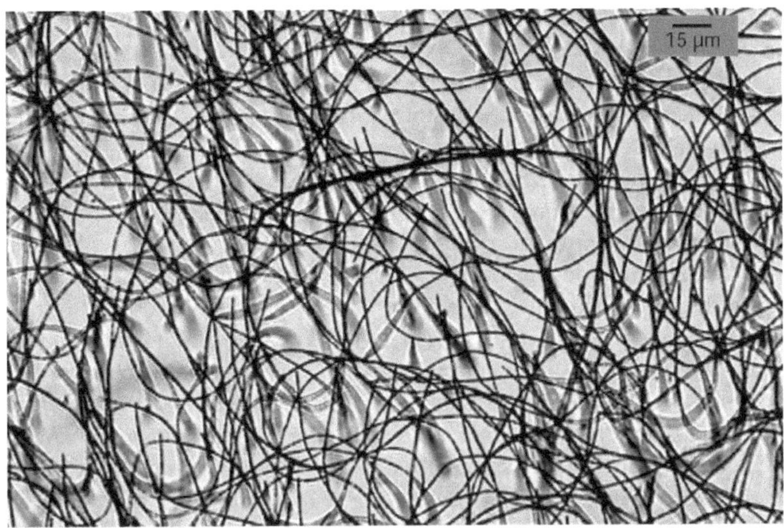

FIGURE 2.16. Long magnetic nanowires made by submitting a mixture of magnetic particles and PAA to a magnetic field. The chains are bent by gravity after field removal. (Reproduced with permission from [47].)

will be residual polymer segment density in the depleted zone. Conformation of regular flexible and nonadsorbing polymers in a good solvent and in a semidilute regime has been calculated by Joanny et al. [48]. For narrowing gaps, the density of polymer segments decreases below some separation and vanishes completely at a critical separation of $\pi\xi$. In the limit of two identical spherical walls of large radius R ($R \gg \pi\xi$), the attractive depletion force resulting from segment exclusion is a linear function of the separation h:

$$F_d \propto -\pi R(\pi\xi - h)k_B T/\xi^3, \quad h \leq \pi\xi \tag{2.29}$$

Because ξ varies with the bulk polymer volume fraction as $\phi_b^{-0.75}$, the range of the depletion interaction should decrease also as $\phi_b^{-0.75}$, while the adhesion $F_d(h = 0)$ should increase as $\phi_b^{1.5}$ when the chain concentration increases. A very different picture is predicted in the case of adsorbing polymers [49]. The layer of adsorbed chains may be partially interpenetrated by free chains in the bulk and therefore the range and strength of the attraction are not determined by the solution concentration. Instead, they are rather sensitive to the coverage and thickness of the adsorbed chains which depend essentially on the solvent quality and on the mean chain length in the dilute regime.

Attempts to measure the depletion force in nonadsorbing polymer medium with an SFA have failed essentially because measurements are hindered by the slow exclusion of the polymer from the narrow gap due to the large viscosity of the polymer solutions. However, depletion forces have been measured in solutions of "living" polymers in a semi-dilute regime by Kékicheff et al. [50]. The measured evolution of the range and strength of the depletion attraction roughly follow the model of Joanny et al. [48]. "Living" polymers are linear and flexible elongated micelles that can break and recombine. Owing to their labile nature, such systems can adapt to release any conformation restriction induced by confinement and so any equilibrium is more rapidly achieved than for regular polymers. The depletion interaction between a sphere and a plane in the presence of nonadsorbing polymers has been measured via AFM [51]. Adhesive energy measurements between lipid bilayer membranes induced by concentrated solutions of nonadsorbing polymer are also reported and show good agreement with mean field theory [52]. By using the TFB technique, Asnacios et al. have studied thin liquid films made from semidilute polyelectrolytes [53]. The TFB method does not allow measuring attraction and therefore no depletion minimum was observed. However, an interesting result is that they have measured oscillatory forces, likely to be structural forces associated with the polymer mesh structure. Indeed, film stratification is observed, with a strata thickness strictly corresponding to the theoretical mesh size ξ of semidilute polyelectrolyte solution. These oscillatory forces are particular to polyelectrolytes and disappear when the electrostatic forces are screened with salt.

2.2.2.3. Interfaces Covered by Polymer–Surfactant Complexes

The ability of polymer–surfactant complexes to alter rheological properties and stability of colloidal formulations has been exploited in oil recovery operations as well as in various industrial products such as paints, detergents, cosmetics, pharmaceuticals, etc. While polyelectrolytes are mainly used for their unique rheological properties, surfactants are usually employed for their ability to adsorb and lower surface energy. Once added to the same solution, polyelectrolytes and charged surfactants can have strong interactions, which affect both bulk and surface properties of the solution. In particular, it is well known that polyelectrolytes and surfactants of opposite charge form hydrophobic complexes at surfactant concentrations lower than the critical micelle concentration, owing to the strong electrostatic interaction between the surfactant head group and the polymer backbone. Bulk interactions and phase behavior [54–56] of polyelectrolytes and surfactants of opposite charge have been studied via several techniques, including turbidimetry [57], fluorescence [56,58], viscometry [59,60], dynamic and static light-scattering [60,61], small-angle neutron scattering [62], small angle X-ray scattering [63], electrophoresis [60], and potentiometric measurements using ion-specific electrodes [64,65]. These techniques have led to estimates of the type of interactions, the aggregation numbers, and the structure of the complexes formed by surfactants and polyelectrolyte molecules. Micelles are bound to the polyelectrolyte chains, their aggregation numbers being in general smaller than that for the free micelles. The concentration of surfactant needed to measure significant aggregation between the polymer and the surfactant is generally called the critical aggregation concentration (CAC).

In addition to the studies in the volume, the adsorption properties of the polyelectrolyte–surfactant complexes at the air–water interface have also been investigated using surface tension measurements [59,66–71] and neutron or X-ray reflectometry [68–71]. Recently, Claesson et al. [72] have introduced the concept of a surface critical aggregation concentration for adsorption onto solid surfaces, to underline the differences seen between bulk and surface behavior. Monteux et al. [73] also show that aggregates can form at the air–water surface at surfactant concentrations different than that seen for solution aggregation. In addition, surface aggregates can form interfacial gels. Surface gelation is identified by a pronounced increase in surface rheological properties which are detected by thin-film drainage studies. Evidence of this phenomenon was obtained by systematic investigation of the air–water surface properties for mixtures of an anionic polyelectrolyte (100% charged sodium polystyrene sulfonate [PSS]) with a cationic surfactant (dodecyltrimethylammonium bromide [DTAB]). By combining surface tension measurements with ellipsometry studies, it was possible to monitor the adsorption at the air–water interface. At the same time, polyelectrolyte–surfactant interactions in the bulk solution were evaluated by measurements of the solution conductivities with ion-specific electrodes.

The formation of complexes is not restricted to mixtures of polyectrolytes and surfactants of opposite charge. Neutral polymers and ionic surfactants can also form bulk and/or surface complexes. Philip et al. [74] have studied the colloidal forces in presence of neutral polymer/ionic surfactant mixtures in the case where both species can adsorb at the interface of oil droplets dispersed in an aqueous phase. The molecules used in their studies are a neutral PVA–Vac copolymer (vinyl alcohol [88%] and vinyl acetate [12%]), with average molecular weight $M_w = 155000$ g/mol, and ionic surfactants such as SDS. The force measurements were performed using MCT. The force profiles were always roughly linear in semilogarithmic scale and were fitted by a simple exponential function:

$$F(h) \propto \exp(-h/\lambda) \tag{2.30}$$

The authors deduce the decay length (λ) from the slope and the first interaction length or onset of repulsion ($2L_0$) defined as the distance at which the magnitude of force is $2\ 10^{-13} N$. One of the main findings of this study is that the force profiles are strongly dependent upon the sequence of adsorption of polymer and surfactant and three cases are envisaged.

Case I (see Fig. 2.17) corresponds to the situation such that the emulsion is initially stabilized with SDS at $8\ 10^{-3}$ mol/l (CMC). The repulsive force as a function of distance between the ferrofluid droplets, stabilized with SDS alone is referred as “0% PVA.” Then, PVA–Vac is introduced at different concentrations varying from 0.002 to 0.5 wt%. After each addition, the emulsion is incubated for 48 h to reach equilibrium. It can be seen that the force profiles remain almost the same as in the case of 0% PVA. As the surfactant concentration is equal to CMC, the expected decay length is 3.4 nm. The experimental value of the decay length obtained from the force profile, 2.9 nm (solid line), is in good agreement with the predicted value. Thus, if the emulsion is preadsorbed with surfactant molecules, the introduction of polymer does not influence the force profile significantly.

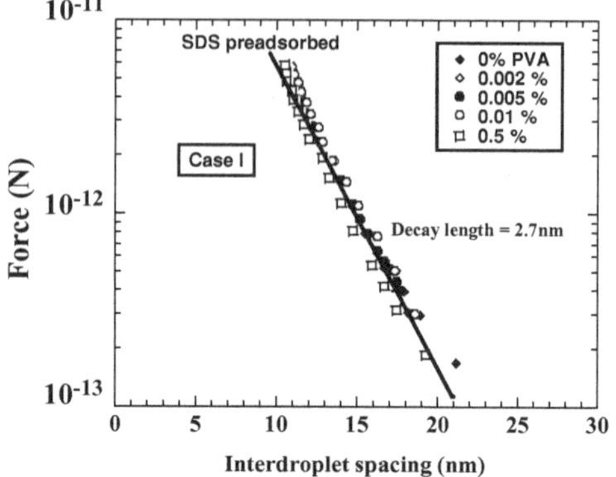

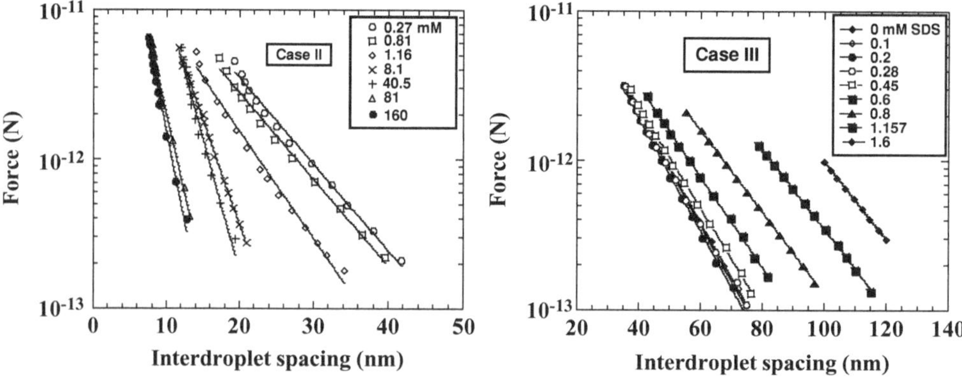

FIGURE 2.17. Forces between the ferrofluid droplets as a function of the interdroplet spacing. The best fits, using Eq. (2.30) are shown by solid lines. Case I: Droplets preadsorbed with sodium dodecyl sulfate (SDS) at 8 10^{-3} mol/l and at various PVA–Vac concentrations. The solid line represents the average value of the best fit. Case II: Droplets preadsorbed with sodium dodecyl sulfate (SDS) at 0.27 10^{-3} mol/l. Premixed PVA–SDS was added to the emulsion. In all the cases, the polymer concentration was 0.6 wt%. The surfactant concentrations are indicated in the inset. Case III: Droplets preadsorbed with PVA–Vac. In all the cases, the polymer concentration was fixed at 0.6 wt%. The surfactant concentrations are indicated in the inset. (Adapted from [75].)

Case II corresponds to an emulsion preadsorbed with surfactant molecules at very low concentration (stabilized with SDS at a concentration of 0.27 10^{-3} mol/l or CMC/30). Polymer and surfactant were premixed separately and later added to the emulsion. In all cases, the polymer concentration was fixed at 0.6 wt%. Premixed polymer/surfactant mixture was incubated sufficiently (>2 h) before adding to the emulsion. The force profiles are again repulsive (Fig. 2.17) and exponentially decaying with a characteristic decay length comparable to the Debye length, corresponding to the equivalent amount of surfactant concentration present in the premixed system.

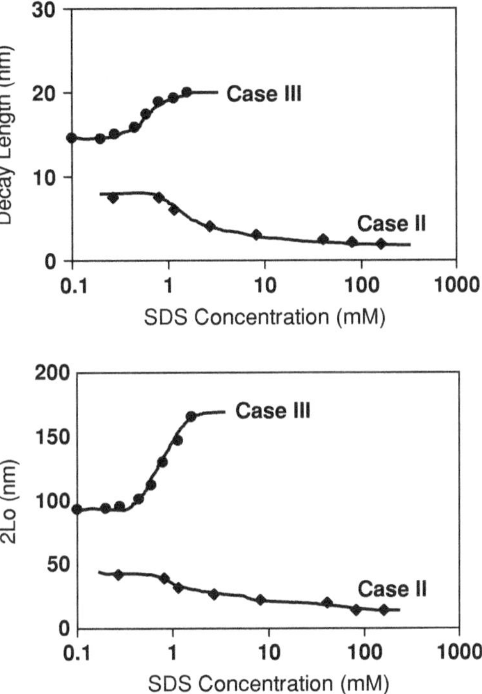

FIGURE 2.18. Decay length and first interaction length ($2L_0$) values deduced from the force curves in Fig. 2.17 (cases II and III), as a function of surfactant concentrations. Solid lines are a visual guide. (Adapted from [75].)

For case III, the polymer is preadsorbed at the emulsion droplet interface. The force profile in the absence of any surfactant is the reference curve (0 mM). The polymer concentration in all the curves was 0.6 wt%, corresponding to the plateau in the adsorption isotherm. The force profiles are repulsive and exponentially decaying with a characteristic decay length comparable to the radius of gyration of the free polymer (16 nm). As the concentration of the surfactant increases, the decay length and the onset of repulsion also increases. Here, the SDS concentration was varied from 0.1 to 1.6 10^{-3} mol/l. The variations of the experimentally obtained decay length and $2L_0$ values for cases II and III are shown in Fig. 2.18. In case II, both decay length and $2L_0$ values decrease with increasing surfactant concentration, while under case III, both values increase with surfactant concentration.

On the basis of the above experimental results, the expected conformations of polymer–surfactant complexes at the oil–water interface are depicted in Fig. 2.19. In case I, the added polymer associates with excess surfactants present in the bulk solution, but the complexes prefer to remain in the bulk phase. Alternately, the polymer–surfactant complexes are unable to displace the adsorbed surfactant molecules from the liquid–liquid interface. Irrespective of the amount of polymer–surfactant concentration in the bulk, the experimental decay length values remain comparable to the Debye lengths, corresponding to the concentration of ion species in the bulk solution (Eq. (2.11)). This means that the force profile is

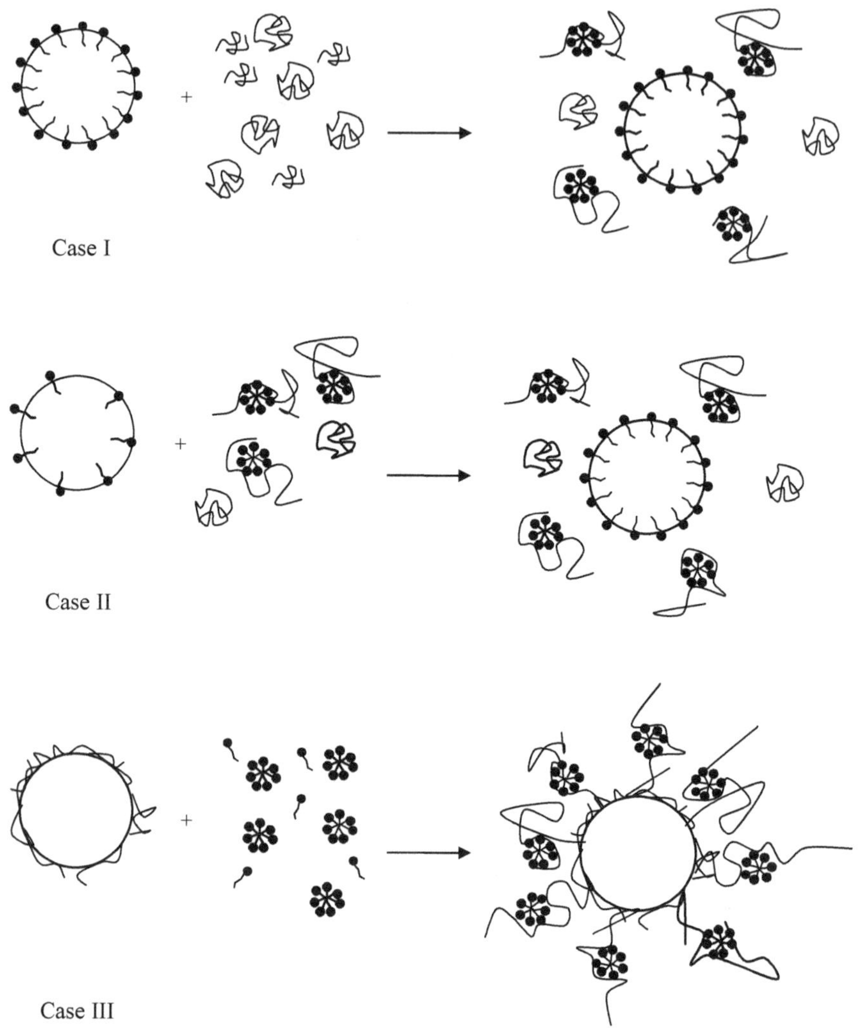

FIGURE 2.19. Schematic diagram of polymer–surfactant complex on emulsions droplets for cases I, II, and III. (Adapted from [74,75].)

dictated by the double-layer forces due to adsorbed surfactant molecules. In case II, the decay length decreases with increasing surfactant concentration. The decay length almost coincides with the Debye length, indicating that the droplet interface is fully adsorbed with surfactant molecules. This shows that polymer–surfactant complexes remain in the bulk solution. In case III, the λ values correspond to the radius of gyration of the polymer, indicating that the polymer is adsorbed at the oil–water interface. The interaction between the surfactant and preadsorbed polymer drastically changes polymer conformation at the interface and hence the repulsive forces, especially the onset of repulsion. It is argued that, as the concentration of SDS increases, more and more surfactant molecules and micelles go into the folded chains (central regime) and stretch the loops and tails. Therefore, at higher surfactant concentrations, the neutral polymer chains at the interface are likely to adopt diblock polyelectrolyte type conformations on the emulsion droplets. The stretching is expected to continue until the coils are saturated with adsorbed surfactant micelles. The critical aggregation concentration CAC for SDS–PVA–Vac ($M_w = 155000$ g/mol) system was measured in bulk solutions (without emulsion) using conductivity and viscometry and was found about 3 10^{-3} mol/l. However, the force experiments with a three-component system show that the onset of repulsion begins at surfactant concentrations of 0.45 10^{-3} mol/l. This suggests that the critical concentrations at which the association occurs for the adsorbed polymer and the free polymer in the bulk solution are different.

To test the generality of their findings, the authors studied the force measurements using polymers of different molecular weights, and different ionic surfactants (cationic, anionic). In all those cases, they observed the same phenomena [74,75].

2.2.2.4. Protein-Covered Interfaces

Protein-stabilized emulsions find a variety of applications in food industry. The main features of these systems arise from the specific interfacial properties of the stabilizing molecules. First, it is noteworthy that proteins do not lower the interfacial tension as much as simple surfactants do. In bulk solutions, protein molecules maintain a tightly packed structure. Their adsorption at fluid–fluid interfaces is accompanied by a gradual unfolding, which means that the biomolecule looses its secondary and higher structure in the adsorbed state. This happens because the presence of a hydrophobic fluid phase (i.e., oil or air) gives the molecule the possibility to minimize the configurational free energy. The adsorbed species are often referred to as denatured or unfolded. This conformation state makes the adsorbed molecules very much different from their nonadsorbed analogues. The process of unfolding uncovers the different segments of adsorbed species and facilitates the lateral interaction between two or more adsorbed molecules. Various interactions are possible, for example, ionic, hydrophobic, covalent (disulfide bridging) or hydrogen bonding. As a result, the adsorbed species form a rigid network and the protein layers, which protects the droplets against coalescence are viscoelastic and almost always tangentially immobile. The large variety of possible interactions as well as their different time scales make the properties of the protein layers largely dependent on the sample history. This feature is frequently termed "aging effects." It is generally not possible to postulate equilibrium adsorption, for at least two reasons. First, the existence of many adsorption sites per molecule makes the hypothesis of instantaneous desorption unlikely. Second, it is an experimental fact that reaching constant interfacial tension does not mean reaching constant adsorption. The protein adsorption can continue via multilayer formation [76,77]. The multilayers comprise reversibly adsorbed molecules and the surface coverage can be augmented up to 5–10 mg/m^2.

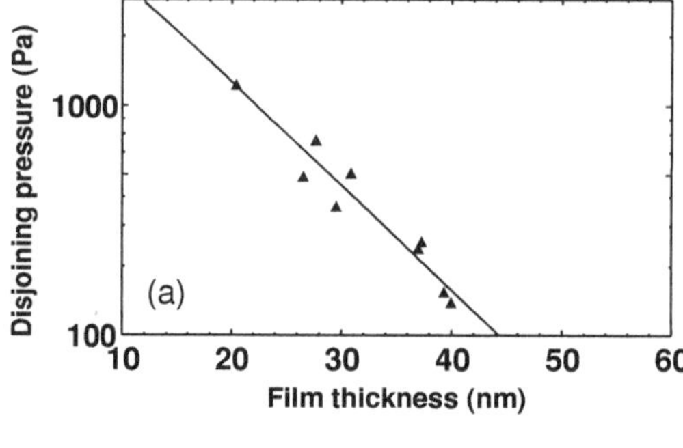

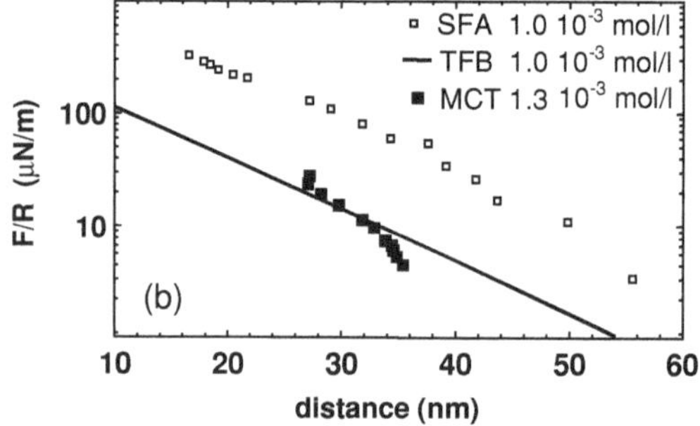

FIGURE 2.20. **(a)** Disjoining pressure vs. thickness isotherm (dots, experimental data; line, double-layer fit) for an emulsion film stabilized by 0.1% β-casein, ionic strength of 10^{-3}mol/l NaCl, oil phase = hexadecane. **(b)** Comparison between the data obtained from TFB, MCT, and SFA. (Adapted from [87].)

There exists relatively little experimental information concerning the surface forces acting in the emulsion films stabilized by proteins. Dimitrova and Leal-Calderon [78] applied two complementary techniques, that is, MCT and TFB to measure the forces acting between protein-stabilized emulsion droplets. Some examples are given below, where typical data sets for bovine serum albumin (BSA), β-lactoglobulin (BLG), and β-casein are shown. These molecules represent two important classes of proteins: the globular (BSA and BLG) and disordered (β-casein) ones. To investigate single foam and emulsions films, a variant of the Mysels–Bergeron thin liquid film set-up was developed [79]. The thin liquid films (TFB) were formed in a porous glass plate immersed in the corresponding oil phase.

i. β-casein

The disjoining pressure vs. thickness isotherms of thin liquid films (TFB) were measured between hexadecane droplets stabilized by 0.1 wt% of β-casein. The profiles obey classical electrostatic behavior. Figure 2.20a shows the experimentally obtained $\Pi(h)$ isotherm (dots) and the best fit using electrostatic standard equations. The Debye length was calculated from the electrolyte concentration using Eq. (2.11). The only free parameter was the surface potential, which was found to be -30 mV. It agrees fairly well with the surface potential deduced from electrophoretic measurements for β-casein-covered particles (-30 to -36 mV).

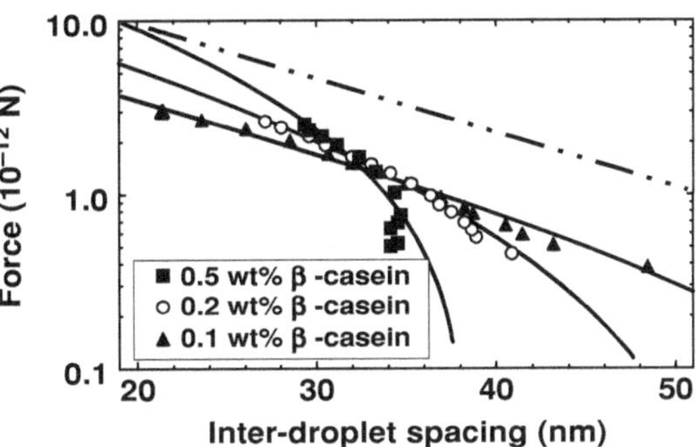

FIGURE 2.21. Force vs. distance profiles for ferrofluid emulsions stabilized with β-casein at different concentrations (points). The ionic strength is 3 10^{-4} mol/l, pH = 6.2. The lines are the best fits (see the text for details). (Adapted from [78].)

Some data obtained from MCT, SFA, and TFB under similar conditions are compared in Fig. 2.20b. The SFA data were taken from reference [80]. They were obtained with hydrophobized mica surfaces, protein concentration of 0.1 wt%, and ionic strength of 10^{-3} mol/l. Data for TFB and MCT led to very similar results. However, comparison with the SFA data demonstrates that the force laws are only qualitatively similar: the curves are parallel but the normalized data for liquid–liquid interfaces (TFB and MCT) lie about one decade below those obtained for solid–liquid interfaces (SFA). This result suggests that the proteins exhibit different adsorption abilities and or adopt different conformations at the two types of surfaces.

The force profiles between β-casein-covered droplets were explored at different protein concentrations in the continuous phase (from 0.1 to 0.5 wt%) using MCT (Fig. 2.21). At high protein content, the measured profiles cannot be interpreted by means of a simple electrostatic force. Instead, the long-range part of the interaction deviates from linearity (in a semilog plot), which suggests the existence of an attractive interaction. The higher the concentration, the stronger the deviation from the classical double-layer repulsion. This evolution strongly suggests the occurrence of depletion attraction due to the presence of β-casein micelles. The lines in Fig. 2.21 represent the fit to the data assuming additivity of the depletion attraction and double-layer forces. The depletion attraction was calculated via Eq. (2.17), where P_{osm} is the osmotic pressure exerted by β-casein micelles, r_m is the radius of the micelles and δ is an extra exclusion distance that accounts for the electrostatic and steric repulsion between micelles and droplet surfaces (Fig. 2.6). For β-casein, the range of the electrostatic repulsion is always longer or equal to the range of the steric repulsion, so δ was supposed to coincide with the Debye length. P_{osm} was calculated assuming the perfect gas approximation. The surface potential, ψ_0, of the droplets and the mean size r_m of the depleting species were used as adjustable parameters. The values deduced from each plot remain constant whatever the ionic strength: $\psi_0 \approx -30$ mV and $r_m \approx 30$ nm. The exclusion volume effect of caseinate micelles has been observed and reported in the literature, where it is found that liquid films stabilized with sodium caseinate at concentrations 0.1–4 wt% undergo stepwise thinning [81]. The height of the stratification steps was 20–25 nm. In the same study, the average size of caseinate submicelles,

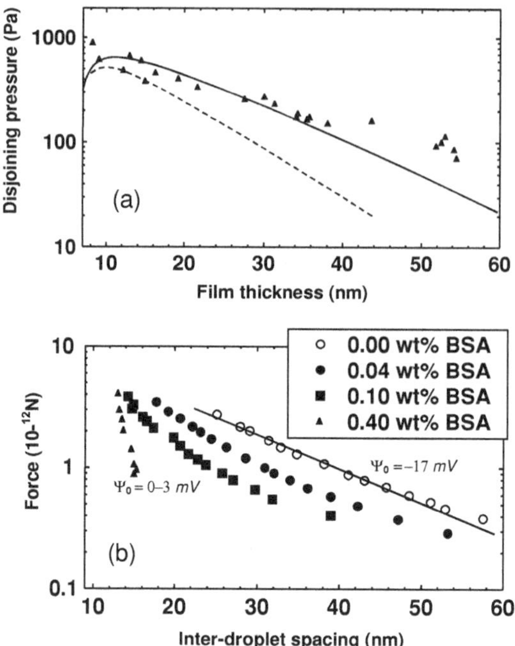

FIGURE 2.22. (**a**) Disjoining pressure vs. thickness isotherm for an emulsion film stabilized by 0.1% BSA, ionic strength of 10^{-3} mol/l NaCl, oil phase = hexadecane. The dots are the experimental data, dashed line is the double-layer contribution to the total disjoining pressure, and the solid line is the best fit done supposing additivity of the double-layer and steric forces. (**b**) Force vs. distance profiles for ferrofluid emulsions stabilized with mixed BSA–Tween-20 adsorption layers. The total concentration of the Tween-20 is kept constant = 5CMC, pH = 5.8. (Adopted from [78].)

measured by dynamic light-scattering, is reported to be about 25 nm, which is also in very good accord with the values deduced from the force profiles. This proposed mechanism is also consistent with the microscopic observation of flocculation in emulsions stabilized with sodium caseinate [82]. Depletion flocculation induced by micellar and/or submicellar protein aggregates/micelles was observed at free protein concentration in the aqueous phase equal (or higher) to about 0.7 wt%. The measurement technique (MCT) allows attaining very low levels of forces, making it possible to detect the depletion attraction *below* the threshold concentration necessary to induce flocculation.

ii. BSA

For TFB experiments, the measurements were carried out with 0.1 wt% BSA solutions. The ionic strength is 10^{-3} mol/l of NaCl and the oil phase is hexadecane. The results from TFB are shown as points in Fig. 2.22a. The value of −10 mV for the surface potential was used to estimate the maximum electrostatic contribution to the total disjoining pressure (dashed line). Electrophoretic measurements of hexadecane droplets, covered by BSA under similar conditions provided values between 0 and −5 mV. It clearly appears that the measured repulsive pressure is longer ranged than the one predicted by a purely electrostatic model. This means that there is an additional repulsive interaction, which is not electrostatic in origin. Generally, BSA layers exhibit strong and long-range steric repulsion [83,84]. For numerous proteins, it is known that after the equilibrium interfacial tension is reached, the adsorption Γ continues to increase due to the formation of additional layer(s) of partially unfolded proteins [76,77]. To take into account the formation of a second protein layer, one may use the approach of Israelachvili and Wennerström [85]. The segment density, $\rho(z)$, in the second layer at a distance z from the interface, can be written:

$$\rho(z) = \rho(z_0)\exp(-h/\lambda^*) \tag{2.31}$$

where λ^* is the characteristic size of the protein species that constitute the second layer. Neglecting the correlations between the moieties in the second layer, the following expression for the repulsive pressure is obtained:

$$\Pi_{repulsion} = \Gamma_2 \frac{k_B T h}{\lambda^{*2}} \exp(-h/\lambda^*) \tag{2.32}$$

where Γ_2 is the adsorption in the second protein layer only. The previous equation is valid for interdroplet spacings h much larger than λ^*. Assuming additivity of the electrostatic contribution and of the repulsive steric pressure given by Eq. (2.32), the experimental data can be fitted using Γ_2 and λ^* as free parameters (solid line in Fig. 2.22a). The values obtained are $\Gamma_2 = 0.23$ mg/m^2 and $\lambda^* = 10.7$ nm. An increase of the total adsorption (after the equilibrium value of the surface tension is reached) of approximately 0.5 mg/m^2 is reported in the literature, in good accord with the Γ_2 value deduced from the fit [86]. The value of λ^* is consistent with the size of the BSA molecule (11.4 × 11.4 × 4.1 nm), as well as with the radius of gyration of a BSA molecule of ~9 nm.

In the presence of BSA alone, the measurements using MCT were impossible because of the flocculated state of the ferrofluid droplets. To prevent flocculation, a nonionic surfactant, polyoxyethylene (20) sorbitan monolaurate (Tween-20, CMC = 2 10^{-5} mol/l) was systematically added to the samples. Typical results are shown in Fig. 2.22b, where the Tween-20 concentration is kept at 5 CMC. For pure Tween-20, the repulsion profile is fitted by adopting a standard electrostatic model, yielding a surface potential of −17 mV. Despite its nonionic nature, this surfactant is known to promote surface ionization [79]. At a constant Tween-20 concentration, when the protein content is increased from 0 to 0.4 wt%, there is a progressive change with respect to classical electrostatic repulsion. The profiles in the presence of the same quantity of BSA at surfactant content of 20CMC and 50CMC are very similar and do not depend on the Tween-20 concentration. This suggests that the presence of the protein dominates the behavior and the properties of these systems. The surface potential deduced from electrophoretic measurements gave values between 0 and −3 mV, meaning that the contribution of the electrostatics to the total repulsive force is negligible.

The disjoining pressure was transformed into force in the same manner as in the case of β-casein and compared with results from MCT and SFA (Fig. 2.23). The data from MCT and TFB show reasonably good agreement. The results from SFA are only qualitatively similar to both MCT and TFB data. The reason is essentially the same as in the case of β-casein-stabilized films, that is, the difference in either

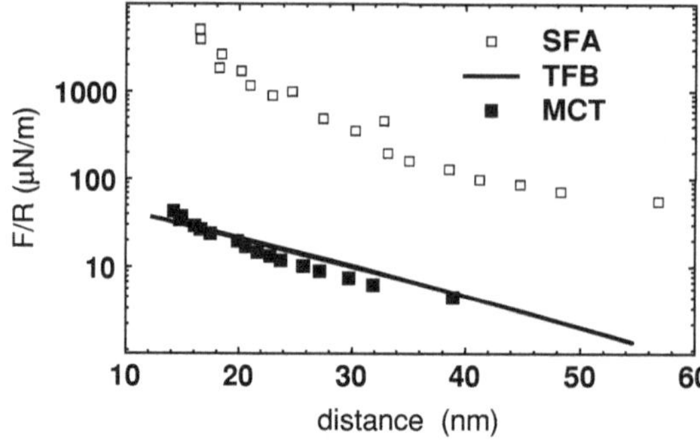

FIGURE 2.23. Comparison between the data obtained from TFB, MCT, and SFA for systems stabilized with BSA. See text for details. (Reproduced from [78], with permission.)

the conformational state of BSA or the amount adsorbed on solid–fluid and on fluid–fluid interfaces.

iii. Threshold Flocculation Force

During the preparation of the samples for MCT, the emulsion properties appeared to be very sensitive to the history of preparation. Flocculation was always observed after centrifugation, that is, when the droplets were brought in contact. For BLG and β-casein, it was possible to produce nonflocculated emulsions stabilized solely by the proteins; however, these systems showed a tendency to flocculate when centrifuged at acceleration larger than 150–200g for more than 30 min. The critical force necessary to induce irreversible flocculation was measured using MCT [87]. The removal of the magnetic field generally leads to instantaneous decomposition of the formed chains. This is observed for ferrofluid oil-in-water emulsions stabilized by low molecular weight surfactants. For proteins, microscopy revealed that above some critical force, the droplets remained irreversibly chained even in the absence of magnetic field. Figure 2.24a,b shows the force profiles for ferrofluid oil-in-water emulsions stabilized by 0.1 wt% protein at various ionic strengths. For both β-casein and BLG, a repulsion is measured up to a force, F^* (indicated by the arrow), above which droplets are held together in flexible chains that persist over time. The dependence of F^* on the ionic strength is shown in Fig. 2.24c. The decrease in the resistance against flocculation when the ionic strength increases, may be due to the screening effect of salt. Under equivalent conditions, the critical threshold flocculation force is lower for BLG than for β-casein-stabilized emulsions. An interesting phenomenon was observed for BLG-stabilized systems. The "irreversible" chains were rigid linear rods that remained unchanged for a period as long as 48 h (Fig. 2.25). The existence of these rods suggests that the contacts between the individual droplets forming the chains are substantially strong and cannot be affected by thermal motion. Instead, the chains formed by β-casein-stabilized particles were rather flexible and tended to decompose for the same period of time.

In Section 2.2.2.2, we described a bridging mechanism in which polymer was laterally squeezed away from the near-contact region upon approach of the droplets.

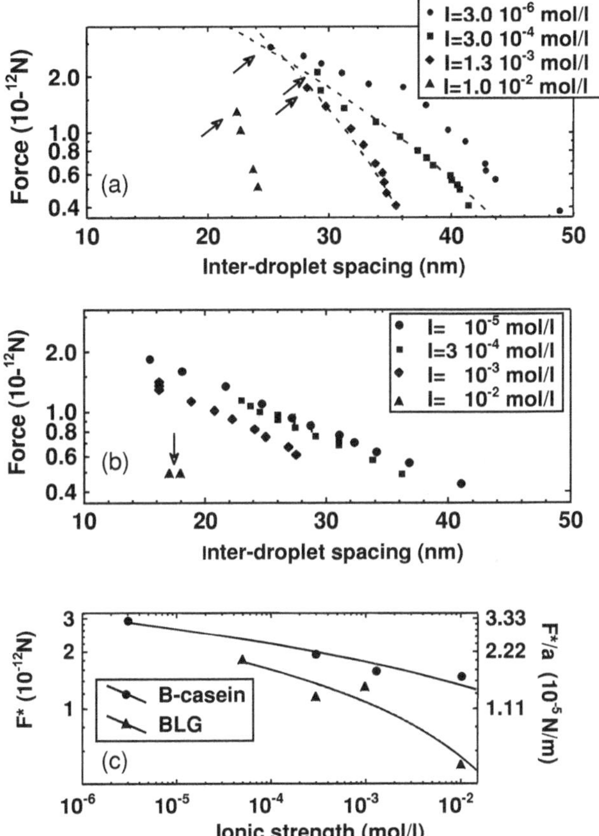

FIGURE 2.24. (a) Force vs. distance profiles for ferrofluid emulsions stabilized with 0.1 wt% β-casein at pH = 6.2 and at different ionic strength (points). The lines (where presented) are double-layer fits. (b) Force vs. distance profiles for ferrofluid emulsions stabilized with 0.1 wt% BLG at pH = 6.0 and at different ionic strengths. (c) Threshold flocculation force for ferrofluid-in-water emulsions stabilized with β-casein and BLG as a function of ionic strength. The lines are visual guides. (Adapted from [87].)

Such mechanism can be ruled out for protein-stabilized droplets whose interfaces are tangentially immobile. When droplets get closer, the approaching protein layers start "touching" each other. This allows some of the adsorbed entities to interact forming a local link. Such links are often termed as "surface aggregates" because in thin liquid film experiments as well as in Brewster angle microscopy studies [88], they appear like lumps of material located on the surface (see Fig. 2.26). These lumps appear as a result of rapid surface distortion. AFM images of β-casein and BLG layers formed at the water–tetradecane interface and further transferred on mica by Langmuir–Blodgett technique, reveal a network formed by the adsorbed species [89]. This is a further evidence for the interaction between the proteins

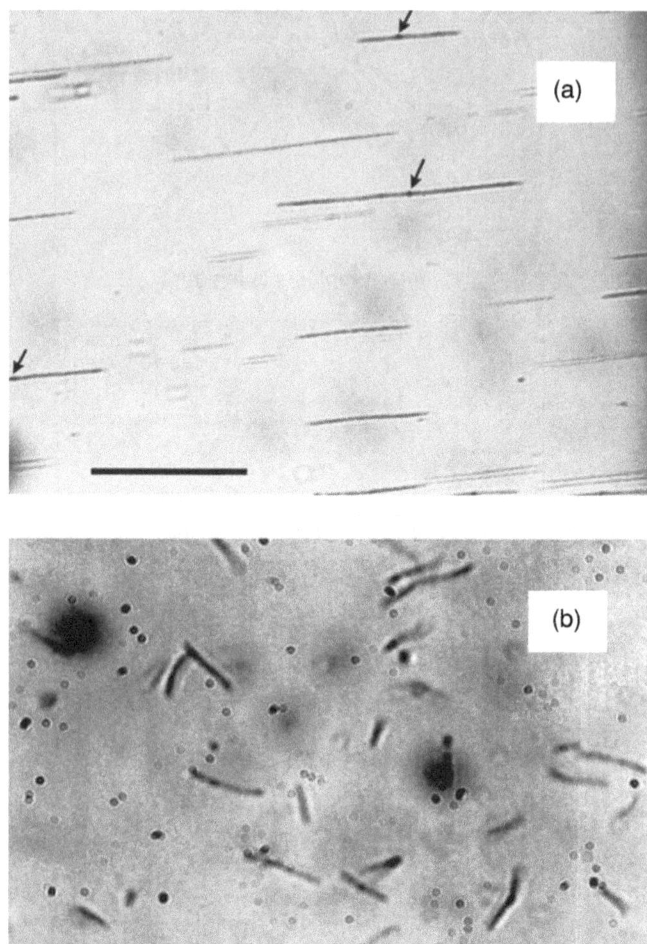

FIGURE 2.25. Optical micrographs of a ferrofluid oil-in-water emulsion stabilized by 0.04 wt% BLG. The bar corresponds to 10 μm. **(a)** On field corresponding to force of 1 pN. **(b)** 30 min after switching off the magnetic field. (Reproduced from [87], with permission.)

in the adsorbed state. The network formation is possible because the adsorbed proteins are unfolded and therefore segments of various types are available for interaction. Provided that two protein-covered droplets are sufficiently close, there is no difference whether the interacting entities belong to the same or to different protein layers. The value of the critical flocculation force, measured as described in the preceding text, has to be considered as the force necessary to bring the droplets at the minimum distance allowing the formation of surface clusters between the two protein layers.

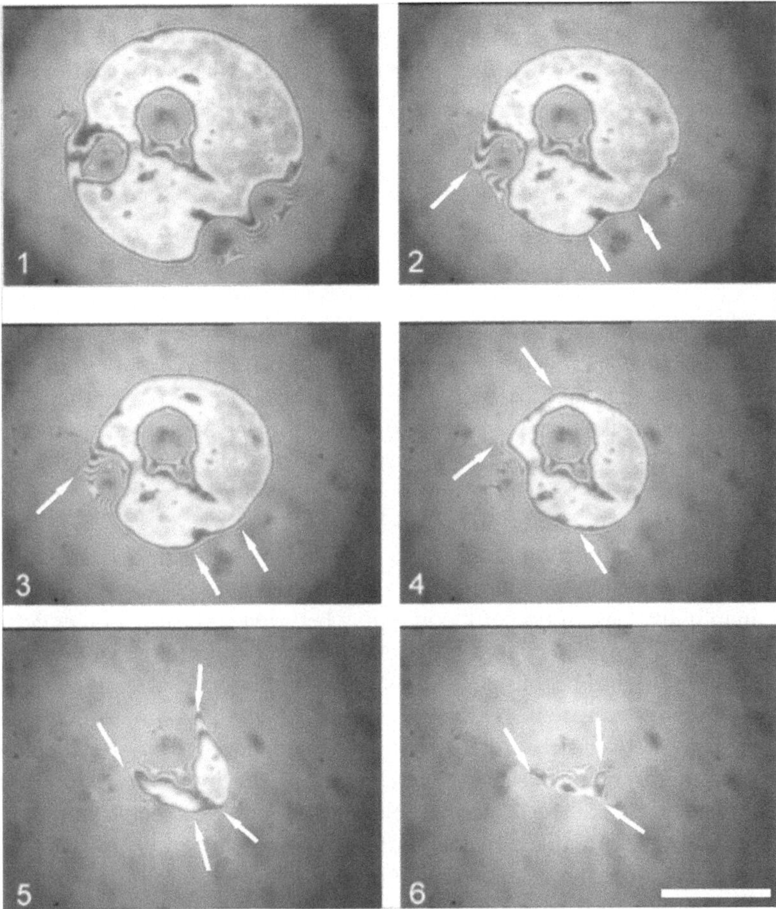

FIGURE 2.26. Six consecutive steps in shrinking of an emulsion film stabilized with 0.1 wt% BSA. Oil phase is hexadecane; the ionic strength is 10^{-3} mol/l. The bar corresponds to 100 μm. The local adhesion on aggregates is evident. Arrows indicate some of the points of adherence at the interfaces. (Adapted from [87].)

2.2.2.5. Role of Droplet Deformability

For nondeformable particles, the theories describing the interaction forces are well advanced. So far, most of the surface force measurements between planar liquid surfaces (TFB) have been conducted under conditions such that the film thickness is always at equilibrium. In the absence of hydrodynamics effects, the forces are correctly accounted considering classical theories valid for planar solid surfaces. When approached at high rate, droplets may deform, which considerably complicates the description; it is well known that when the two droplets are sufficiently large, hydrodynamic forces result in the formation of a dimple that flattens prior to film thinning. Along with the hydrodynamic interactions, the direct

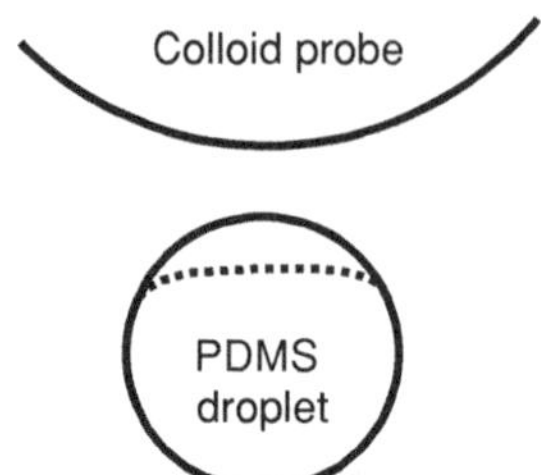

FIGURE 2.27. Schematic representation of colloid probe–PDMS droplet interaction during the AFM experiment. Solid line depicts the undeformed profile of the PDMS droplet and the rigid colloid probe. Dashed line shows the deformed profile of the PDMS droplet.

interactions due to surface forces can be strongly affected by the deformation as well. A theoretical approach for calculation of the different contributions (van der Waals, electrostatics, steric, depletion) has been developed [90]. The shapes of two deformed droplets in contact are approximated by truncated spheres separated by a planar film. The contribution of the surface extension energy and/or bending elasticity is also included. The extension of the drop surface upon the deformation is equivalent to a soft interdroplet repulsion.

Experimental interaction studies on droplets of colloidal size are still in a pioneering phase and much further work is required before the development of a fully quantitative description of the interplay between emulsion droplet deformation and interaction. The colloid probe AFM technique, originally developed for hard particles [6], has recently been applied to bubbles and soft polymer colloids [91–96], and is proving to be a valuable tool in characterising droplet deformation. The interplay between the Laplace pressure and internal droplet rheology in controlling deformation has been well demonstrated by Gillies and Prestidge [97,98]. They used AFM to determine the interaction forces between a spherical silica probe and immobilized colloidal droplets of polydimethylsiloxane (PDMS) in aqueous solution, the characteristic size of both objects being of the order of some tens of microns (Fig. 2.27). Under alkaline conditions where PDMS droplets and the silica probe are both negatively charged, repulsive forces are evidenced (Fig. 2.28). At large separations the repulsive forces are well characterized by the electrostatic interactions of rigid colloids, while at smaller separations, droplet deformation results in deviation from this behavior: the force increases less rapidly on surface approach than expected for electrostatic interaction of rigid particles. The departure from hard-sphere behavior enables the extent of droplet deformation to be determined and by varying the extent of internal cross-linking within the PDMS droplets, the influence of internal droplet rheology and interfacial tension on droplet deformation has been explored. For highly cross-linked droplets, the extent of deformation is controlled by the internal droplet rheology rather than the interfacial properties. On retraction of the surfaces, force curve hysteresis was observed and was due to the viscoelastic response of the PDMS. Instead, for liquid-like droplets, with a low level of cross-linking, no force curve hysteresis was observed; in this limit, the deformation was dominated by interfacial effects and not by the bulk rheology.

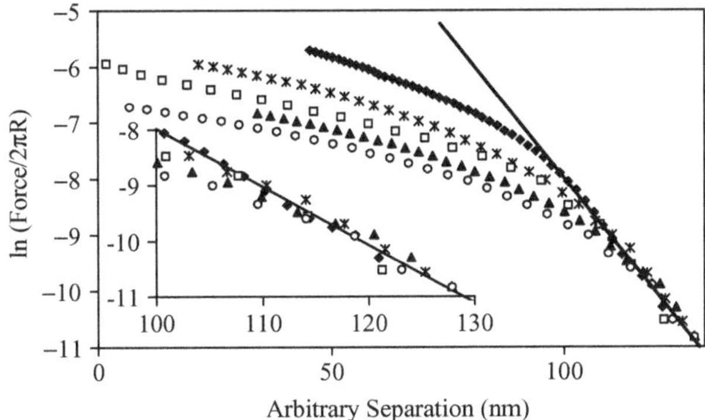

FIGURE 2.28. Log force vs. arbitrary separation curves for the approach of a glass probe and a droplet from emulsion. The symbols correspond to different levels of cross-linking of PDMS droplets: (♦) 50%; (✱) 45%; (□) 40%; (▲) 35%; (○) 30%. The background electrolyte solution contains KNO_3 at 10^{-3} mol/l and pH = 9.5. Solid line represents the theoretical electrostatic repulsion with a Debye length of 9.6 nm. All forces were determined at 1 μm/s. (Reproduced from [97], with permission.)

2.3. Short-Range Forces and Adhesion Between Emulsion Droplets

As shown in the previous section, long-range forces between interfaces in colloidal suspensions and emulsions systems are now rather well understood. Emulsion droplets can also experience short-range attractions or repulsions, whose range and amplitude is more difficult to determine. Emulsion droplets exhibit "adhesion" when they experience pronounced attraction. A strong, steep repulsion at short range is then necessary to stabilize the droplets against coalescence. Under such conditions, emulsion droplets form large contact angles as they adhere to one another or as they adhere onto a substrate. As in classical wetting phenomena, the contact angles are imposed by the energy of adhesion. However, in contrast to classical wetting, a thin liquid film of the continuous phase persists between the interfaces. This film is stabilized by the surfactant molecules adsorbed at the interfaces. The adhesion is governed by the properties of the surfactant film, rather than by the nature of the phases on contact. That is why adhesion in emulsion systems is closely related to adhesion in other surfactant systems such as soap films [99–102] and biological membranes [103].

2.3.1. Energy of Adhesion and Contact Angles

The thermodynamics of thin liquid films and adhesion is well documented in the literature [15,17,100–103]. The first theoretical approaches were developed mainly for thin soap films. Most of the results of the thermodynamics of soap

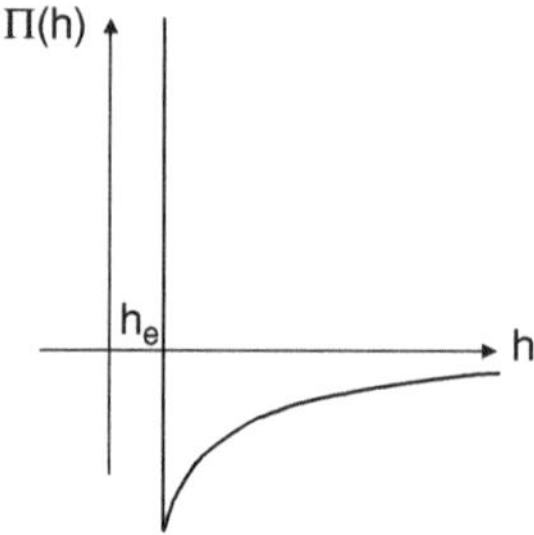

FIGURE 2.29. Disjoining pressure between interfaces of a liquid film.

films can be equivalently used for emulsion systems. A thin liquid film can be modeled by two interfaces at a distance h from each other. The excess pressure in the film due to the interactions between the interfaces has been already defined as the disjoining pressure, $\Pi(h)$. As mentioned earlier, adhesion occurs when the interfaces experience a strong attraction and a steep repulsion at short range. A typical $\Pi(h)$ for adhesive systems is shown in Fig. 2.29. The pressure is negative at longer range whereas it sharply rises at shorter range, ensuring stabilization against coalescence. The short-range repulsion can be often considered as a hard wall because it is usually due to steric or solvation forces [17]. Under these conditions, the interfaces spontaneously approach each other up to the equilibrium distance h_e. This distance is imposed by the sharp rise of the repulsion. The resultant work of adhesion, $E(h_e)$, is given by:

$$E(h_e) = -\int_{\infty}^{h_e} \Pi(h)\, dh \tag{2.33}$$

$E(h_e)$ is negative for adhesive systems and $-E(h_e)$ is traditionally known as the free energy of adhesion.

An important consequence resulting from the approach of the interfaces is that γ_f, the surface energy of the film, is lowered by the energy of adhesion. When the interfaces are far apart, $E(h)$ is equal to zero and γ_f is simply equal to $2\gamma_{\text{int}}$, where γ_{int} is the tension of a single interface. At equilibrium, the surface energy of the film is $\gamma_f = 2\gamma_{\text{int}} + E(h_e)$.

As the tension of the film is different from the tension of two single isolated interfaces, a contact angle, θ, is expected at the junction between adherent and free interfaces. This is depicted in Fig. 2.30. The mechanical equilibrium of the contact line at the junction between free and adherent interfaces dictates the value of the contact angle:

$$\gamma_f = 2\gamma_{\text{int}} \cos(\theta) \tag{2.34}$$

The contact angle can be written as a function of $E(h_e)$:

$$E(h_e) = 2\gamma_{\text{int}} [\cos(\theta) - 1] \tag{2.35}$$

This relation is known as the Young–Dupré equation. It shows that the energy of adhesion can be determined by measuring the contact angle and the surface

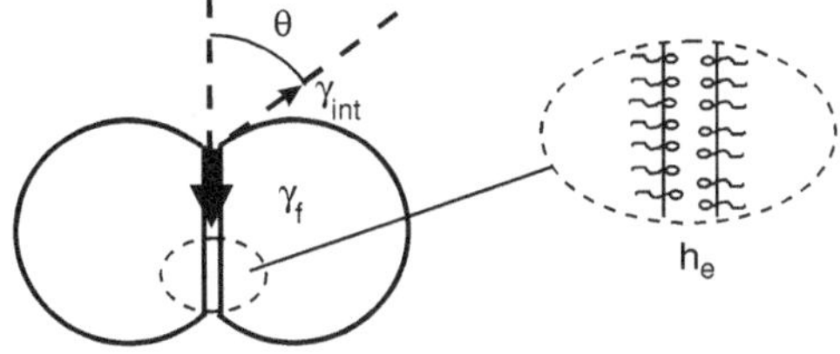

FIGURE 2.30. Two adhesive emulsion droplets. A flat liquid film stabilized by the surfactant layers is located between the droplets. This film being very thin, it can be usually considered as a surfactant bilayer. γ_f is the tension of the film and γ_{int} the tension of single isolated interface.

tension of a single interface. The pair interaction energy, u, between two droplets separated by an adhesive film of radius r (Fig. 2.30) is given by:

$$u = \gamma_{\text{int}} \pi r^2 [\cos(\theta) - 1] \tag{2.36}$$

where u is equal to half of the adhesion energy times the surface area of the flat adhesive film.

As shown in Fig. 2.31, a contact angle can be experimentally measured by looking at two adhesive droplets. However, a direct determination from side views is rather difficult unless the contact angle is large enough, as in Fig. 2.31. A more convenient way to achieve measurements of contact angles consists of measuring the radius of two adhesive droplets, a_1 and a_2, and the radius of the adhesive film between the droplets, r (Fig. 2.32) [104–107]. One has:

$$2\theta = Arc\sin\left(\frac{r}{a_1}\right) + Arc\sin\left(\frac{r}{a_2}\right) \tag{2.37}$$

in the case of two droplets with different radii.

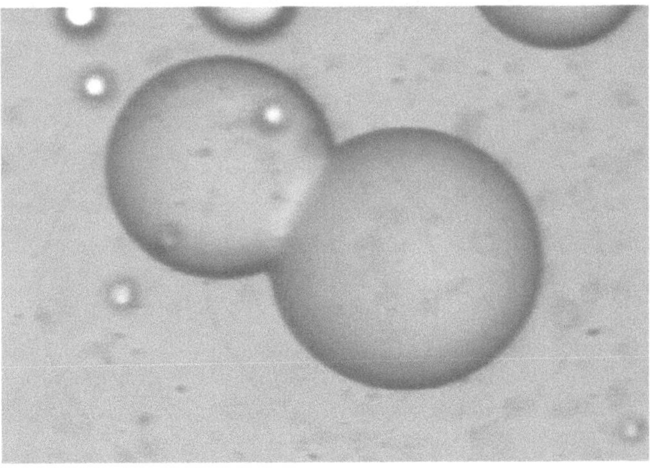

FIGURE 2.31. Two adhesive emulsion droplets of few tens of microns. The adhesion induces the formation of a large contact angle of about 40°. (Reproduced from [113], with permission.)

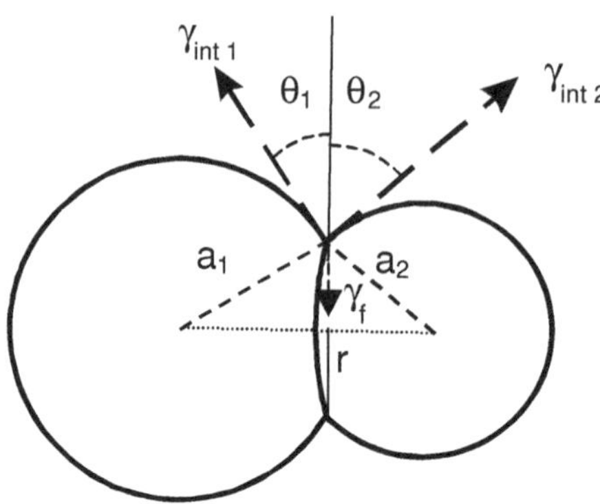

FIGURE 2.32. Two adhesive droplets of size a_1 and a_2. If the droplets are of the same nature $\gamma_{\text{int}1} = \gamma_{\text{int}2} = \gamma_{\text{int}}$ and $\theta_1 = \theta_2$.

Moreover, an accurate measurement of r can be easily performed by looking at droplets from above or in perspective as sketched in Fig. 2.33. As shown in the experimental picture of Fig. 2.34, the adhesive film looks brighter than the surrounding part of droplets. This comes from the minute thickness of the film. It does not reflect light as do the surrounding oil–water interfaces. For the same reason, thin soap films look black when they are observed with reflected light, thus their name. With transmitted light, as for optical microscopy observations, the "black films" look brighter. The r corresponds to the radius of the white circle viewed from above or to the larger axis of the white ellipse when observed in perspective. This way of measurement allows small contact angles to be accurately determined and also the onset of formation of nonzero contact angle to be characterized by the onset of apparition of a white spot between the droplets.

The contact line is considered as a sharp boundary in the classical picture presented above. In this picture, there is a discontinuous transition from the flat

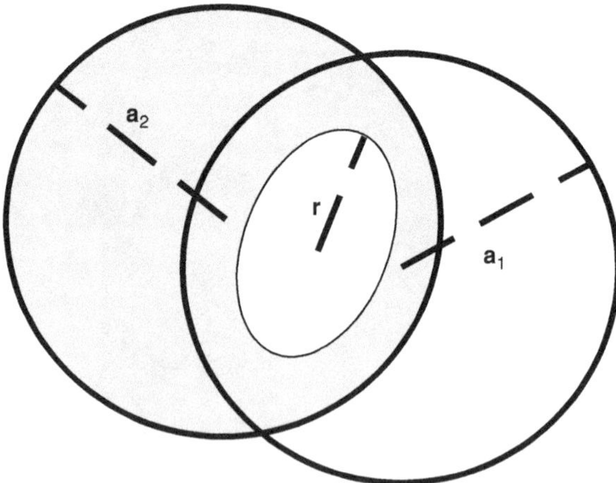

FIGURE 2.33. Scheme of two adhesive droplets viewed in perspective. The brighter ellipse between the droplets is the adhesive film. (Adapted from [113].)

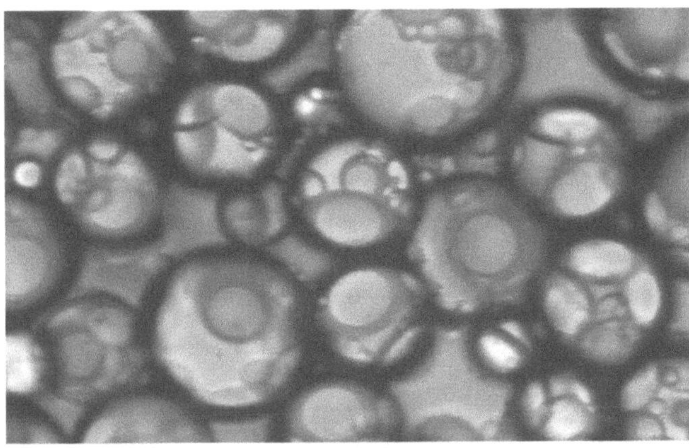

FIGURE 2.34. Collection of adhesive emulsion droplets (of a few tens of microns). A white ellipsoid can be seen between superimposed droplets. This ellipsoid is the projection of the adhesive film between the droplets. The contact angle is about 29°. (Reproduced from [113], with permission.)

adhesive film to the curved isolated interfaces that are considered at infinity, as they are not in the film. However, in practice, there is a transition region where the curvature changes continuously as the separation between the interfaces increases. At short range, in the adhesive film, the interactions are attractive whereas they can become repulsive in the region of separation. A phenomenological approach can be used to account for the additional energy due to the contact line. It consists in assigning a line tension to the transition region [107,108]. If the interactions are repulsive this tension is expected to be positive and unfavorable to adhesion. Its competition with surface and adhesion phenomena depends on the size a_0 of the droplet. Indeed, the involved surface energy scales as a_0^2, whereas the energy of the transition region scales as a_0. The consequences of the contact line are thus expected to be important for smaller droplets. However, it was shown for soap films and air bubbles that the line tension is in fact very small [107,108]. Indirect experiments lead to the same conclusion for emulsion systems [109]. This is due to the shortness of the range of the interactions involved in these systems. The long-range interactions are too weak to significantly affect the adhesion.

2.3.2. Experimental Measurements of the Adhesive Energy

2.3.2.1. Oil-in-Water Emulsions

Observations of large contact angles in emulsions were first reported by Aronson and Princen [105,106]. The authors have studied oil-in-water droplets stabilized by anionic surfactant in the presence of various salts. Similar systems were studied by Poulin [110]. Anionic surfactants such as sulfate, sulfonate, or carboxylate surfactants [106,110] exhibit a good stability and a strong adhesion in the presence

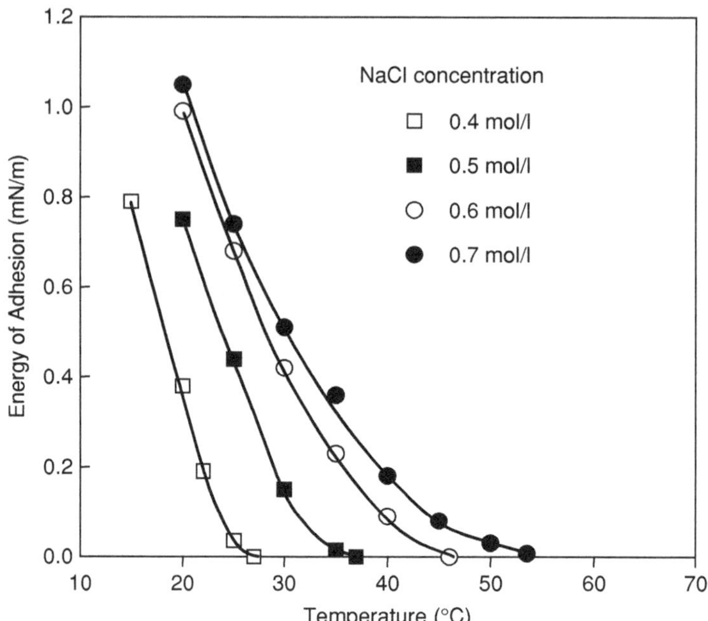

FIGURE 2.35. Energy of adhesion between hexadecane droplets stabilized in water by SDS, at various NaCl concentrations. (Adapted from [111].)

of monovalent salts such as NaCl and KCl. Systems made with cationic surfactants such as alkyl ammonium are less stable when salt is added. However, some systems can exhibit adhesion in well-defined conditions without coalescing too quickly [110]. For example, highly viscous silicon oil droplets stabilized in water by tetradecyl trimethyl ammonium bromide (TTAB) exhibit adhesion in the presence of KI. However, the droplets coalesce in a few minutes after the salt is added. Finally, strong adhesion with pure nonionic surfactants in water was not observed to our knowledge. However, mixtures of nonionic and ionic surfactant can lead to adhesion [111].

The energy of adhesion between hexadecane droplets stabilized in water by SDS in the presence of NaCl is shown in Fig. 2.35. It is observed that the adhesion depends strongly on the temperature and on the salt concentration. For a given salt concentration, there is a well defined temperature, T^*, above which there is no adhesion. As the behavior of the surface energy changes at T^*, this temperature can be referred to as a wetting transition temperature [109]. The dependence of T^* versus the salt concentration is plotted on Fig. 2.36.

Moreover, the adhesion depends strongly on the nature of the salt [105]. A pronounced difference is noted in Fig. 2.37, where the energy of adhesion with KCl, NaCl and LiCl are compared [110]. The system is much more adhesive with KCl. This is the contrary with a carboxylate surfactant [106,110]. This behavior is important because it suggests that the adhesion is linked to the solubility of the

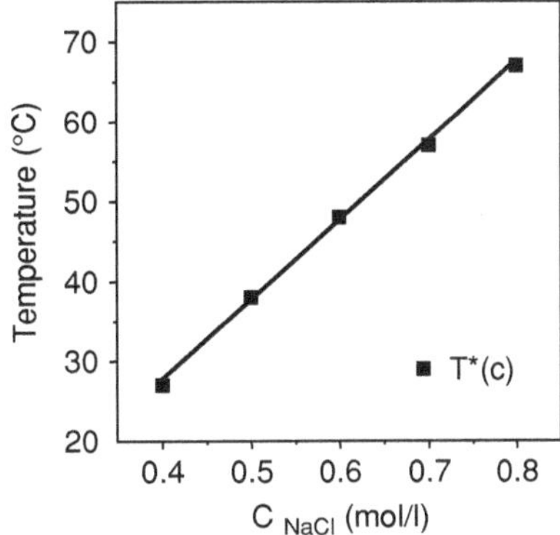

FIGURE 2.36. Onset of adhesion, T^*, between hexadecane droplets stabilized in water by SDS as a function of the NaCl concentration. (Adapted from [111].)

surfactant. Indeed, sulfate sodium salts are more soluble than sulfate potassium salts whereas carboxylate sodium salts are less soluble than carboxylate potassium salts. Moreover, cooling down the system and increasing the salt concentration favors surfactant insolubility and emulsion adhesion at the same time.

2.3.2.2. Water-in-Oil Emulsions

The study of inverse adhesive emulsions has revealed the same features as direct emulsions [112,113]. Here again, it was shown that adhesion is favored when the surfactant becomes less soluble in the continuous phase [113]. This can be tested experimentally by using binary mixtures of oils, one in which the surfactant is soluble and another one in which the surfactant is insoluble. For example, water droplets can be stabilized in mineral oil by sorbitan monooleate (Span 80). This surfactant is soluble in dodecane whereas it is not in silicon oil. The affinity of the surfactant for the organic solvent can be tuned by mixing dodecane and silicon oil. As shown in Fig. 2.38, the energy of adhesion between water droplets strongly varies as the ratio of the mixture is changed. A sharp rise is noted as the surfactant

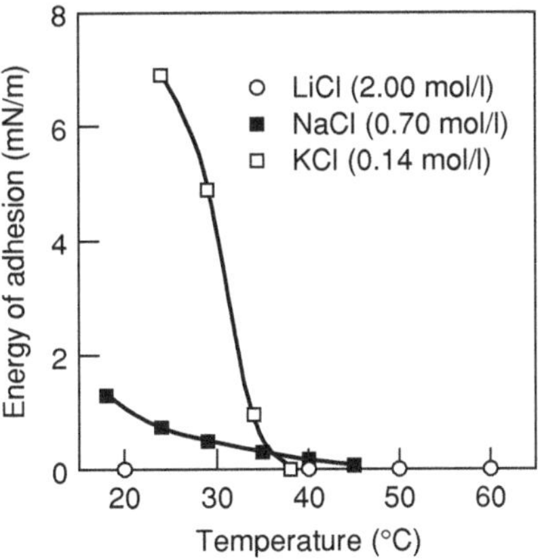

FIGURE 2.37. Energy of adhesion between hexadecane droplets stabilized in water by sodium dodecyl sulfate (SDS). The energy of adhesion is much greater with KCl than with NaCl or LiCl, although the KCl concentration is lower. (Adapted from [110].)

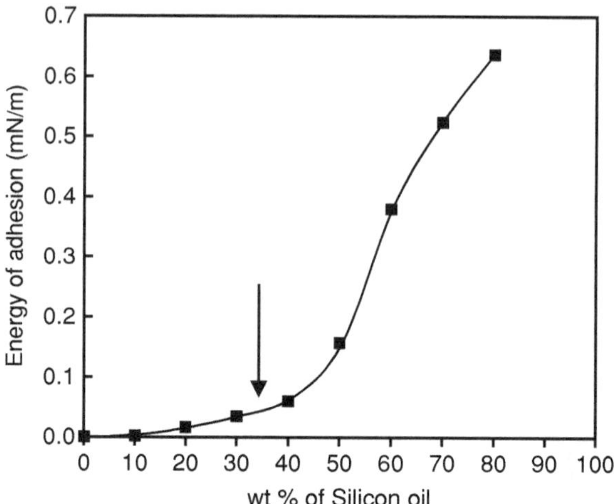

FIGURE 2.38. Energy of adhesion between water droplets stabilized with sorbitan monooleate (Span 80) in a silicon oil–dodecane mixture. The arrow indicates the insolubility threshold of the amphiphile. (Adapted from [113].)

becomes insoluble in the solvent due to the large ratio of silicone oil. Similar behavior was observed for inverted emulsions stabilized by phospholipids [114].

2.3.2.3. Influence of Droplet Size

For weakly adhesive systems, the pair energy between two identical droplets, u, can be written as [109]:

$$u = -\frac{\pi a_0^2 E(h_e)^2}{2\gamma_{\text{int}}} \tag{2.38}$$

where a_0 is the droplet's undeformed radius. The energy u scales as a_0^2. This means that for a given energy of adhesion $E(h_e)$, the pair energy u can strongly vary as a function of the droplet size. If the net energy is much larger than $k_B T$, the droplets are expected to be adhesive. However, if u is of the order of $k_B T$, or even weaker, the droplets can be dispersed. Because u depends on the size of the particles, small droplets can be homogeneously dispersed in conditions where large droplets are adhesive. This effect was observed by Poulin et al. in oil-in-water emulsions stabilized by SDS in the presence of salt [109]. The onset of adhesion of very large droplets for such systems is given in Fig. 2.39. T^* defines the temperature below which macroscopic interfaces become adhesive. It was observed that below T^* small droplets can remain totally dispersed. However, below a critical temperature T_g, lower than T^*, aggregation and adhesion of the drops occur. T_g is plotted in Fig. 2.39 for various salt concentrations and different drop sizes. The temperature shift of adhesion, from macroscopic to colloidal scale, increases as the size of the drops decreases. This effect was quantitatively explained by a very simple model

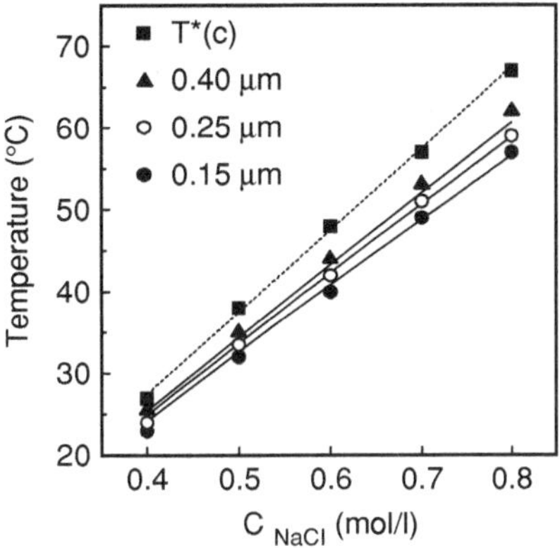

FIGURE 2.39. Evolution of the wetting transition temperature $T^*(c)$ for macroscopic interfaces and gelation temperature $T_g(c)$ of emulsions of various droplet sizes. (Adapted from [113].)

that takes into account the entropy of the drops and the net energy of adhesion between them [109].

2.3.2.4. Film Thickness Measurements

Film thickness measurements can be achieved between macroscopic solid or liquid interfaces via optical techniques [17,90,99–102,115]. For example, an early estimation of the thickness of Newton black soap films stabilized by SDS was about 40 Å [99–101]. More accurate measurements were performed using X-ray reflectivity [116]. It was shown that the thickness of Newton black soap films is, in fact, about 30 Å. This means that such film is essentially a surfactant bilayer standing in air. Neutron-scattering from concentrated emulsions was used to measure the thickness of adhesive films between small emulsion droplets suspended in a liquid phase [111]. This technique takes advantage of the large surface area that is involved in a concentrated emulsion of small emulsion droplets. It is found that the global film thickness is 29 Å. Moreover, this thickness is independent of the temperature. The thickness in emulsion systems is thus similar to the thickness of Newton black soap films [116]. Emulsion and foam films stabilized with the same surfactant exhibit similar adhesive behavior [117], although the Hamaker constants for air and oil systems are significantly different [17]. This shows that the van der Waals interactions between oil or air media do not play an important role. Moreover, the order of magnitude of the energy of adhesion in surfactant films can reach a few mN/m. Such value reflects the molecular origin of the adhesion in surfactant films. Indeed, a few mN/m corresponds to about a few tenths of $k_B T$ per surfactant molecule. The cohesion of adhesive liquid films is thus usually arising from short-range interactions between the surfactant layers.

2.4. Conclusion

Several techniques (SFA, TFB, AFM, MCT, etc.) have allowed very important advances in the field of surface forces between solid and liquid surfaces. Force measurements offer insight into interaction of emulsion droplets and have implications when considering the stability and processing of emulsions. So far most of the studies have concerned equilibrium forces. Additional effects occur when the droplets move relative to one another. Two kinds of dynamical effects can be distinguished. The most obvious occurs when droplets move so rapidly that there is not enough time for equilibrating. For example, the ions of the electric double-layer can be left slightly behind when a charged droplet moves. Identically, for droplets covered with adsorbed polymer, the rearrangements at the surface are often slow, causing the force between two such surfaces to depend on the rate of the relative motion. The second, dynamical effect arises when droplets or aggregates move in the liquid medium. As a big unit moves, it creates a flow in the surrounding incompressible liquid. The presence of a second particle will influence the flow patterns in the liquid and will result in the two particles "seeing" each other trough the flow of the intervening fluid. This effect is called hydrodynamic interaction. So far it is difficult to state general rules concerning dynamic effects and it seems likely that dynamic measurements of hydrodynamic and viscous forces will receive increased attention in the near future.

References

[1] D. Tabor and R.H.S. Wintertor: "Direct Measurement of Normal and Retarded Van Der Waals Forces." Proc. R. Soc. Lond. **A312**, 435 (1969).

[2] J.N. Israelachvili and D. Tabor: "Measurement of Van Der Waals Dispersion Forces in the Range 1.5 to 130 Nm." Proc. R. Soc. Lond. **A331**, 19 (1972).

[3] J.N. Israelachvili and G.E. Adams: "Measurement of Forces between Two Mica Surfaces in Aqueous Electrolyte Solutions in the Range 0-100 Nm." J. Chem. Soc. Faraday Trans. I **74**, 975 (1978).

[4] J.L. Parker: "A Novel Method for Measuring the Force Between Two Surfaces in a Surface Force Apparatus." Langmuir **8**, 551 (1992).

[5] J.L. Parker: "Surface Force Measurements in Surfactant Systems." Prog. Surf. Sci. **47**, 205 (1994).

[6] W.A. Ducker, T.S. Senden, and R.M. Pashley: "Direct Measurement of Colloidal Forces Using an Atomic Force Microscope." Nature **353**, 239 (1991).

[7] D.C. Prieve and N.A. Frej: "Total Internal Reflection Microscopy: A Quantitative Tool for the Measurement of Colloidal Forces." Langmuir **6**, 396 (1990).

[8] A. Scheludko and D. Exerowa: "Flow of Liquids from Soap Films." Kolloids Z. **155**, 39 (1957).

[9] R. Aveyard, B.P. Binks, W.G. Cho, L.R. Fisher, P.D.I. Fletcher, and F. Klinkhammer: "Investigation of the Force-Distance Relationship for a Small Liquid Drop Approaching a Liquid-Liquid Interface." Langmuir **12**, 6561 (1996).

[10] F. Leal-Calderon, T. Stora, O. Mondain Monval, P. Poulin, and J. Bibette: "Direct Measurement of Colloidal Forces." Phys. Rev. Lett. **72**, 2959 (1994).

[11] D.M. LeNeveu, R.P. Rand, and V.A. Parsegian: "Measurement of Forces Between Lecithin Bilayers." Nature **259**, 601 (1976).

[12] V. Bergeron: "Stability of Emulsion Films." In Proceedings of the Second World Congress of Emulsions **4**, 247, Bordeaux, France (1997).

[13] V. Bergeron, M.E. Fagan, and C.J. Radke: "Criterion for Foam Stability Against Oil in Porous Media." Langmuir **9**, 1704 (1993).
[14] T. Tchaliovska, P. Herder, R. Pugh, P. Stenius, and J.C. Ericksson: "Studies of the Contact Interaction Between an Air Bubble and a Mica Surface Submerged in Dodecylammonium Chloride Solution." Langmuir **6**, 1535 (1990).
[15] B.V. Derjaguin, N.V. Churaev, and V.M. Muller: *Surface Forces*. Consultants Bureau, New York (1987).
[16] H. Zhang and M. Widom: "Field-Induced Forces in Colloidal Particle Chains." Phys. Rev. E **51**, 2099 (1995).
[17] J.N. Israelachvili: *Intermolecular and Surface Forces*. Academic Press, San Diego (1985).
[18] E.J.W. Verwey and J.T.H. Overbeek: *Theory of the Stability of Lypophobic Colloids*. Elsevier, Amsterdam (1948).
[19] S. Levine and A. Suddaby: "Simplified Forms for Free Energy of the Double-Layers of Two Plates in a Symmetrical Electrolyte." Proc. Phys. Soc. A **64**, 287 (1951).
[20] S. Asakura and J. Oosawa: "Interaction Between Particles Suspended in Solutions of Macromolecules." J. Polym. Sci. **32**, 183 (1958).
[21] P. Richetti and P. Kékicheff: "Direct Measurement of Depletion and Structural Forces in a Micellar System." Phys. Rev. Lett. **68**, 1951 (1992).
[22] O. Mondain-Monval, F. Leal-Calderon, J. Phillip, and J. Bibette: "Depletion Forces in the Presence of Electrostatic Double-Layer Repulsion." Phys. Rev. lett. **75**, 3364 (1995).
[23] E.D. Manev, S.V. Sazdanova, and D.T. Wasan: "Stratification in Emulsion Films." J. Dispersion Sci. Technol. **5**, 111 (1984).
[24] P.A. Kralchevsky and N.D. Denkov: "Analytical Expression for the Oscillatory Structural Surface Force." Chem. Phys. Lett. **240**, 385 (1995).
[25] V. Bergeron and C.J. Radke: "Equilibrium Measurements of Oscillatory Disjoining Pressures in Aqueous Foam Films." Langmuir **8**, 3020 (1992).
[26] P.G. De Gennes: *Scaling Concepts in Polymer Physics*. Cornell University Press London (1979).
[27] G. Fleer, M. Cohen Stuart, J. Scheutjens, T. Cosgrove, and B. Vincent: *Polymers at Interfaces*. Chapman and Hall, London (1993).
[28] S.S. Patel and M. Tirrell: "Measurement of Forces Between Surfaces in Polymer Fluids." Annu. Rev. Phys. Chem. **40**, 597 (1989).
[29] J. Lyklema and T. Van Vliet: "Polymer-Stabilized Free Liquid Films." Faraday Disc. Chem. Soc. **65**, 26 (1978).
[30] O. Mondain-Monval, A. Espert, P. Omarjee, J. Bibette, F. Leal-Calderon, J. Phillip, and J.F. Joanny: "Polymer Induced Repulsive Forces: An Exponential Scaling." Phys. Rev. Lett. **80**, 1778 (1998).
[31] P. Omarjee: "Direct Force Measurements Between Polymer Stabilized Colloids." Ph.D thesis, Bordeaux I University (1999).
[32] A.N. Semenov, J. Bonet-Avalos, A. Johner, and J.F. Joanny: "Adsorption of Polymer Solutions onto a Flat Surface." Macromolecules **29**, 2179 (1996).
[33] H. Sonntag, B. Unterberger, and S. Zimontkowski: "Experimental Investigation of the Steric Stabilization of Emulsions by (Poly)Vinyl Alcohol." Colloid Polym. Sci. **257**, 286 (1979).
[34] P. Omarjee, A. Espert, and O. Mondain Monval: "Polymer-Induced Repulsive Forces at Solid-Liquid and Liquid-Liquid Interfaces." Langmuir **17**, 5693 (2001).

[35] J. Klein and P.F. Luckham: "Long-Range Attractive Forces Between Two Mica Surfaces in an Aqueous Polymer Solution." Nature **308**, 836 (1984).
[36] J. Klein and G. Rossi: "Analysis of the Experimental Implications of the Scaling Theory of Polymer Adsorption." Macromolecules **31**, 1979 (1998).
[37] L. Cohen-Tannoudji, E. Bertrand, L. Bressy, C. Goubault, J. Baudry, J. Klein, J.-F. Joanny, and J. Bibette: "Polymer Bridging Probed by Magnetic Colloids." Phys. Rev. Lett. **94**, 038301 (2005).
[38] H. Vogel: "The Law of the Relation Between the Viscosity of Liquids and the Temperature." Physik. Z. **22**, 645 (1921).
[39] G.S. Fulcher: "Analysis of Recent Measurements of the Viscosity of Glasses. II." J. Am. Ceram. Soc. **8**, 789 (1925).
[40] G. Tammann and W. Hesse: "The Dependence of Viscosity Upon the Temperature of Supercooled Liquids." Z. Anorg. Allg. Chem. **156**, 245 (1926).
[41] M.L. Williams, R.F. Landel, and J.D. Ferry: "The Temperature Dependence of Relaxation Mechanisms in Amorphous Polymers and Other Glass-Forming Liquids." J. Am. Chem. Soc. **77**, 3701 (1955).
[42] M.D. Ediger, C.A. Angell, and S.R. Nagel: "Supercooled Liquids and Glasses." J. Phys. Chem. **100**, 13200 (1996).
[43] K. Kremer: "Glassy States of Adsorbed Flexible Polymers and Spread Polymer "Monolayers." J. Phys. (Paris) **47**, 1269 (1986).
[44] J.B. Avalos, J.-F. Joanny, A. Johner, and A.N. Semenov: "Equilibrium Interaction Between Adsorbed Polymer Layers." Europhys. Lett. **35**, 97 (1996).
[45] G.D.J. Phillies, D. Rostcheck, and S. Ahmed: "Probe Diffusion in Intermediate Molecular Weight Polyelectrolytes: Temperature Dependence." Macromolecules **25**, 3689 (1992).
[46] D.W. Krevelen: *Properties of Polymers*. Elsevier, Philadelphia (1997).
[47] C. Goubault, F. Leal-Calderon, J.-L. Viovy, and J. Bibette: "Self-Assembled Magnetic Nanowires Made Irreversible by Polymer Bridging." Langmuir **21**, 3725 (2005).
[48] J.-F. Joanny, L. Leibler, and P.G. de Gennes: "Effects of Polymer Solutions and Colloid Stability." J. Polym. Sci. Polym. Phys. **17**, 1073 (1979).
[49] J.M. Scheutjens and G.J. Fleer: "Interaction Between Two Adsorbed Polymer Layers." Macromolecules **18**, 1882 (1985).
[50] P. Kékicheff, F. Nallet, and P. Richetti: "Measurement of Depletion Interaction in Semi-Dilute Solutions of Worm-Like Surfactant Aggregates." J. Phys. II France **4**, 735 (1994).
[51] A. Milling and S. Biggs: "Direct Measurement of the Depletion Force Using an Atomic Force Microscope." J. Colloid Interface Sci. **170**, 604 (1995).
[52] E. Evans and D. Needham: "Attraction Between Lipid Bilayer Membranes in Concentrated Solutions of Nonadsorbing Polymers: Comparison of Mean-Field Theory with Measurements of Adhesion Energy." Macromolecules **21**, 1822 (1988).
[53] A. Asnacios, A. Espert, A. Colin, and D. Langevin: "Structural Forces in Thin Films Made from Polyelectrolyte Solutions." Phys. Rev. Lett. **78**, 4974 (1997).
[54] K. Thalberg, B. Lindman, and G. Karlstrom: "Phase Behavior of Cationic Surfactant and Anionic Polyelectrolyte: Influence of Surfactant Chain Length and Polyelectrolyte Molecular Weight." J. Phys. Chem. **95**, 3370 (1991).
[55] K. Thalberg, B. Lindman, and G. Karlstrom: "Phase Behavior of a System of Cationic Surfactant and Anionic Polyelectrolyte: The Effect of Salt." J. Phys. Chem. **95**, 6004 (1991).

[56] P. Hansson and M. Almgren: "Interaction of Alkyltrimethylammonium Surfactants with Polyacrylate and Poly(Styrenesulfonate) in Aqueous Solution: Phase Behavior and Surfactant Aggregation Numbers." Langmuir **10**, 2115 (1994).

[57] P.L. Dubin, D.R. Rigsbee, L.M. Gan, and M.A. Fallon: "Equilibrium Binding of Mixed Micelles to Oppositely Charged Polyelectrolytes." Macromolecules **21**, 2555 (1988).

[58] M. Almgren, P. Hansson, E. Mukhtar, and S. van Stam: "Aggregation of Alkyltrimethylammonium Surfactants in Aqueous Poly(Styrenesulfonate) Solutions." Langmuir **8**, 2405 (1992).

[59] A. Asnacios, D. Langevin, and J.-F. Argillier: "Complexation of Cationic Surfactant and Anionic Polymer at the Air-Water Interface." Macromolecules **29**, 7412 (1996).

[60] A.V. Gorelov, E.D. Kudryashov, J.C. Jacquier, D.M. McLoughlin, and K.A. Dawson: "Complex Formation Between DNA and Cationic Surfactant." Physica A **249**, 216 (1998).

[61] Y. Li, J. Xia, and P.L. Dubin: "Complex Formation Between Polyelectrolyte and Oppositely Charged Mixed Micelles: Static and Dynamic Light Scattering. Study of the Effect of Polyelectrolyte Molecular Weight and Concentration." Macromolecules **27**, 7049 (1994).

[62] P.M. Claesson, M. Bergstrom, A. Dedinaite, M. Kjellin, J.F. Legrand, and I. Grillo: "Mixtures of Cationic Polyelectrolyte and Anionic Surfactant Studied with Small Angle Neutron Scattering." J. Phys. Chem. B **104**, 11689 (2000).

[63] A. Svensson, L. Picullel, B. Cabane, and P.A. Ilekti: "New Approach to the Phase Behavior of Oppositely Charged Polymers and Surfactants." J. Phys. Chem. B **106**, 1013 (2002).

[64] K. Hayakawa, J.P. Santerre, and J.C.T. Kwak: "Study of Surfactant-Polyelectrolyte Interactions. Binding of Dodecyl-and Tetradecyltrimethylammonium Bromide by Some Carboxylic Polyelectrolytes." Macromolecules **16**, 1642 (1983).

[65] K. Hayakawa and J.C.T. Kwak: "Surfacatant-Polyelectrolyte Interactions. 4. Surfactant Chain Length Dependence on the Binding of Alkylpyridinium Cations to Dextran Sulfate." J. Phys. Chem. **88**, 1930 (1984).

[66] J. Lucassen, F. Hollway, and J.H. Buckingham: "Surface Properties of Mixed Solutions of Poly-Lysine and Sodium Dodecyl Sulfate. 2. Dynamic Surface Properties." J. Colloid Interface Sci. **67**, 432 (1978).

[67] A. Asnacios, R. Kitzling, and D. Langevin: "Mixed Monolayers of Polyelectrolytes and Surfactants at the Air-Water Interface." Colloid Surf. A **167**, 189 (2000).

[68] D.J.F. Taylor, R.K. Thomas, and J. Penfold: "The Adsorption of Oppositely Charged Polyelectrolyte/Surfactant Mixtures: Neutron Reflection from Dodecyl Trimethylammonium Bromide and Sodium Poly(Styrene Sulfonate) at the Air/Water Interface." Langmuir **18**, 4748 (2002).

[69] E. Staples, I. Tucker, J. Penfold, R.K. Thomas, and D.J.F. Taylor: "Organization of Polymer-Surfactant Mixtures at the Air-Water Interface: Sodium Dodecyl Sulfate and Poly Dimethyldiallylammonium Chloride." Langmuir **18**, 5147 (2002).

[70] C. Stubenrauch, P.A. Albouy, R. Kitzling, and D. Langevin: "Polymer/Surfactant Complexes at the Water/Air Interface: A Surface Tension and X-Ray Reflectivity Study." Langmuir **16**, 3206 (2000).

[71] H. Ritacco, P.A. Albouy, A. Bhattacharyya, and D. Langevin: "Influence of the Polymer Backbone Rigidity on Polyelectrolyte-Surfactant Complexes at the Air/Water Interface." Phys. Chem. Chem. Phys. **2**, 5243 (2000).

[72] P.M. Claesson, A. Dedinaite, M. Fielden, M. Kjellin, and R. Audebert: "Polyelectrolyte-Surfactant Interactions at Interfaces." Prog. Colloid Polym. Sci. **106**, 24 (1997).
[73] C. Monteux, C.E. Williams, J. Meunier, O. Anthony, and V. Bergeron: "Adsorption of Oppositely Charged Polyelectrolyte/Surfactant Complexes at the Air/Water Interface : Formation of Interfacial Gels." Langmuir **20**, 57 (2004).
[74] J. Philip, G. Gnana Prakash, T. Jayakumar, P. Kalyanasundaram, and B. Raj: "Stretching and Collapse of Neutral Polymer Layers Under Association with Ionic Surfactants." Phys. Rev. Lett. **89**, 268301 (2002).
[75] J. Philip, G. Gnana Prakash, T. Jayakumar, P. Kalyanasundaram, and B. Raj: "Three Distinct Scenarios under Polymer, Surfactant and Colloidal Interaction." Macromolecules **36**, 9230 (2003).
[76] D.E. Graham and M.C. Phillips: "Proteins at Liquid Interfaces: II. Adsorption Isotherms." J. Colloid Interface Sci. **70**, 415 (1979).
[77] T. Sengupta, L. Razumovsky, and S. Damodaran: "Energetics of Protein-Interface Interactions and its Effect on Protein Adsorption." Langmuir **15**, 6991 (1999).
[78] T.D. Dimitrova and F. Leal-Calderon: "Forces Between Emulsion Droplets Stabilized with Tween 20 and Proteins." Langmuir **15**, 8813 (1999).
[79] T.D. Dimitrova, F. Leal-Calderon, T.D. Gurkov, and B. Campbell "Disjoining Pressure vs. Thickness Isotherms of Thin Emulsion Films Stabilized by Proteins." Langmuir **17**, 8069 (2001).
[80] T. Nylander and M.N. Wahlgren: "Forces Between Adsorbed Layers of β-Casein." Langmuir **13**, 6219 (1997).
[81] K. Kozco, A.D. Nikolov, D.T. Wasan, R.P. Borwankar, and A. Gonsalves: "Layering of Sodium Caseinate Submicelles in Thin Liquid Films–A New Stability Mechanism for Food Dispersions." J. Colloid Interface Sci. **178**, 694 (1996).
[82] E. Dickinson, M. Golding, and M.J.W. Povey: "Creaming and Flocculation of Oil-in-Water Emulsions Containing Sodium Caseinate." J. Colloid Interface Sci. **185**, 515 (1997).
[83] J.P. Gallinet and B. Gauthier-Manuel: "Adsorption–Desorption of Serum Albumin on Bare Mica Surfaces." Colloid Surfaces **68**, 189 (1992).
[84] G. Narsimham: "Maximum Disjoining Pressure in Protein-Stabilized Concentrated Oil-in-Water Emulsions." Colloid Surfaces **62**, 31 (1992).
[85] J.N. Israelachvili and H. Wennerström: "Entropic Forces Between Amphiphilic Surfaces in Liquids." J. Phys. Chem. **96**, 520 (1992).
[86] J.R. Lu, T.J. Su, and R.K. Thomas: "Structural Conformation of Bovine Serum Albumin Layers at the Air–Water Interface Studied by Neutron Reflection." J. Colloid Interface Sci. **213**, 426 (1999).
[87] T.D. Dimitrova, F. Leal-Calderon, T.D. Gurkov, and B. Campbell: "Surface Forces in Model Oil-in-Water Emulsions Stabilized by Proteins." Adv. Colloid Interface Sci. **108–109**, 73 (2004).
[88] B.S. Murray, B. Cattin, E. Schuler, and Z.O. Sonmez: "Response of Adsorbed Protein Films to Rapid Expansion." Langmuir **18**, 9476 (2002).
[89] A.R. Mackie, P. Gunning, P. Wilde, and V. Morris: "Orogenic Displacement of Protein from the Oil/Water Interface." Langmuir **16**, 2242 (2000).
[90] N.D. Denkov, N.D. Petsev, and K.D. Danov: "Interaction Between Deformable Brownian Droplets." Phys. Rev. Lett. **71**, (1993).
[91] W.A. Ducker, Z. Xu, and J.N. Israelachvili: "Measurements of Hydrophobic and DLVO Forces in Bubble-Surface Interactions in Aqueous Solutions." Langmuir **10**, 3279 (1994).

[92] M.L. Fielden, R.A. Hayes, and J. Ralston: "Surface and Capillary Forces Affecting Air Bubble-Particle Interactions in Aqueous Electrolyte." Langmuir **12**, 3721 (1996).

[93] M. Preuss and H.-J. Butt: "Direct Measurement of Particle-Bubble Interactions in Aqueous Electrolyte: Dependence on Surfactant." Langmuir **14**, 3164 (1998).

[94] G. Gillies, C.A. Prestidge, and P. Attard: "An AFM Study of the Deformation and Nanorheology of Cross-Linked PDMS Droplets." Langmuir **18**, 1674 (2002).

[95] G. Gillies, C.A. Prestidge, and P. Attard: "Determination of the Separation in Colloid Probe Atomic Force Microscopy of Deformable Bodies." Langmuir **17**, 7955 (2001).

[96] I. Vakarelski, A. Toritani, M. Nakayama, and Higashitani: "Effects of Particle Deformability on Interaction Between Surfaces in Solutions." Langmuir **19**, 110 (2003).

[97] G. Gillies and C.A. Prestidge: "Interaction Forces, Deformation and Nano-Rheology of Emulsion Droplets as Determined by Colloid Probe AFM." Adv. Colloid Interface Sci. **108–109**, 197 (2004).

[98] G. Gillies and C.A. Prestidge: "Colloid Probe AFM Investigation of the Influence of Cross-Linking on the Interaction Behavior and Nano-Rheology of Colloidal Droplets." Langmuir **21**, 12342 (2005).

[99] J.A. De Feijter, B. Rijnbout, and A. Vrij: "Contact Angles in Thin Films. I. Thermodynamic Description." J. Colloid Interface Sci. **64**, 258 (1978).

[100] J.A. De Feijter and A. Vrij: "Contact Angles in Thin Films. II. Contact Angle Measurements in Newton Black Soap Films." J. Colloid Interface Sci. **64**, 269 (1978).

[101] J.A. De Feijter and A. Vrij: "Contact Angles in Thin Films. III. Interaction Forces in Newton Black Soap Films." J. Colloid Interface Sci. **70**, 456 (1979).

[102] I.B. Ivanov (ed): "*Thin Liquid Films. Fundamentals and Applications*." Marcel Dekker, New York (1988).

[103] R.B. Gennis: *Biomembranes*. Springer-Verlag, New York (1989).

[104] M.P. Aronson and H.M. Princen: "Contact Angles in Oil-in-Water Emulsions Stabilized by Ionic Surfactants." Nature **286**, 370 (1980).

[105] M.P. Aronson and H.M. Princen: "Geometry of Clusters of Strongly Coagulated Fluid Drops and the Occurence of Collapsed Plateau Borders." Colloids Surfaces **4**, 173 (1982).

[106] H. Princen: "Geometry of Clusters of Strongly Coagulated Fluid Drops and the Occurence of Collapsed Plateau Borders." Colloids Surfaces **9**, 47 (1984).

[107] I.B. Ivanov, A.S. Dimitrov, A.D. Nikolov, N.D. Denkov, and P.A. Kralchevsky: "Contact Angle Film and Line Tension of Foam Films. II. Film and Line Tension Measurements." J. Colloid Interface Sci. **151**, 446 (1992).

[108] A.S. Dimitrov, A.D. Nikolov, P.A. Kralchevsky, and I.B. Ivanov: "Wetting of Emulsions Droplets: From Macroscopic to Colloidal Scale." J. Colloid Interface Sci. **151**, 462 (1992).

[109] P. Poulin and J. Bibette: "Interaction Between Deformable Brownian Droplets." Phys. Rev. Lett. **79**, 3290 (1997).

[110] P. Poulin: "Adhésion d'interfaces Fluides et Agrégation Colloïdale dans les Émulsions." Ph.D thesis, Bordeaux I University (1995).

[111] P. Poulin and J. Bibette: "Influence of the Alkyl Surfactant Tail on the Adhesion Between Emulsion Drops." Langmuir **15**, 4731 (1999).

[112] F.A.M. Leermarkers, Y.S. Sdranis, J. Lyklema, and R.D. Groot: "Adhesion of Water Droplets in Organic Solvent." Colloids Surfaces **85**, 135 (1994).

[113] P. Poulin and J. Bibette: "Stability and Permeability of Amphiphile Bilayers." Langmuir **14**, 6341 (1998).

[114] P. Poulin, W. Essafi, and J. Bibette: "On the Colloidal Stability of Water-in-Oil Emulsions. A Self-Consistent Field Approach." J. Chem. Phys. B **103**, 5157 (1999).

[115] D. Exerowa, D. Kashiev, and D. Platikanov: "Structural Properties of Soap Black Films Investigated by X-Ray Reflectivity." Adv. Colloid Interface Sci. **40**, 201 (1992).

[116] O. Bélorgey, and J. Benattar: "Wetting: Statistics and Dynamics." Phys. Rev. Lett. **66**, (1991).

[117] P. Poulin, F. Nallet, B. Cabane, and J. Bibette: "Evidence for Newton Black Films Between Adhesive Emulsion Droplets." Phys. Rev. Lett. **77**, 3248 (1996).

3
Phase Transitions

3.1. Introduction

Attractive interactions between droplets may induce flocculation and, depending on the respective density of the solvent and the droplets, sedimentation or creaming occurs. Flocculation is the process in which emulsion droplets aggregate, without rupture of the stabilizing film between them. Flocculation occurs when the total pairwise interaction between droplets becomes appreciably attractive at a given separation. If the depth of the attractive well is of the order of the thermal energy, $k_B T$, then the droplets form aggregates that coexist with single Brownian droplets. The thermal energy allows an equilibrium state to be reached with permanent exchange between aggregated and free droplets. At the macroscopic scale, after a few hours of settling, the emulsions undergo a phase separation between a concentrated cream/sediment and a dilute phase. When the depth of the attractive well is large compared to $k_B T$ (i.e., more than 10 $k_B T$), the droplets are strongly bound to the aggregates and cannot be redispersed by thermal motion. After a short period of time, all the droplets are entrapped within large and tenuous clusters that fill the whole space. The contraction of this gel-like network may be very long, especially in concentrated emulsions. In Section 3.2, we describe flocculation phenomena produced by weak attractive interactions. In this limit, flocculation can be regarded as a reversible phase transition, exactly as in molecular fluids. To illustrate the analogy, we shall present a method that allows partitioning of a polydisperse emulsion into monodisperse fractions (fractionated crystallization) based on reversible aggregation. Finally, irreversible flocculation phenomena and their associated gel structures are presented in Section 3.3.

3.2. Weak Attractive Interactions and Equilibrium Phase Transitions

3.2.1. Experimental Observations

The droplet structure in attractive emulsions can be directly observed under a microscope. Besides direct observations, a precise determination of the structure and a quantitative determination of the interdroplet attraction generally requires the use of scattering techniques.

3.2.1.1. Scattering Techniques

Scattering methods, such as light scattering, are generally useful for studying emulsions owing to the characteristic visible wavelength which is of the same order as the droplet diameter. In a typical scattering experiment, a monochromatic beam of light is focused on the liquid sample and the intensity of the scattered beam is measured via a detector, as illustrated in Fig. 3.1. Collisions between photons and the molecules of the sample are elastic (with no energy exchange), but a transfer of momentum takes place. The larger the momentum transfer, the higher the scattering angle θ. The momentum transfer is determined by the scattering vector $\vec{q}$ (Fig. 3.1) such that:

$$q = |\vec{q}| = \frac{4\pi n_r}{\lambda_0} \sin \frac{\theta}{2} \tag{3.1}$$

where λ_0 is the light wavelength in vacuum and n_r is the average refractive index of the sample. A straightforward scattering experiment consists of measuring the intensity of the scattered beam as a function of the scattering vector. In colloidal systems, light scattering is caused by local variations in the refractive index. A light beam hitting a colloidal system will "see" the dispersed phase as particles of refractive index n_d, more or less randomly distributed in a solvent of refractive index n_s. Provided $n_d \neq n_s$, the refractive index averaged over a volume of size λ_0^3 will vary in accordance with the local concentrations of the dispersed objects. The average intensity of the scattered light depends on four factors: an instrumental constant, a contrast factor, a concentration factor, and a sample-dependent scattering function. The three first factors can be measured independently, and the relevant information is found in the scattering function, which is termed $I(q)$. If

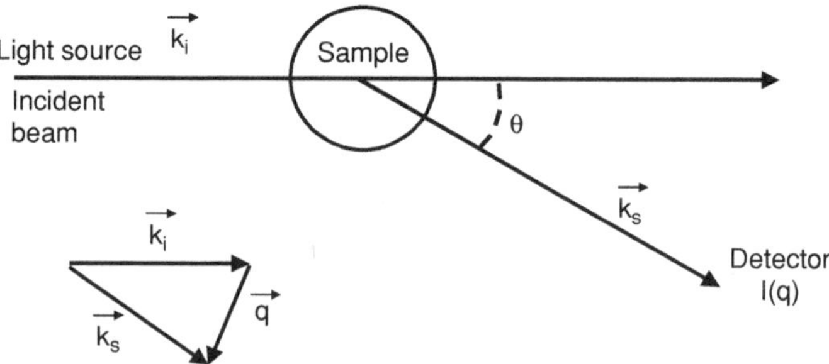

FIGURE 3.1. Scheme of a light-scattering experiment. For elastic scattering, the magnitude of the wave-vector is conserved: $|\vec{k}_i| = |\vec{k}_s| = 2\pi n_r/\lambda_0$. The magnitude of the scattering vector $\vec{q} = \vec{k}_s - \vec{k}_i$ is given by Eq. (3.1).

the intra- and interparticle interferences are independent, the scattering function becomes a product of a form factor $P(q)$ and a structure factor $S(q)$:

$$I(q) = P(q)\mathrm{S(q)} \tag{3.2}$$

The form factor depends only on intraparticle interferences and is independent of concentration as long as the particles remain unchanged. For example, for a uniform sphere of radius a comparable to λ_0:

$$P(q) = \left[\frac{3\,(\sin qa - qa\cos qa)}{(qa)^3}\right]^2 \tag{3.3}$$

The structure factor accounts for interparticle interferences. It relates to the so-called pair distribution function, $g(r)$, of particles through:

$$S(q) = 1 + \frac{3\phi}{qa^3}\int_0^\infty [g(r) - 1]r\sin(qr)\,dr \tag{3.4}$$

where ϕ is the particle volume fraction. The function $g(r)$ describes the probability to find one particle at center-to-center distance r from a test particle (at the origin). Thus, if we know the particle size, we can calculate $P(q)$, measure $I(q)$, and then determine the particle distribution $g(r)$. Knowing the radial distribution function, one may calculate all the macroscopic properties of suspended colloidal particles as the osmotic pressure, the isothermal compressibility, and so forth [1]. The radial distribution function is influenced by the many-body interactions that exist between the dispersed particles. Describing the radial distribution function for a multiparticle system is not simple. The reason is that the interactions between each two particles is affected by the presence of all others in the ensemble. This is why all equations for determining the radial distribution function introduce assumptions and approximations [1]. For example, for very dilute colloidal dispersions with short-range interparticle interactions, many-body correlations (i.e., more than two particles) can be neglected, and $g(r)$ can be expressed as a function of the two-body (pair) interaction energy, $u(r)$:

$$g(r) \approx \exp[-u(r)/k_BT] \tag{3.5}$$

Hence, by adopting the appropriate approximation, it is possible to obtain quantitative information about the pair interaction energy from scattering experiments, over a wide variety of systems and particle concentrations.

3.2.1.2. Oil-in-Water Emulsions

We first consider emulsion droplets submitted to attractive interactions of the order of k_BT. Reversible flocculation may be simply produced by adding excess surfactant in the continuous phase of emulsions. As already mentioned in Chapter 2, micelles may induce an attractive depletion interaction between the dispersed droplets. For equal spheres of radius a at center-to-center separation r, the depletion energy may be written as [2]:

$$u_d = \int_{\infty}^{r} F_d(r')dr'$$
$$= -\frac{4\pi}{3}(a+\Delta)^3 P_{osm}\left[1 - \frac{3r}{4(a+\Delta)} + \frac{r^3}{16(a+\Delta)^3}\right] \quad (3.6)$$

where F_d is the depletion force defined in Chapter 2, Δ is the exclusion characteristic length which is comparable to the micellar radius, and P_{osm} is the micellar osmotic pressure. This potential decreases monotonically, from zero at $r = 2(a+\Delta)$, to a minimum value u_{dc} at contact ($r = 2a$):

$$u_{dc} = -\frac{4\pi\Delta^3}{3}\left(1 + \frac{3a}{2\Delta}\right)P_{osm} \quad (3.7)$$

The first observation of depletion flocculation by surfactant micelles was reported by Aronson [3]. Bibette et al. [4] have studied the behavior of silicone-in-water emulsions stabilized by sodium dodecyl sulfate (SDS). They have exploited the attractive depletion interaction to size fractionate a crude polydisperse emulsion [5]. Because the surfactant volume fraction necessary to induce flocculation is always lower than 5%, the micelle osmotic pressure can be taken to be the ideal-gas value:

$$P_{osm} \approx n_m k_B T = \frac{\phi_m}{\frac{4}{3}\pi\Delta^3} k_B T \quad (3.8)$$

where n_m and ϕ_m are the micelle concentration and volume fraction, respectively. Moreover, because silicone oil droplets have a size ranging from 0.1 μm to a few microns, the ratio a/Δ is very large and Eq. (3.7) can be approximated by:

$$u_{dc} \approx -\frac{3}{2}k_B T\phi_m\frac{a}{\Delta} \quad (3.9)$$

This equation predicts a simple linear dependence of the contact potential on the micelle volume fraction and on the ratio a/Δ. For example, for $\phi_m = 2\%$ and $\Delta = 5$ nm (typical effective micelle radius), Eq. (3.9) predicts that the pair contact energy between droplets having a radius of 1 μm is equal to $-6\,k_B T$, an attraction that is in principle sufficient to produce flocculation. By adjusting the amount of surfactant above its critical micellar concentration (CMC), the volume fraction of micelles can be easily controlled. This type of interaction can be exploited to size fractionate a primary polydisperse emulsion [5]. To produce monodisperse fractions, a crude polydisperse emulsion is diluted to a droplet volume fraction of $\phi \approx 10\%$ According to Eq. (3.9), for or any given micelle volume fraction ϕ_m, the absolute value of the depletion attraction will be greater than $k_B T$ for oil droplets larger than some radius and will cause these larger droplets to flocculate. The density mismatch between the oil and the water will cause these flocs to cream after approximately 12 h of settling. The creamed droplets can be removed from the initial emulsion and redispersed at $\phi \approx 10\%$ in a separate suspension. By this means, the droplet distribution has been divided into two parts; droplets larger than some size will be preferentially in one part, while smaller droplets will be in the other. For the second fractionation step, the surfactant concentration of the emulsion containing the large droplets is reduced in order to flocculate only the largest droplets. In the emulsion containing the smaller droplets, the surfactant concentration is raised to flocculate still smaller droplets. By repeating this separation procedure five or six times, monodisperse emulsions having different droplet radii can be fractionated from a single polydisperse emulsion.

The uniformity of the droplet size distribution can be appreciated by comparing an initial polydisperse emulsion, shown in the optical microscopy image in Fig. 3.2a, with a monodisperse emulsion at the same volume fraction ($\approx$60%) obtained after six fractionation steps, shown in Fig. 3.2b. The polydisperse emulsion is highly disordered and possesses a wide range of droplet sizes. By contrast, the droplet size of the fractionated emulsion is very uniform, allowing the development of ordered packing. This fractionation method can produce monodisperse emulsions with droplet radii that range from 0.1 to 1 μm. The degree of polydispersity defined as the ratio of the size distribution's width to its average (see Chapter 1, eq 1.4 for the exact definition), may be as low as 10% after six purification steps. Owing to the entropic origin of the depletion attraction, this method is applicable to any kind of emulsion stabilized by any kind of surfactant provided the emulsion is stable against coalescence and Ostwald ripening.

Once produced, monodisperse emulsions can be used to study the physics of the phase separation occurring in the presence of excess surfactant [6]. As previously described, adding surfactant leads to the formation of aggregates coexisting with free droplets. Likewise, the two coexisting phases, the dense one containing aggregates and the dilute one containing free droplets, exchange continuously particles as can be observed under a microscope (Fig. 3.3). This strongly suggests the existence of a fluid–solid thermodynamic equilibrium as defined in the field of conventional molecular fluids. Macroscopic samples of the dense phase may be obtained after creaming. Iridescence is observed when a 1–5 μm thick film of the dense phase is deposited between two glass slides. This property gives fairly good certitude that the dense phase is solid-like with long range ordering of the oil droplets. This is confirmed by the photomicrograph in Fig. 3.2b in which one can easily distinguish oil droplets forming randomly oriented crystallites and by light diffraction experiments, the results of which are consistent with a face-centered crystalline structure [6]. Because a thermodynamic equilibrium is expected, one can draw a phase diagram for various droplet sizes as a function of surfactant concentration and oil droplet volume fraction [6] (Fig. 3.4). The experimental points separate the one-phase region where only free droplets are observed from the two-phase region (above the curve) where aggregated and free droplets coexist. Creaming is systematically observed at the macroscopic level when the surfactant concentration threshold is reached. Qualitatively, the droplet size dependence of the experimental phase diagrams is in accord with the above described depletion mechanism. Indeed, the colloidal aggregation at low droplet volume fraction can be considered as a gas–solid phase transition and may be simply described by

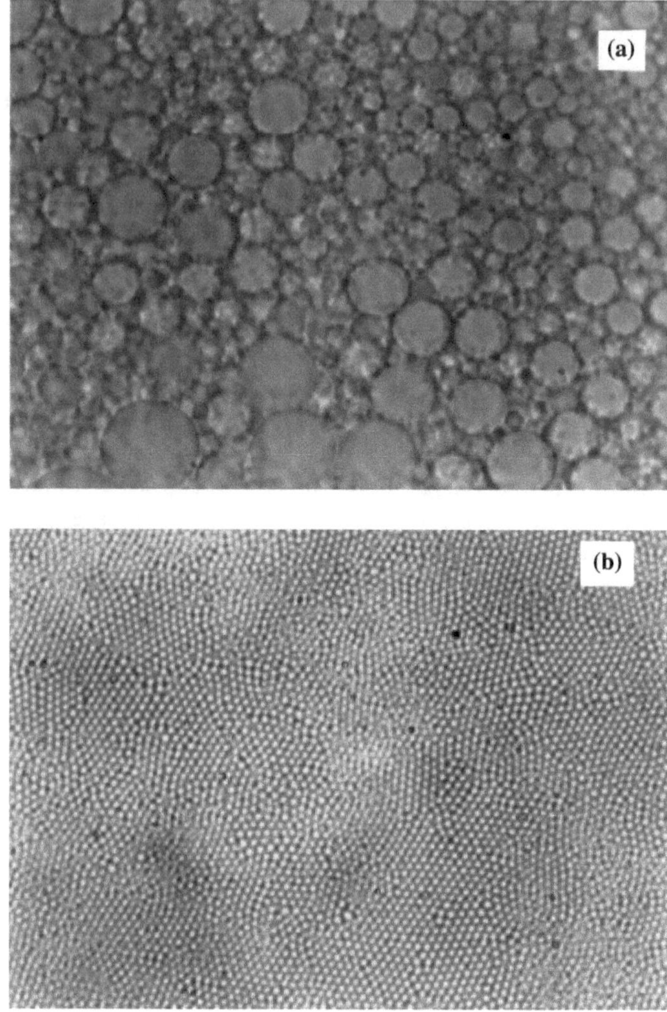

FIGURE 3.2. Microscopic image of a polydisperse emulsion **(a)** and of the emulsion obtained after six fractionation steps **(b)**. The droplet volume fraction in both pictures is around 60%. (Reproduced with permission from [5].)

equating both the chemical potentials and pressures of an ideal gas and a purely incompressible dense phase involving only nearest neighbors [6]:

$$\ln \phi = \frac{z}{2k_B T}(u_c + \Delta\mu^0) \tag{3.10}$$

where z is the coordination number of a droplet within the dense phase ($z = 12$ in a compact structure), u_c is the pair energy at each contact, ϕ is the droplet volume fraction, and $\Delta\mu^0$ is the reference chemical potential difference between

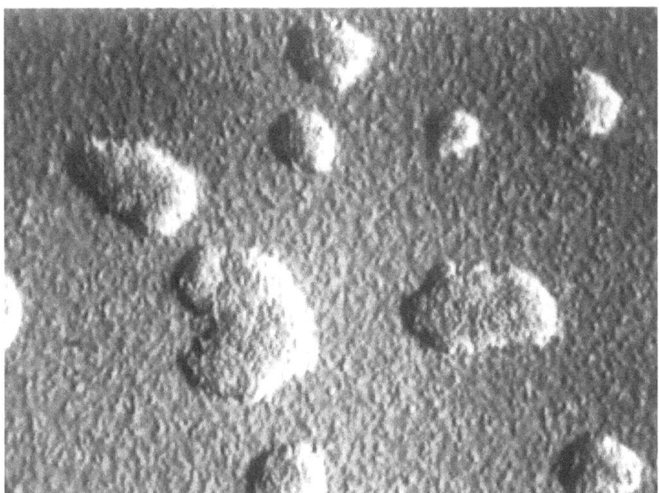

FIGURE 3.3. Microscopic image showing the effect of adding excess surfactant in the continuous phase of the emulsions: flocs separate from a coexisting fluid phase.

fluid and dense phases. Hence, at constant droplet volume fraction, the boundary corresponds to a constant pair interaction energy, u_c, within the aggregates. Following Eq. (3.9), higher ϕ_m values are required for smaller droplets to get phase separation ($u_c = u_{dc}$), as observed experimentally. The colloidal structure of dilute phases has been investigated by means of static light-scattering experiments [6]. The dilute phase exhibits a correlation peak indicating hard-sphere-like interactions between emulsion droplets at low surfactant concentration, replaced by intense small-angle scattering indicative of attractive interactions at higher surfactant concentration. The attractive interactions deduced from the scattered intensity profiles confirm the depletion mechanism.

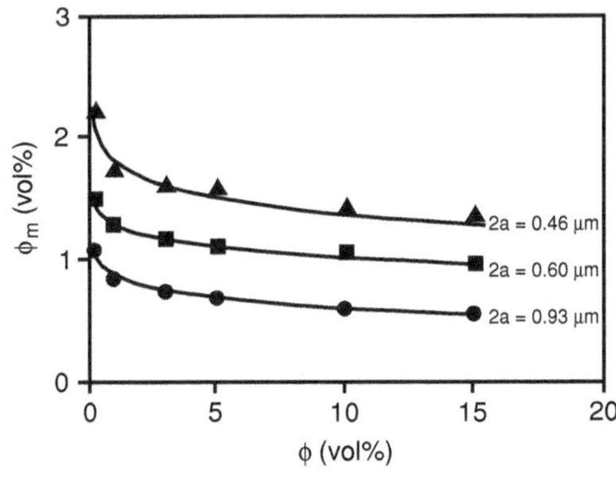

FIGURE 3.4. Experimental phase diagrams in the micelle volume fraction ϕ_m/oil volume fraction ϕ plane. The lines are visual guides that delimitate the one-phase region (fluid) from the two-phase region (fluid + solid). (Adapted from [6].)

As described in Chapter 2, depletion attraction may also arise from the presence of free polymer coils in the continuous phase of emulsions. Meller and Stavans have investigated the effect of nonadsorbing hydrophilic polymers (poly[ethylene oxide]) on the stability of monodisperse oil-in-water emulsions [7]. Above some threshold polymer concentration, the polymer coils induce a fluid–solid equilibrium due to depletion effects. Indeed, in the dilute regime, macromolecules may be regarded as hard spheres whose radius is comparable to the radius of gyration R_g. Polymer coils are thus excluded when droplet surface separation becomes lower than $\Delta = R_g$. Steiner et al. have investigated the phase behavior of mixtures of emulsion droplets, covering a large range of relative compositions and size ratios [8]. Their results confirm the general phenomenology of phase separation induced by osmotic depletion forces due to the smaller droplets. From their study it may be concluded that polydispersity in itself may induce aggregation.

Many different types of interaction can induce reversible phase transitions. For instance, weak flocculation has been observed in emulsions stabilized by nonionic surfactants by increasing the temperature. It is well known that many nonionic surfactants dissolved in water undergo a phase separation: above a critical temperature, an initially homogeneous surfactant solution separates into two micellar phases of different composition. This demixtion is generally termed as "cloud point transition." Identically, oil droplets covered by the same surfactants molecules become attractive within the same temperature range and undergo a reversible fluid–solid phase separation [9].

3.2.1.3. Water-in-Oil Emulsions

Inverse emulsions are generally made of water droplets covered by a layer of short, aliphatic chains adsorbed at the oil–water interface. Because the droplets are not charged and ions do not form easily in the low-dielectric-constant organic solvents, electrostatic interaction can be neglected. Hence, the interaction between the droplets depends essentially on the London-van-der Waals forces between the droplet cores and on the quality of the solvent with regard to the stabilizing chains. A rough approximation of the van der Waals potential between spherical particles at contact is given by [10]:

$$u_{VdW} = -\frac{Aa}{24\delta} \tag{3.11}$$

where A is the Hamaker constant and δ is the thickness of the surfactant stabilizing layer. To impart stability against van der Waals attraction, the absolute value of this interaction potential must be of the order or lower than the thermal energy, $k_B T$. For this to be the case, the stabilizing thickness δ must satisfy the condition:

$$\frac{\delta}{a} \geq \frac{A}{24 k_B T} \tag{3.12}$$

The interaction that arises from the overlapping of the surfactant chains is due to a complex interplay between enthalpic and entropic effects involving surfactant chain segments (monomer units) and solvent molecules. The enthalpic part of the interaction is determined by the balance between segment–segment and solvent–segment interactions. If the latter are highly favorable, the chains are solvated by solvent molecules and a slight interpenetration of two stabilizing layers leads to a strong repulsion, which is modeled as a "hard sphere" interaction. This repulsion is very strong and short ranged because the chains are short, uniform in length, and cover the surface densely. If the segment–solvent interaction is unfavorable, then in addition to the usual steric repulsion, an attractive interaction may appear between the stabilizing chains.

Leal-Calderon et al. [11] have studied the flocculation of water-in-oil emulsions stabilized with sorbitan monooleate (SMO), an oil-soluble surfactant that possesses an average unsaturated C_{18} hydrophobic tail. Above the critical micellar concentration, inverted micelles are present within the continuous oil phase, thus allowing micellar depletion forces to induce reversible flocculation and further fractionation of crude polydisperse emulsions. Within several decantation steps, different monodisperse emulsion droplets were obtained via the same procedure as described for direct emulsions. The monodisperse samples were then exploited to characterize the role of the oil continuous phase in inducing aggregation. It appeared that the droplets were dispersed in some oils referred as "good solvents," such as dodecane, and remained aggregated in many other different oils referred as "bad solvents," such as vegetable or silicone oils. Hence, it is interesting to identify and characterize the aggregation threshold as a function of the continuous phase composition made of a mixture of a so-called good solvent (dodecane) and bad solvent (isopropyl myristate [IPM]). Figure 3.5a,b, shows the phase diagram obtained when both the surfactant and the bad solvent contents are varied at constant water droplet volume fraction ($\phi = 5\%$). The system changes continuously from a Brownian emulsion to a gel on addition of IPM or excess surfactant. For intermediate solvent compositions, coexisting states (fluid–solid equilibrium) were observed. It is also noteworthy that this threshold depends on the droplet size. Indeed aggregation was found at lower surfactant or IPM contents as the droplet diameter was increased (Fig. 3.5a). In other words, a smaller amount of surfactant or of the so-called bad solvent was required for larger droplets to phase separate. Figure 3.6 shows a photomicrograph of an emulsion in which the solvent composition is much above the aggregation threshold. The clusters are very reminiscent of those observed in direct emulsions undergoing gelation on addition of electrolytes. (see Figure 3.11). Such an observation suggests that the attractive interaction between water droplets induced by the addition of the second solvent may become much larger than k_BT (in absolute value).

As for direct emulsions, the presence of excess surfactant induces depletion interaction followed by phase separation. Such a mechanism was proposed by Binks et al. [12] to explain the flocculation of inverse emulsion droplets in the presence of microemulsion-swollen micelles. The microscopic origin of the interaction driven by the presence of the "bad solvent" is more speculative. From empirical considerations, it can be deduced that surfactant chains mix more easily with alkanes than with vegetable, silicone, and some functionalized oils. The size dependence of such a mechanism, reflected by the shifts in the phase transition thresholds, is

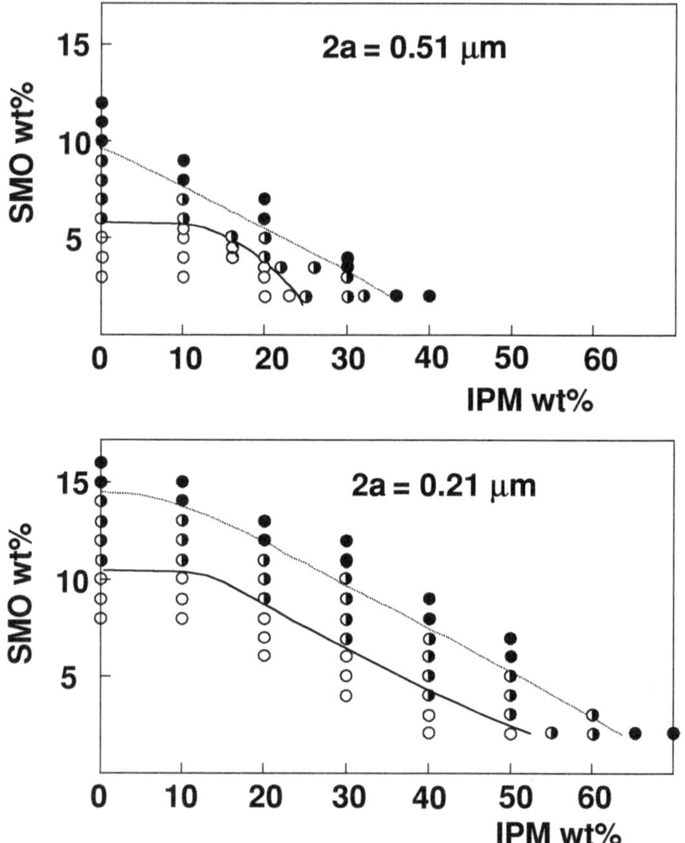

FIGURE 3.5. Phase diagrams obtained for two monodisperse water-in-oil emulsions stabilized with SMO. (○) totally dispersed; (◑) flocculated (fluid–solid); (●) totally flocculated (gel). (Adapted from [11].)

certainly due to the increasing contact surface between larger droplets. As previously mentioned, the interaction between the stabilizing chains are quite complex and rather sensitive to many different microscopic parameters such as segment–segment and chain–solvent molecular interactions, surfactant coverage, thickness of the adsorbed layer, solvent chain length, and so forth.

Leal-Calderon et al. [13] have proposed some basic ideas that control the colloidal interactions induced by solvent or a mixture of solvent and solute, when varying their length from molecular to colloidal scale. They have investigated the behavior of water- and glycerol-in oil emulsions in the presence of linear flexible chains of various masses. Figure 3.7 shows the phase behavior of both water and glycerol droplets of diameter 0.4 μm when dispersed in a linear aliphatic solvent of formula C_nH_{2n+2}, from $n = 5$ to $n = 30$. Because, for n larger than 16, solvent crystallization occurs at room temperature, a second series of experiments

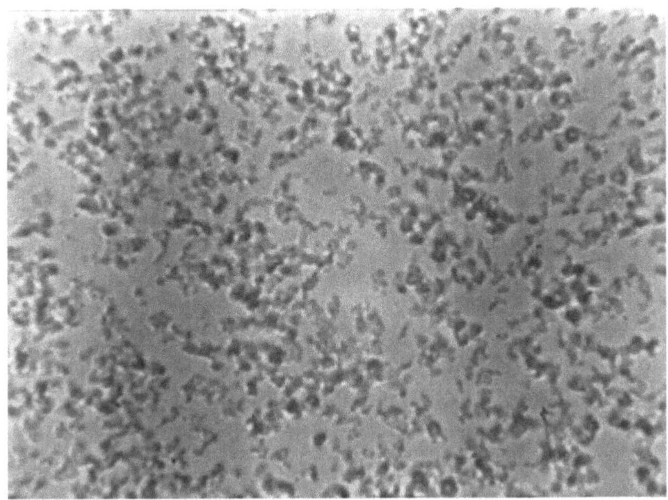

FIGURE 3.6. Microscopic image of a totally flocculated emulsion (water droplets of diameter 0.51 μm in a mixture of dodecane and IPM).

was performed at 65°C. The absolute value of the refractive index mismatch Δn_r between oil and water or glycerol is plotted as a function of n. The state of aggregation is reported for each sample (dark symbols, aggregated; empty symbols, dispersed). For water droplets, increasing n leads to increase $|\Delta n_r|$ and an aggregation threshold appears at large n. Below this threshold occurring at $n = 24$,

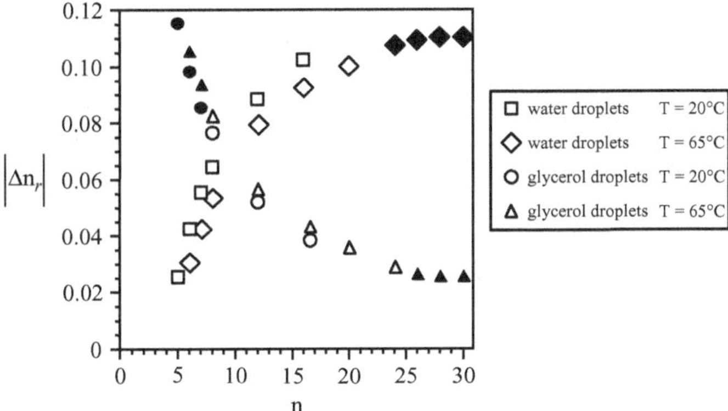

FIGURE 3.7. State of aggregation of water and glycerol droplets in different oils (C_nH_{2n+2}) as a function of n and of the absolute value refractive index mismatch Δn_r between the dispersed and the continuous phase. The surfactant concentration (SMO) is equal to 1 wt%. The droplet volume fraction is set at 5%. Water and glycerol droplets have a diameter close to 0.4 μm. Black symbols, aggregated droplets; empty symbols, dispersed droplets. (Adapted from [13].)

droplets are perfectly Brownian while above the threshold, the emulsions turn into a completely aggregated system. In contrast, for glycerol droplets, increasing n reduces $|\Delta n_r|$ and, instead, two distinct thresholds at $n = 7$ and $n = 26$ are observed. The droplets are first aggregated ($|\Delta n_r|$ large), become dispersed as $|\Delta n_r|$ decreases, and are finally aggregated again as $|\Delta n_r|$ decreases further. Note that this second aggregation threshold occurs in the limit of large n.

From these results, it can be concluded that two distinct and independent mechanisms affect the colloidal aggregation and therefore the contact pair interaction. One is controlled by the refractive index mismatch, which reflects the magnitude of the van der Waals forces [10]. The hydrocarbon–water Hamaker constant lies in the range 3–7 10^{-21} J when n varies from 5 to 30 [10]. Assuming a layer thickness δ of about 2 nm, Eq. (3.11) gives a potential energy between -2 and $-6\ k_B T$, a range in which phase separation is expected to occur, as found experimentally for the water droplets. However, for the glycerol droplets, the Hamaker constant is continuously decreasing as n increases and the flocculation observed for $n > 26$ suggests that the attraction in the presence of long alkanes is not produced only by van der Waals forces. The interaction that arises from the overlapping of the surfactant chains covering the droplets is controlled by the chain length of the solvent, which certainly suggests an entropically driven mechanism. Indeed, enthalpic contributions should be weakly changing with n because the chemical nature of the solvent remains the same (alkanes), and the surfactant chain is also kept the same.

To test this idea, the coupling between the length parameter n and the volume fraction φ of flexible chains when mixed with dodecane was investigated. Because the variable φ allows one to continuously vary the magnitude of the attraction, it is possible to induce a transition from the dispersed to the totally aggregated state with a coexisting state in between (fluid–solid equilibrium). Figure 3.8 shows the aggregation threshold φ^* of glycerol droplets ($\phi = 5\%$) within the φ/n plane from $n = 25$ to $n = 40$. φ^* is defined as the volume fraction of linear alkane C_nH_{2n+2} required to reach the coexisting state. Observations are performed at 65°C in order to avoid crystallization of the dodecane/C_nH_{2n+2} mixture. As expected, there is a strong coupling between φ and n: the longer the chain is, the fewer are required

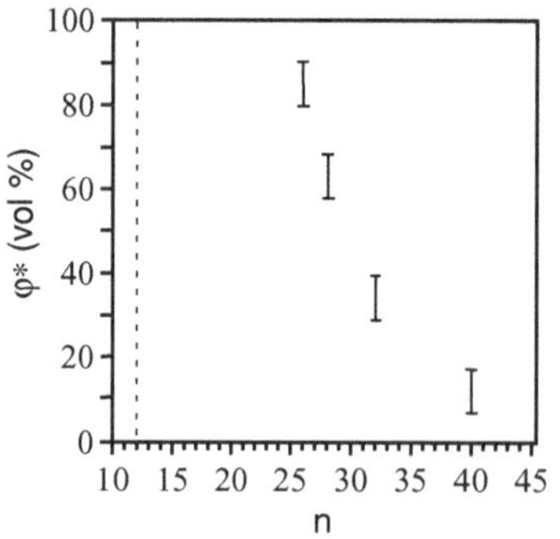

FIGURE 3.8. Volume fraction of n-alkane at the onset of flocculation as a function of n-alkane chain length (C_nH_{2n+2}). The continuous phase is made of a mixture of n-alkane, dodecane, and SMO (1 wt%). Glycerol droplets (5% in volume) have a diameter of 0.38 μm. $T = 65$°C. (Adapted from [13].)

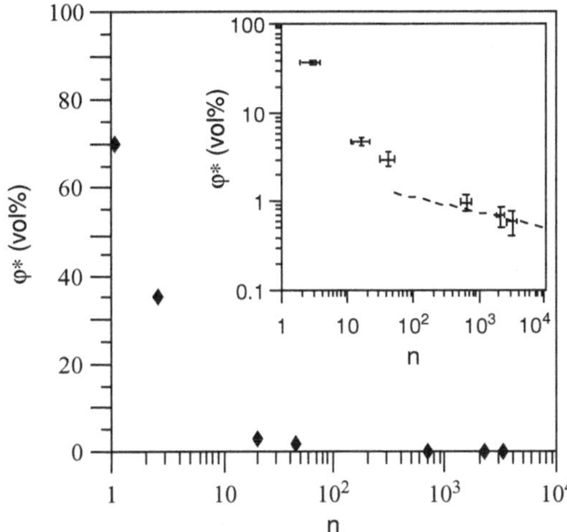

FIGURE 3.9. PDMS volume fraction vs. average number of unit segments at the onset of flocculation. The dashed line corresponds to the scaling $n^{-0.1}$ ($T = 20\,°\mathrm{C}$; water droplet diameter $= 0.28\ \mu\mathrm{m}$; droplet volume fraction $=$ 1%; 1 wt% of SMO). Inset: same plot in a log–log scale. (Adapted from [13].)

for reaching the coexisting state. Note that $\varphi^*(n)$ dramatically increases when n approaches the number of unit segments that compose the adsorbed surfactant tail ($n = 18$).

To explore the form of the function $\varphi^*(n)$ on a larger scale, the same authors studied the behavior of water droplets dispersed in a mixture of polydimethylsiloxane (PDMS) chains and dodecane (with 1 wt% of SMO). PDMS chains conform to the general formula $(CH_3)_3\,Si - [O - Si\,(CH_3)_2]_n - CH_3$. In Fig. 3.9 the concentration threshold φ^* of PDMS oil necessary to reach the coexisting state is plotted as a function of the average number n of monomers per chain (at 20°C and fixed water volume fraction ($\phi = 1\%$)). The boundary is clearly governed by two distinct regimes: a sharp drop that occurs for low n, as for alkanes, and a smoother decrease at large n. As in the case of normal alkanes, the aggregation is totally reversible because it disappears on simple dilution with pure dodecane.

The colloidal aggregation at low droplet volume fraction can be considered as a gas–solid phase transition. Following Eq. (3.10), it can also be assumed that the φ/n boundary at which the colloidal aggregation occurs corresponds to a constant contact energy, u_c, of the order of k_BT between droplets within the dense phase (at constant ϕ). In the limit of large n ($n > 500$), a hard-core depletion mechanism may govern the evolution of the φ/n boundary of Fig. 3.9. Indeed, the van der Waals force should remain constant, owing to the very small amount of polymer ($\varphi^* < 1\%$) at which the aggregation occurs. Therefore, because polymer chains are required to induce colloidal aggregation, van der Waals interactions are obviously not sufficient. However, they certainly do contribute as a constant background. The simplest description of the depletion force consists of assuming that for a droplet surface separation lower than twice the radius of gyration R_g, the polymer coils are totally excluded. Following equation 3.7 (with $\Delta = R_g$), such a mechanism leads to a contact energy given by:

$$u_c = -2\pi P_{osm} a R_g^2 \tag{3.13}$$

where P_{osm} is the osmotic pressure of the polymer solution. This relationship assumes that $a/R_g \gg 1$. Because the PDMS concentration remains very low (below the semidilute critical concentration), the perfect-gas approximation may be assumed for P_{osm}. For the larger n value ($n \approx 3400$), $P_{osm} \approx 60$Pa at the aggregation threshold. Using viscosimetric measurements, a hydrodynamic radius (which is assumed to be equal to R_g) of 147 Å has been measured at 20°C. At such a temperature, the solvent behaves roughly as a theta solvent (the hydrodynamic radius is found to scale as n^{α}, where $\alpha = 0.53 \pm 0.05$ [14]). Therefore, Eq. (3.13) gives $u_c \approx -2.7\, k_B T$, which perfectly agrees with the initial assumption: the hard-core depletion mechanism might be responsible for the evolution of the φ/n boundary, at least for such a large value of n. In a theta solvent, R_g scales as $n^{0.5}$ and P_{osm} as φ/n, and therefore the aggregation boundary in that limit should become essentially independent of n. In a good solvent, R_g scales as $n^{0.6}$, and therefore φ^* scales as $n^{-0.2}$. So, if the depletion interaction governs the experimental φ/n dependence, φ^* is expected to exhibit a very weak dependence on n, which is clearly the case in the limit of large n. In the inset of Fig. 3.9, the data are plotted in a log–log plot and confirm that the slope is becoming comparable to the expected one (between 0 and –0.2; as a guide a line of slope –0.1 has been drawn). In the limit of small n ($n < 100$), the simple depletion mechanism which assumes a total exclusion of polymer chains is unrealistic. Indeed, the polymer is small enough to swell the adsorbed surfactant brush and to be only partially excluded when the two droplets approach. As an example, at the aggregation threshold corresponding to $n \approx 40$ ($\varphi^* \approx 3\%$), the ideal gas osmotic pressure is 22 10^3 Pa. For this system, the radius of gyration is about 15 Å, leading to a depletion contact potential of about $-10\, k_B T$. As seen in Fig. 3.9 (inset) for $n \approx 40$ the data do not agree with the previous scaling and accordingly, the deduced absolute value of the contact potential at the threshold is much larger than $k_B T$. This suggests that the hard-core depletion mechanism is not realistic any longer. Indeed, such a mechanism overestimates the pairwise interaction at the precipitation threshold and this overestimation becomes more dramatic as n decreases. For the same reasons, the slope of $\varphi^*(n)$ for $n < 100$ deviates from the prediction based on that mechanism.

3.2.2. *Models for Phase Transitions*

An analogy may be drawn between the phase behavior of weakly attractive monodisperse dispersions and that of conventional molecular systems provided coalescence and Ostwald ripening do not occur. The similarity arises from the common form of the pair potential, whose dominant feature in both cases is the presence of a shallow minimum. The equilibrium statistical mechanics of such systems have been extensively explored. As previously explained, the primary difficulty in predicting equilibrium phase behavior lies in the many-body interactions intrinsic to any condensed phase. Fortunately, the synthesis of several methods (integral equation approaches, perturbation theories, virial expansions, and computer simulations) now provides accurate predictions of thermodynamic properties and phase behavior of dense molecular fluids or colloidal fluids [1].

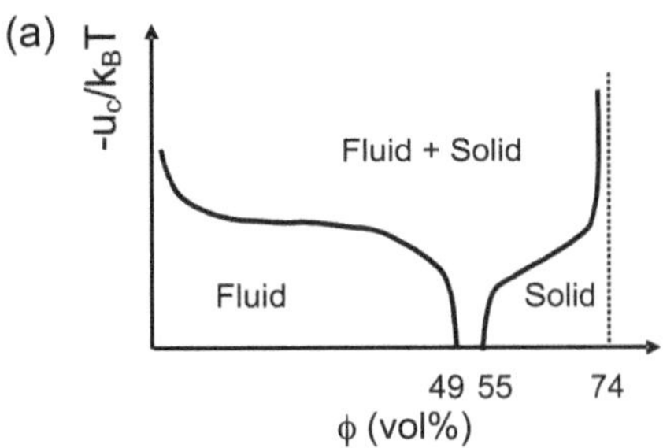

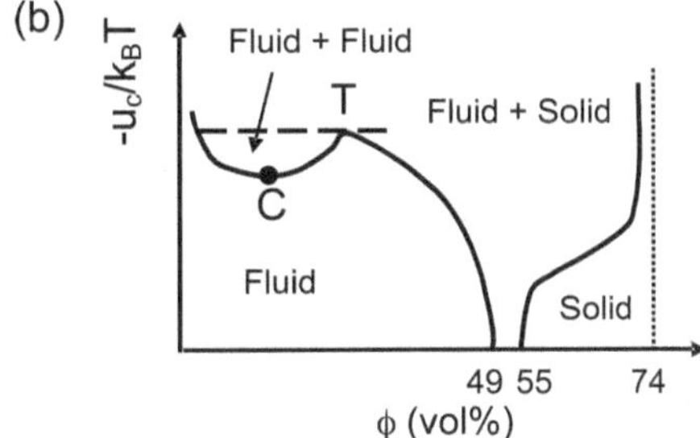

FIGURE 3.10. Phase diagrams of attractive monodisperse dispersions. u_c is the contact pair potential and ϕ is the particle volume fraction. For $u_c/k_BT = 0$, the only accessible one-phase transition is the hard sphere transition. If $u_c/k_BT \neq 0$, two distinct scenarios are possible according to the value of the ratio ξ (range of the pair potential over particle radius). For $\xi < 0.3$ **(a),** only fluid–solid equilibrium is predicted. For $\xi > 0.3$ **(b),** in addition to fluid–solid equilibrium, a fluid–fluid (liquid–gas) coexistence is predicted with a critical point (C) and a triple point (T).

Freezing and crystallization are generally driven by potential minima that overcome the entropic tendency toward disorder. In some cases however, entropy can actually induce order and cause a simple liquid to freeze. While this seems counterintuitive, such entropy driven freezing transition has been observed in computer simulations of spheres interacting through a purely repulsive hard sphere potential [1]. This freezing transition was observed experimentally in monodisperse, hard-sphere colloidal suspensions [15]. For liquids and suspensions made of single-size spheres, freezing is observed when the volume fraction ϕ of spheres exceeds 49.4% and melting at volume fractions lower than 54.5%. In addition, it is now well established, from theory and numerical simulations, that the topology of the phase diagram depends on the ratio ξ of the interparticle attraction range (Δ) to the interparticle hard-core repulsion range (a) : $\xi = \frac{\Delta}{a}$ [16–19]. In the case of sufficiently small ξ, theory predicts that the only effect of the interparticle attraction is to expand the above-mentioned fluid-crystal coexistence region ($49.4\% < \phi < 54.5\%$) (Fig. 3.10a). For ξ values larger than approximately 0.3, in addition to the fluid–solid coexistence region, a fluid–fluid coexistence (gas–liquid) is also predicted (Fig. 3.10b). Experimental evidence for the effect of the range of the interparticle attraction on equilibrium phase behavior has been provided by studies on colloid–polymer and colloid–colloid mixtures [20–24], thus confirming the predicted topologies as a function of ξ. So far, in the field of emulsions, the only type of equilibrium that has been recognized is fluid–solid like, which suggests that only small ξ values are accessible. This is due to the fact that emulsion droplets have a characteristic diameter that generally spans between 0.1 and 1 μm, a hard-core size that is much larger than the range of classical interactions.

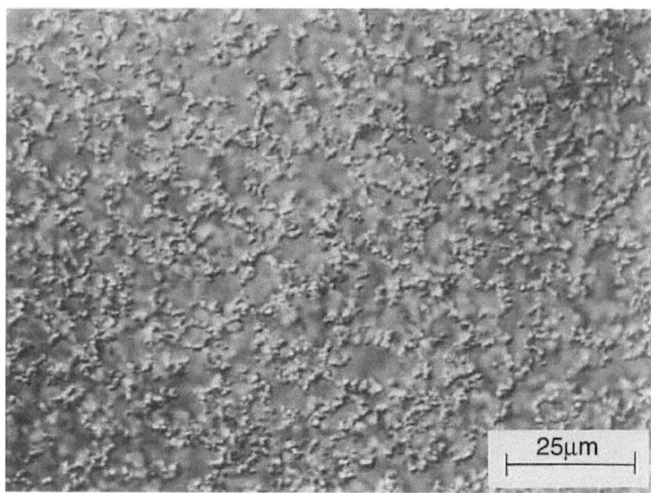

FIGURE 3.11. Emulsion gel. The droplets are about 0.5 μm in diameter. They form large and ramified aggregates.

3.3. Gelation and Kinetically Induced Ordering

If the attractive energy between the droplets is around $-10\ k_B T$, equilibrium phase behavior is no longer possible. Strongly attractive emulsion droplets irreversibly stick to one another, as they collide owing to their diffusive motion. As shown in Fig. 3.11, the resultant bonds can be sufficiently strong to prevent further rearrangements, leading to highly disordered and tenuous clusters, whose structure can be well described as fractal [25]. In contrast to a compact aggregate in which the mass M and the radius R are related as $M \propto R^3$, the radius of a fractal object depends more strongly on M. By definition, if M and R are related through:

$$M \propto R^{d_f} \tag{3.14}$$

we are dealing with a fractal object. The exponent d_f represents the fractal dimension, which is a rational number less than 3; the smaller d_f, the less compact the aggregate. If there is no repulsive barrier preventing clusters from sticking upon collision, the aggregation is driven solely by the diffusion-induced collisions between the growing clusters. This regime, called diffusion-limited cluster aggregation (DLCA), is amenable to detailed theoretical analysis, and has become an important base case on which understanding of other kinetic growth processes can be built. At long times, DLCA must produce a gel that spans the system, preventing further coarsening. DLCA is completely random; nevertheless, experiments performed with concentrated suspensions, have shown that a surprising order can develop, manifested by a pronounced peak in the small angle scattering intensity $I(q)$ as a function of wave vector, which reflects the development of a characteristic length scale in the suspension [26,27].

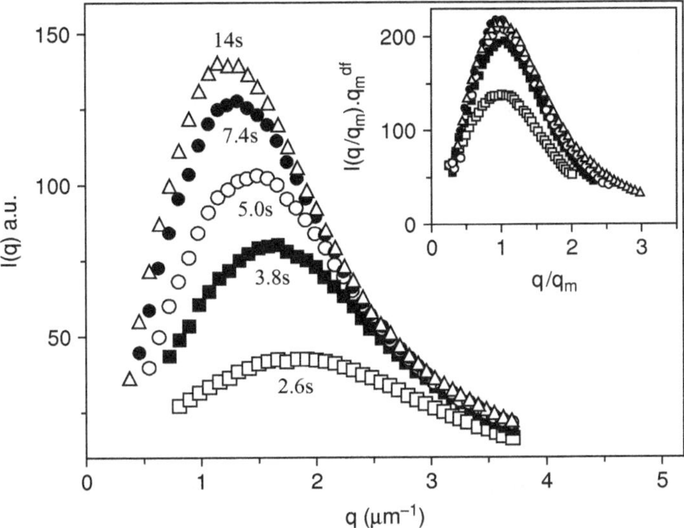

FIGURE 3.12. Light-scattering intensity as a function of the wave vector for an emulsion with $\phi = 4\%$ at a series of times. The inset shows that, except at the earliest times, the data can be scaled onto a single curve. (Adapted from [27].)

The structure of adhesive emulsions stabilized with ionic surfactant in the presence of salt was studied using light-scattering [27,28]. These emulsions are ideal systems to reach the limit of DLCA as the droplets can be made unstable toward aggregation by temperature quenches. Before the quench, at high temperature, the droplets are homogeneously dispersed whereas they stick to one another just after the quench (see Chapter 2, Section 2.3). Figure 3.12 shows a series of $I(q)$ taken at different times after the aggregation begins, using a sample with droplet volume fraction $\phi = 4\%$; $t = 0$ is chosen as the time when the first perceptible change in the scattering intensity is observed. There is a pronounced peak in $I(q)$ that grows in intensity and moves to lower q with time; ultimately the system gels, and there are no further changes in $I(q)$. The scattering wave vector of the maximum intensity defines a characteristic length scale of the system, q_m^{-1}; q_g is defined as the asymptotic limit of q_m at long times, that is, when gelation is achieved. As previously observed [26], after a brief initial stage, all the scattering curves can be scaled onto a master curve by plotting $q_m^{d_f} I(q/q_m)$ as a function of q/q_m. This scaling is illustrated in the inset of Fig. 3.12, where the fractal dimension $d_f = 1.9$ has been used. This value is close to 1.8, the expected fractal dimension for DLCA [25]. Such scaling behavior has also been observed in other three-dimensional [26] and in two-dimensional [29] colloidal systems. Several authors have pointed out the intriguing similarity between DLCA and spinodal decomposition where scaling behavior is also observed [26,29–32]. Numerical simulations [30,31] and theoretical analysis [33] agree with the experimental results obtained with different systems. The origin of the peak is mainly due to a depletion shell around the larger

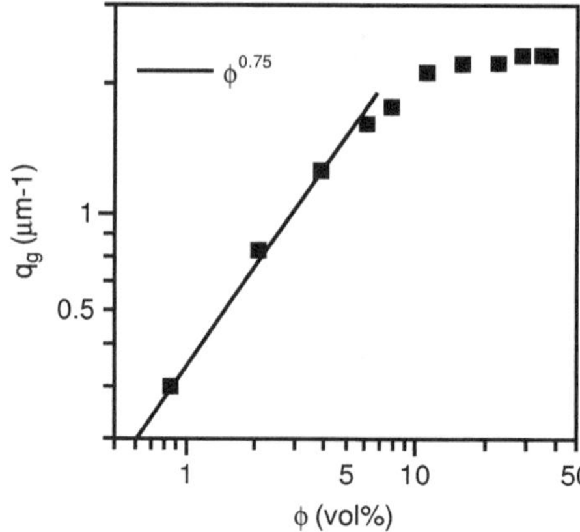

FIGURE 3.13. Dependence of q_g vs. ϕ. For $\phi < 10\%$, the data have a $\phi^{0.75}$ dependence, consistent with DLCA clusters with a fractal dimension of $d_f = 1.9$. (Adapted from [28].)

clusters as aggregation proceeds [26]. However, at higher volume fraction and later times, the peak originates mainly from correlations between the growing clusters.

For diffusion-controlled gelation and low droplet fraction, q_g follows a power-law dependence [27]:

$$q_g \sim \phi^{1/(3-d_f)} \tag{3.15}$$

This is confirmed in Fig. 3.13. For $\phi < 10\%$, the exponent is 0.75, close to the expected value, 0.83 for $d_f = 1.8$. However, for $\phi > 10\%$, there is a sharp change, and very little variation of q_g with ϕ is observed. The origin of this behavior and the mechanisms of colloidal aggregation at very high volume fractions are still the subject of current research. It has been shown that the similarity with spinodal decomposition tends to be more pronounced in these regimes of highly concentrated droplets [34]. There is still a pronounced peak in $I(q)$; however, q_m is independent of t, as shown in Fig. 3.14 for $\phi = 23\%$. Moreover, the intensity at q_m grows exponentially with time until gelation occurs, typically in less than 2 s;

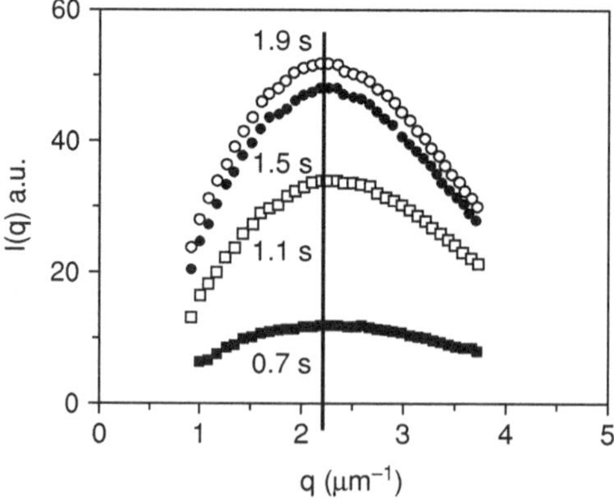

FIGURE 3.14. Time evolution of $I(q)$ for $\phi = 23\%$. The peak position does not change. (Adapted from [28].)

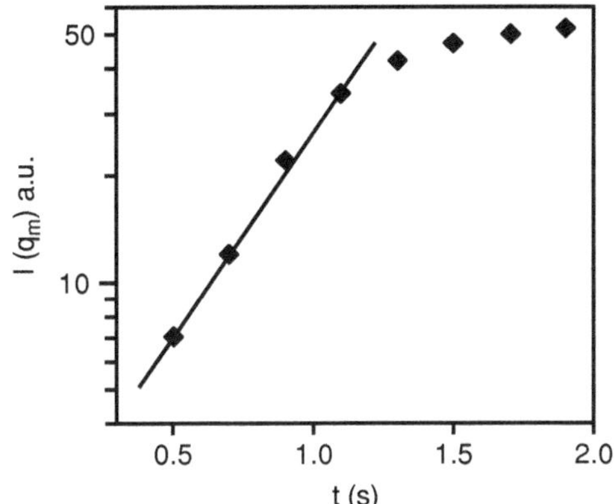

FIGURE 3.15. Time evolution of the peak of $I(q)$ for $\phi = 23\%$. The intensity increases exponentially with time until gelation is reached, typically in less than 2 s. (Adapted from [28].)

this is shown in Fig. 3.15. Similar behavior is observed for all $\phi > 10\%$, although the peak in $I(q)$ becomes broader as ϕ_d is increased. This broadening is noticeably more pronounced when $\phi > 30\%$. For $\phi > 40\%$, a peak is no longer observed. The behavior in this second scenario is similar to that expected for the early stages of a spinodal decomposition within the framework of the Cahn–Hilliard linear theory [35,36].

Finally, an important feature of gels made of adhesive emulsions arises from the deformation of the droplets. Indeed, as the temperature is lowered the contact angles between the droplets increase [27,28] (see Chapter 2, Section 2.3). Consequently, the structure of the final flocs depends on the time evolution of the strength of the adhesion. Initially, the adhesion results in the formation of a random, solid gel network in the emulsion. Further increase of adhesion causes massive fracturing of the gel, disrupting the rigidity of the structure and leading to well separated, and more compact flocs [27,28].

3.4. Conclusion

Many practical applications of emulsion science revolve around the problem of controlling the interaction between droplets or between droplets and surfaces. Some consequences of weak or strong attractions between emulsions droplets have been examined and we hope that the concepts presented in this chapter will provide a useful guide for controlling aggregation phenomena. Such phenomena have a strong impact on optical and rheological properties of emulsions. For instance, aggregation may induce viscoelastic behavior at relatively low droplet volume fraction.

In numerous applications, emulsions are used to coat a surface and adhesion of the droplets on a substrate is required. So far, droplet–substrate interactions have been much less examined and it is clear that any progress in coating technology will require further investigation. Among the most exciting challenges, we can mention the specific deposition or "targeting" of emulsion droplets on biological substrates (skin, living cells, etc.) for medical purposes. Targeting requires the formation of specific complexes between ligand–receptor couples over a wide variety of hydrodynamic and compositional conditions. Some promising biotechnological applications of emulsions are reviewed in Chapter 7.

References

[1] J.P. Hansen and I.R. McDonald: *Theory of Simple Liquids*, 3rd ed. Boston Elsevier 2006.

[2] S. Asakura and J. Oosawa: "Interaction Between Particles Suspended in Solutions of Macromolecules." J. Polym. Sci. **32**, 183 (1958).

[3] M.P. Aronson: "The Role of Free Surfactant in Destabilizing Oil-in-Water Emulsions." Langmuir **5**, 494 (1989).

[4] J. Bibette, D. Roux, and F. Nallet: "Depletion Interactions and Fluid-Solid Equilibrium in Emulsions." Phys. Rev. Lett. **65**, 2470 (1990).

[5] J. Bibette: "Depletion Interactions and Fractionated Crystallization for Polydisperse Emulsion Purification." J. Colloid Interface Sci. **147**, 474 (1991).

[6] J. Bibette, D. Roux, and B. Pouligny: "Creaming of Emulsions: The Role of Depletion Forces Induced by Surfactant." J. Phys. II France **2**, 401 (1992).

[7] A. Meller and J. Stavans: "Stability of Emulsions with Nonadsorbing Polymers." Langmuir **12**, 301 (1996).

[8] U. Steiner, A. Meller, and J. Stavans: "Entropy Driven Phase Separation in Binary Emulsions." Phys. Rev. Lett. **74**, 4750 (1995).

[9] V. Digiorgio, R. Piazza, M. Corti, and C. Minero: "Critical Properties of Nonionic Micellar Solutions." J. Chem. Phys. **82**, 1025 (1984).

[10] J.N. Israelachvili: *Intermolecular and Surface Forces*." Academic Press New York (1992).

[11] F. Leal-Calderon, B. Gerhardi, A. Espert, F. Brossard, V. Alard, J.F. Tranchant, T. Stora, and J. Bibette: "Aggregation Phenomena in Water-in-Oil Emulsions." Langmuir **12**, 872 (1996).

[12] B.P. Binks, P.D.I. Fletcher, and D.I. Horsup: "Effect of Microemulsified Surfactant in Destabilizing Water-in-Oil Emulsions Containing C12E4." Colloids Surfaces **61**, 291 (1991).

[13] F. Leal-Calderon, O. Mondain-Monval, K. Pays, N. Royer, and J. Bibette: "Water-in-Oil Emulsions: Role of the Solvent Molecular Size on Droplet Interactions." Langmuir **13**, 7008 (1997).

[14] P.G. De Gennes: *Scaling Concepts in Polymer Physics*. Cornell University Press London (1979).

[15] P.N. Pusey and W. van Megen: "Phase Behaviour of Concentrated Suspensions of Nearly Hard Colloidal Spheres." Nature (Lond.) **320**, 340 (1986).

[16] A.P. Gast, C.K. Hall, and W.B. Russel: "Polymer-Induced Phase Separations in Nonaqueous Colloidal Suspensions." J. Colloid Interface Sci. **96**, 251 (1983).

[17] B. Vincent, J. Edwards, S. Emmett, and R. Croot: "Phase Separation in Dispersions of Weakly Interacting Particles in Solutions of Non Adsorbing Polymer." Colloids Surfaces **31**, 267 (1988).

[18] H.N.W. Lekkerkerker, W.C.K. Poon, P.N. Pusey, A. Stroobants, and P.B. Warren: "Phase Behaviour of Colloid + Polymer Mixtures." Europhys. Lett. **20**, 59 (1992).
[19] M.H.J. Hagen and D. Frenkel: "Determination of Phase Diagrams for the Hard-Core Attractive Yukawa System." J. Chem. Phys. **101**, 4093 (1994).
[20] P.R. Sperry: "Morphology and Mechanism in Latex Flocculated by Volume Restriction." J. Colloid Interface Sci. **99**, 97 (1984).
[21] B. Vincent: "The Stability of Non-Aqueous Dispersions of Weakly Interacting Particles." Colloids Surfaces **24**, 269 (1987).
[22] F. Leal-Calderon, J. Bibette, and J. Biais: "Experimental Phase Diagrams of Polymer and Colloid Mixtures." Europhys. Lett. **23**, 653 (1993).
[23] S.M. Illet, A. Orrock, W.C.K. Poon, and P.N. Pusey: "Phase Behavior of a Model Colloid-Polymer Mixture." Phys. Rev. E **51**, 1344 (1995).
[24] S. Sanyal, N. Easwear, S. Ramaswamy, and A.K. Sood: "Phase Separation in Binary Nearly-Hard-Sphere Colloids: Evidence for the Depletion Force." Europhys. Lett. **18**, 107 (1993).
[25] R. Jullien and R. Botet: *Aggregation and Fractal Aggregates*. World Scientific, ISBN 9971502488, Singapore (1987).
[26] M. Carpineti and M. Giglio: "Spinodal-Type Dynamics in Fractal Aggregation of Colloidal Clusters." Phys. Rev. Lett **68**, 3327 (1992).
[27] J. Bibette, T.G. Mason, H. Gang, and D.A. Weitz: "Kinetically Induced Ordering in Gelation of Emulsions." Phys. Rev. Lett **69**, 981 (1992).
[28] J. Bibette, T.G. Mason, H. Gang, D.A. Weitz, and P. Poulin: "Structure of Adhesive Emulsions." Langmuir **9**, 3352 (1993).
[29] D.J. Robinson and J.C. Earnshaw: "Long Range Order in Two Dimensional Fractal Aggregation." Phys. Rev. Lett **71**, 715 (1993).
[30] A. Hasmy, E. Anglaret, M. Foret, J. Pellous, and R. Jullien: "Small-Angle Neutron-Scattering Investigation of Long-Range Correlations in Silica Aerogels: Simulations and Experiments." Phys. Rev. B **50**, 6006 (1994).
[31] A. Hasmy and R. Jullien: "Sol-Gel Process Simulation by Cluster-Cluster Aggregation." J. Non-Crystalline Solids **186**, 352 (1995).
[32] A.E. Gonzalez and G. Ramirez-Santiago: "Spatial Ordering and Structure Factor Scaling in the Simulations of Colloid Aggregation." Phys. Rev. Lett **74**, 1238 (1995).
[33] F. Sciortino and P. Tartaglia: "Structure Factor Scaling During Irreversible Cluster-Cluster Aggregation." Phys. Rev. Lett **74**, 282 (1995).
[34] P. Poulin, J. Bibette, and D.A. Weitz: "From Colloidal Aggregation to Spinodal Decomposition in Sticky Emulsions." Eur. J. Physics B **9**, 3352 (1999).
[35] Domb C. and Lebowitz J.L. (eds.): *Phase Transition and Critical Phenomena*. vol. 8. Academic Press, London (1983).
[36] J.W. Cahn and J.E. Hilliard: "Free Energy of a Nonuniform System. I. Interfacial Free Energy." J. Chem. Phys. **28**, 258 (1958).

4
Compressibility and Elasticity of Concentrated Emulsions

4.1. Introduction

Emulsions are dispersions of deformable droplets that can therefore span droplet volume fractions from zero to almost one. Much effort has been spent in studying the rheology of relatively diluted emulsions (oil volume fraction ϕ below 40%) ([1], and references therein, [2,3]). At low volume fraction, nonadhesive emulsions consist of unpacked spherical droplets; such samples generally show a Newtonian behavior, while flocculated emulsions show shear-thinning behavior. A significant change in the rheological behavior is observed if the droplets are concentrated up to volume fractions higher than the volume fraction ϕ^* corresponding to the random close packing of hard spheres. For randomly packed monodisperse spheres, $\phi^* = 64\%$. Above ϕ^*, the droplets can no longer pack without deforming; although being composed of fluids only, emulsions resemble a solid. The elasticity exists only because the repulsive droplets have been concentrated up to a sufficiently large volume fraction, ϕ, which permits the storage of interfacial energy. As pointed out by Princen and Kiss [4,5], the considerable elasticity of concentrated emulsions exists because the repulsive droplets have been compressed by an external osmotic pressure. Two compressed droplets will begin to deform before their interfaces actually touch, owing to the intrinsic repulsive interactions between them. Emulsions minimize their total free energy by reducing the repulsion (which may have different origins) at the expense of creating some additional surface area by deforming the droplet interfaces. The necessary work to deform the droplets arises from the application (by any means) of an external osmotic pressure, Π, and the excess surface area of the droplets determines the equilibrium elastic energy stored at a given osmotic pressure. The additional excess surface area created by shear deformation determines the elastic shear modulus, $G(\phi)$. Although Π and G represent fundamentally different properties, they both depend on the degree of droplet deformation and therefore on ϕ. This chapter aims to describe the basic physics of compressibility and elasticity of concentrated emulsions.

4.2. Basic Concepts

Let us first consider interfaces at equilibrium. Any stress (osmotic or shear stress) imposed to the emulsion increases the amount of interface, leading to a modification of the free energy. For a monodisperse collection of N droplets of radius a, the total interfacial area of the undeformed droplets is $S_0 = 4\pi Na^2$. If the emulsion is compressed up to $\phi > \phi^*$, each droplet is pressing against its neighbors through flat facets. As a consequence, the total surface area, S, becomes larger than S_0. The osmotic pressure is defined as the derivative of the free energy F with respect to the total volume V, at constant number of droplets:

$$\Pi = -\left(\frac{\partial F}{\partial V}\right)_N = -\left(\frac{\partial F}{\partial S}\right)_N \left(\frac{\partial S}{\partial V}\right)_N \tag{4.1}$$

The derivative of F with respect to S,

$$\sigma = \left(\frac{\partial F}{\partial S}\right)_N \tag{4.2}$$

characterizes the mechanical behavior of the surface and is equal to the interfacial tension, γ_{int}, in the case of a surfactant-covered interface. $\left(\frac{\partial S}{\partial V}\right)_N$ is a purely geometrical term that represents the excess surface due to compression.

A similar approach can be adopted for the bulk shear modulus. When a small strain is applied to a solid, the latter is stressed, and one can measure the resulting stress. At low deformation, the bulk shear stress, τ, is proportional to the strain, Γ, following Hooke's law:

$$\tau = G\Gamma \tag{4.3}$$

where G is the elastic shear modulus. For materials that do not store perfectly the elastic energy (i.e., materials that exhibit some viscous loss), G can be generalized considering that the stress varies linearly with the strain. Thus G is transformed into a complex value that is frequency-dependent. If the emulsion—already compressed to a surface S—experiences a small shear strain Γ, the total interface increases quadratically with the strain, as shown by Princen [6]. The total droplet surface area thus increases as:

$$\Delta S = \frac{1}{2}\Gamma^2 \left(\frac{\partial^2 S}{\partial \Gamma^2}\right) \tag{4.4}$$

The bulk stress can be expressed as:

$$\tau = \frac{1}{V}\left(\frac{\partial F}{\partial \Gamma}\right)_N = \frac{\Gamma}{V}\left(\frac{\partial F}{\partial S}\right)_N \left(\frac{\partial^2 S}{\partial \Gamma^2}\right) \tag{4.5}$$

The shear modulus is then given by:

$$G = \frac{\tau}{\Gamma} = \frac{1}{V}\left(\frac{\partial F}{\partial S}\right)_N \left(\frac{\partial^2 S}{\partial \Gamma^2}\right) \tag{4.6}$$

4.3. Experimental Techniques

4.3.1. *Elasticity Measurements*

G is measured by means of a rheometer. The shear modulus is a complex number that has a phase and a magnitude, because the viscous dissipation causes a lag (i.e., a phase shift) between the resulting stress and the applied strain. The real part of G is related to the energy storage, while the imaginary part is related to the viscous loss. In a real measurement, a sinusoidal strain or stress is applied at a given frequency. The response of the system (i.e., the time-dependent stress or strain) is measured. In the linear regime, the response of the studied material is also sinusoidal. Considering a periodic strain, with an amplitude Γ_0 and pulsation ω, one writes:

$$\begin{aligned}\Gamma(t) &= \Gamma_0 \cos(\omega t)\\ \tau(t) &= \tau_0 \cos(\omega t + \varphi)\end{aligned} \tag{4.7}$$

where φ is the phase shift due to dissipation. Equations (4.6) and (4.7) yield:

$$\begin{aligned}G' &= \frac{\tau_0}{\Gamma_0} \cos \varphi(\omega)\\ G'' &= \frac{\tau_0}{\Gamma_0} \sin \varphi(\omega)\end{aligned} \tag{4.8}$$

where G' and G'' are the real and imaginary part, respectively, of G. Hereafter, G' is called elastic modulus, and G'' loss modulus.

4.3.2. *Compressibility Measurements*

At least two different techniques are available to compress an emulsion at a given osmotic pressure Π. One technique consists of introducing the emulsion into a semipermeable dialysis bag and to immerse it into a large reservoir filled with a stressing polymer solution. This latter sets the osmotic pressure Π. The permeability of the dialysis membrane is such that only solvent molecules from the continuous phase and surfactant are exchanged across the membrane until the osmotic pressure in the emulsion becomes equal to that of the reservoir. The dialysis bag is then removed and the droplet volume fraction at equilibrium is measured.

Another technique consists of submitting the emulsion to centrifugation and determining the droplet volume fraction ϕ_f at the top (bottom) of the cream (sediment). The centrifugation typically takes several hours until the equilibrium volume fraction is achieved. After equilibration, if the droplets occupy a distance much less than that of the centrifuge lever arm, the spatial gradient in the acceleration can be neglected, and the osmotic pressure can be determined (see Fig. 4.1):

$$\Pi = |\Delta \rho|\, \Omega^2 \phi_f \left(d_c H \pm \frac{H^2}{2} \right) \tag{4.9}$$

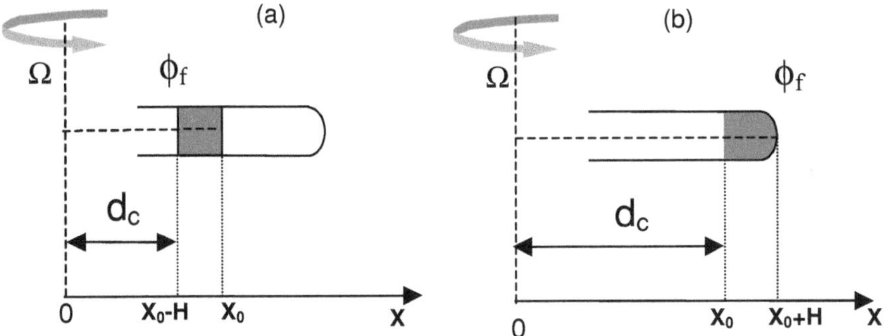

FIGURE 4.1. Scheme of the centrifugation experiment for a) a creaming emulsion, b) a sedimenting emulsion. See text for details.

where H is the final height of the cream (sediment), ϕ_f is the final oil (water) volume fraction, $\Delta\rho$ is the density mismatch between the dispersed and the continuous phase, d_c is the length of the lever arm, and Ω is the rotation speed of the centrifuge. In the right-hand term of Eq. (4.9), the operator + applies to creaming and – to sedimentation. This maximum osmotic pressure reflects the stress exerted by the droplet layers below (above) the top (bottom) of the cream (sediment).

4.4. Compressibility and Elasticity of Surfactant-Stabilized Emulsions

In this section, we shall describe the bulk properties of emulsions that are due to surface tension only, as generally observed with surfactant-stabilized systems: $\sigma = \gamma_{\text{int}}$.

4.4.1. Experimental Results

The first quantitative study on the elastic properties of monodisperse emulsions was carried out by Mason et al. [7,8], following the pioneering work of Princen [4,5], performed on polydisperse systems. Mason et al. prepared concentrated monodisperse silicon oil-in-water emulsions, stabilized by sodium dodecyl sulfate (SDS). Typical results for both G' and G'', as a function of the applied oscillatory strain, Γ_0, are shown in Fig. 4.2, for several volume fractions of an emulsion with mean radius $a = 0.53$ μm. The elastic modulus increases by nearly four decades as ϕ increases. For low strain values (in the linear regime), G' is greater than G'', reflecting the elastic nature of the emulsions. However, at larger strains, there is a slight but gradual drop in the storage modulus, while the loss modulus begins to rise, indicating the approach to the nonlinear yielding behavior and plastic flow. At very large strains, beyond the yield strain marked by the drop of G', the apparent G'' dominates, reflecting the dominance of the energy loss due to the nonlinear

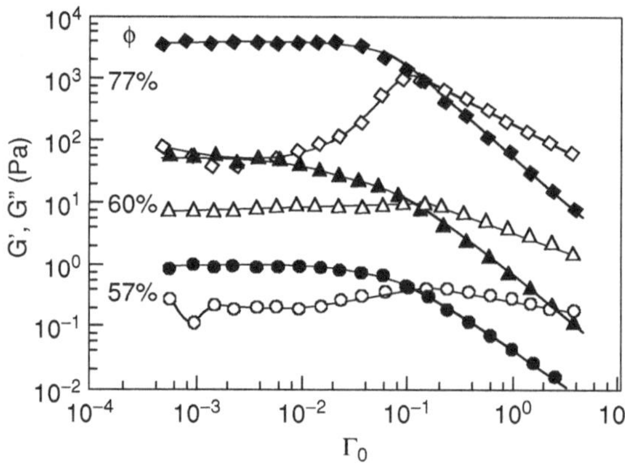

FIGURE 4.2. The Γ dependence of the storage G' (solid symbols) and loss G'' (open symbols) moduli of a monodisperse silicon oil-in-water emulsion stabilized with SDS, with radius $a = 0.53$ μm, for three volume fractions; from top to bottom: $\phi = 77\%, 60\%$, and 57%. The frequency is 1 rad/s; the lines are visual guides. (Adapted from [10].)

flow. Hébraud et al. [9] have investigated the evolutions of droplet motions as the emulsions undergo the transition from the linear to the nonlinear regime. They used diffusing-wave spectroscopy as a tool to estimate the fraction of moving drops in concentrated emulsions subjected to a periodic shear strain. They showed that the strain gives rise to periodic echoes in the correlation function, which decays with increasing strain amplitude Γ_0. For a given Γ_0, the decay of the echoes implies that a finite fraction of the emulsion droplets never rearranges while the remaining fraction of droplets repeatedly rearranges. Yielding occurs when about 4–5% of the droplets rearrange.

The frequency dependence of the moduli was measured by Mason et al. [7,8] and is shown in Fig. 4.3 for several values of ϕ. In all cases, there is a plateau in $G'(\omega)$; at high ϕ, this extends over the full four decades of explored frequency, while for the lower ϕ, the plateau is no longer strictly independent on ω but reduces to an inflection point at G'_p. In contrast, for all ϕ, $G''(\omega)$ exhibits a shallow minimum, G''_m. Mason et al. used G'_p to characterize the elasticity and G''_m to characterize the viscous loss.

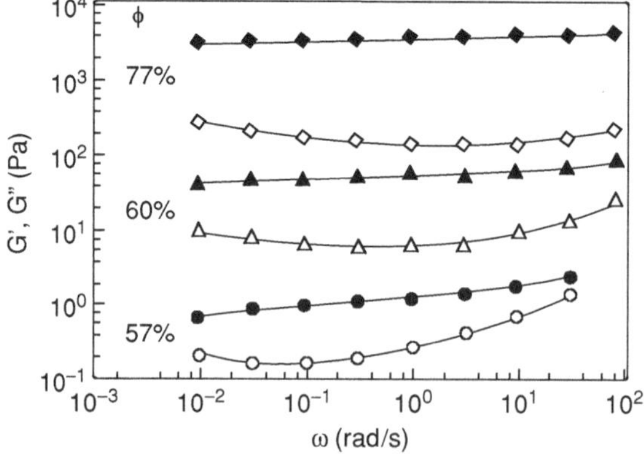

FIGURE 4.3. The frequency dependence of the storage G' (solid symbols) and loss G'' (open symbols) moduli of a monodisperse silicon oil-in-water emulsion stabilized with SDS, with radius $a = 0.53$ μm, for three volume fractions; from top to bottom: $\phi = 77\%, 60\%$, and 57%. The strain is 0.005 in all cases. The lines are visual guides. (Adapted from [10].)

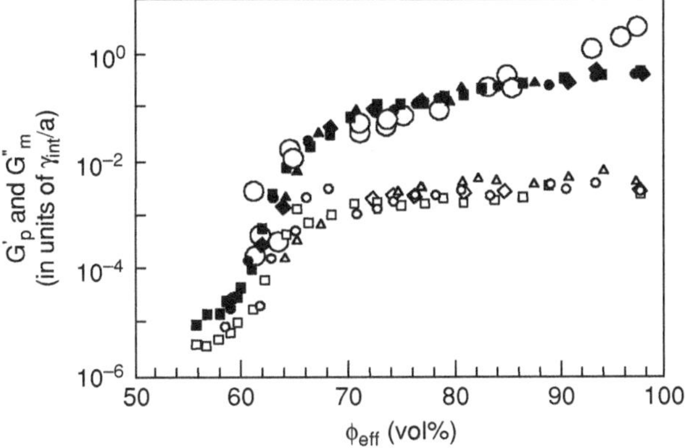

FIGURE 4.4. The plateau storage modulus G'_p (small solid symbols) and the minimum of the loss modulus G'''_m (small open symbols) as a function of the effective oil volume fraction. $a = 0.25$ μm (circles), $a = 0.37$ μm (triangles), $a = 0.53$ μm (squares), and $a = 0.74$ μm (diamonds). The large circles are the measured values for the osmotic pressure. All data are normalized by γ_{int}/a (Adapted from [10].)

Mason et al. [7,8,10] measured the ϕ dependence of G'_p, G''_m, and Π, for emulsions of different radii and they normalized the results by the Laplace pressure, γ_{int}/a; the results are reported in Fig. 4.4. Measurements of G', G'', and Π were performed on droplets of various sizes. Osmotic pressure data were obtained, in the regime of large ϕ, from a dialysis technique, while in the regime of low ϕ, gentle centrifugation was more accurate. The repulsive electrostatic interaction induced by adsorbed SDS molecules was accounted by assuming that the droplets behave as deformable spheres with an effective radius larger than the real one, the difference being of the order of the Debye length. Indeed, as initially pointed out by Princen [11], it is the effective oil volume fraction that governs the rheological properties. ϕ_{eff} is calculated from:

$$\phi_{eff} = \phi \left[1 + \frac{1}{2} \frac{h(\phi)}{a} \right]^3 \tag{4.10}$$

where $h(\phi)$ is the film thickness between the droplets at volume fraction ϕ. A linear variation of h on ϕ can be assumed. Hence one needs to evaluate the film thickness at two oil volume fractions (above the close packing). For $\phi = \phi^*$, h has been taken equal to 175 Å, consistent with force–distance measurements for emulsions coated with the same surfactant [12] (see Chapter 2, Fig. 2.5). Near $\phi = 100\%$, h has been taken equal to 50 Å; this is consistent with the Debye length of the double-layer repulsion between highly deformed interfaces and a disjoining pressure of γ_{int}/a [13].

The rescaled data of G'_p and G''_m all fall onto single curves as shown in Fig. 4.4; G'_p rises by about four decades, as ϕ_{eff} increases from 60% to 85%. The scaling with γ_{int}/a confirms that the origin of elasticity results from the storage of energy at droplet interfaces. Moreover, the scaling with ϕ_{eff} indicates that the elasticity of these compressed droplets depends only on the packing geometry. Like the elasticity, the osmotic pressure also reflects the energy storage at the interfaces as they are deformed with increasing ϕ. Remarkably, $\Pi/(\gamma_{int}/a)$ is nearly the same as $G'/(\gamma_{int}/a)$, until it diverges at high ϕ_{eff}. When the droplets are highly compressed ($\phi \to 100\%$), the emulsion's elasticity resembles that of a dry foam and it is determined by γ_{int}/a. A random dry foam is predicted to have $G'_p = 0.55\,(\gamma_{int}/a)$ [14], in excellent agreement with both the results of Mason et al. [7,8,10] and Princen's earlier data on polydisperse systems. In contrast, the measured Π exhibits a pronounced increasing slope, as $\phi \to 100\%$.

4.4.2. Theoretical Approaches

The first theoretical considerations concerning $\Pi\,(\phi)$ and $G\,(\phi)$ of concentrated 3-D emulsions and foams were based on perfectly ordered crystals of droplets [4,5,15–18]. In such models, at a given volume fraction and applied shear strain, all droplets are assumed to be equally compressed, that is, to deform affinely under the applied shear; thus all of them should have the same shape. Princen [15,16] initially analyzed an ordered monodisperse 2-D array of deformable cylinders and concluded that $G = 0$ for $\phi < \phi^*$, and that G jumps to nearly the 2-D Laplace pressure of the cylinders at the approach of $\phi = 100\%$, following a $(\phi - \phi^*)^{1/2}$ dependence.

Three-dimensional problems are much more elaborate. The behavior of 3-D concentrated emulsions can be inferred by considering a single droplet confined in a box, whose dimensions are decreased below $2a$, thus deforming the droplet and forming flat facets at the walls. Hence, within this picture, the description of $\Pi\,(\phi)$ and $G(\phi)$ reduces to the problem of a single droplet within a unit cell, each facet behaving as a spring that repels the walls. This picture can be generalized to describe the bulk emulsion by assuming the neighboring droplets to form the box. Therefore, soft spheres interact with their nearest neighbors through central repulsive potentials that reflect the "spring-like" behavior of each of the facets. The repulsive pair elastic energy is $U\,(\xi)$, with $\xi = 1 - d/2a$, d being the distance between the centers of the interacting deformed spheres. The interaction is said to be harmonic when $U\,(\xi)$ varies as ξ^2, and nonharmonic when the exponent is larger than 2. Morse and Witten [19] have found from analytical considerations that, when ξ is close to zero (small deformation), the pair elastic potential $U(\xi)$ varies as $\xi^2/\ln\xi$ (the "effective" exponent is thus larger than 2). This was the first demonstration that the elastic interaction between deformable droplets is nonharmonic. This claim was the first important step in understanding elasticity of dense emulsions, because it raised the possibility of nonaffine droplet displacements under shear strain. Numerical calculations proposed by Lacasse et al. [20] have extended the range of applicability of the elastic pair interaction toward larger degree of deformation, also including the nature and the role of the unit cell. In an extensive numerical study of the response of a single droplet to compression by various Wigner–Seitz cells, it was shown that, for moderately compressed emulsions, the interaction

$U(\xi)$ can be approximated by a power law:

$$U(\xi) = K\gamma_{\text{int}}a^2\xi^\alpha \tag{4.11}$$

where K is a constant and α is a power exponent larger than 2. Note that K and α depend on the number of interacting neighbors. A better fit over a wider range of the numerical data was obtained via the expression:

$$U(\xi) = K\gamma_{\text{int}}a^2[(1-\xi)^{-3} - 1]^\alpha \tag{4.12}$$

Again, K and α are depending on the number of interacting neighbors. As an example, $\alpha = 2.4$ for a face cubic-centered lattice.

We now derive an expression for the equation of state $\Pi(\phi)$ from $U(\xi)$ as defined in the preceding text. By combining Eq. (4.1) and

$$\phi = \frac{4\pi Na^3}{3V} \tag{4.13}$$

we obtain:

$$\Pi = \frac{3\gamma_{\text{int}}\phi^2}{4\pi a^3}\left(\frac{\partial s}{\partial\phi}\right)_N \tag{4.14}$$

where s is the single droplet surface. Since

$$\left(\frac{d}{2a}\right)^3 = \frac{\phi^*}{\phi} \tag{4.15}$$

ξ can be rewritten as:

$$\xi = 1 - \left(\frac{\phi^*}{\phi}\right)^{1/3} \approx \frac{\phi - \phi^*}{3\phi^*} \tag{4.16}$$

Thus, if we take the limit in which ξ tends to zero (small deformations), ξ is proportional to $(\phi - \phi^*)$, and $U(\xi)$ can be transformed into $U(\phi)$. $U(\phi)$ is related to $s(\phi)$ by:

$$(s(\phi) - s_0)\gamma_{\text{int}} = zU(\phi) \tag{4.17}$$

where z is the number of nearest neighbors and s_0 is the undeformed droplet surface area. We can thus derive an expression (for small deformations) for the equation of state $\Pi(\phi)$ by differentiating $U(\phi)$ in Eq. (4.14). A harmonic potential is sufficient to describe the main features of the equation of state, at least in the regime of small deformations, so $U(\xi)$ can be taken as proportional to ξ^2, leading to $U(\phi)$ proportional to $(\phi - \phi^*)^2$. Therefore Π is proportional to $\phi^2(\phi - \phi^*)$, in agreement with the experimental data of Mason et al. [7] which can be fitted by the following empirical equation for $\phi^* < \phi < 95\%$:

$$\Pi \approx 1.7\frac{\gamma_{\text{int}}}{a}\phi^2(\phi - \phi^*) \tag{4.18}$$

An expression for the regime of very large compressions, where the emulsion is almost dry ($\phi \to 100\%$) can be derived following Princen's argument [4,5]. In highly concentrated emulsions, the droplets are forced to deform, that is, they flatten in those areas where they make "contact" and adopt the shape of polyhedra with rounded edges. The so-called Plateau borders are confined zones that appear at the intersection of three films. The surfaces delimiting Plateau borders are the rounded edges of the polyhedra. By assuming that the residual continuous phase is contained only within Plateau borders, whose shapes can be approximated by cylinders of radius r and length a, the water volume fraction $1-\phi$ is proportional to r^2a/a^3. Because in the limit of ϕ close to 100% the osmotic pressure is equal to the Laplace pressure, $2\gamma_{\text{int}}/r$, of the deformed droplets, where r is the plateau border curvature, we find that Π diverges as $(1-\phi^*)^{-1/2}$.

Similarly to the osmotic pressure, the static shear modulus $G(\phi)$ can be calculated within a mean field picture. Models based on ordered structures and harmonic potential predict a sudden raise of G at ϕ^* and a continuous increase with $(\frac{\partial^2 G}{\partial \phi^2}) \leq 0$. Such models do not account properly for the experimental data that reveal that G exhibits a smooth raise at ϕ^* followed by an increase with $(\frac{\partial^2 G}{\partial \phi^2}) \geq 0$ (G roughly follows the same evolution as Π for $\phi < 95\%$). Using the nonharmonic potential obtained by Morse and Witten [19] described earlier, and assuming an emulsion that obeys simple cubic packing, Buzza et al. [17,18] predicted a sharp but continuous rise of the shear modulus at ϕ^* and then a continuous increase with $(\frac{\partial^2 G}{\partial \phi^2}) \leq 0$. For more details concerning the calculation of $G(\phi)$, the reader is directed to [17,18,21]. The mean field approach, although including a more realistic elastic interaction, is not adequate for describing the experimental dependence of G on ϕ.

Lacasse et al. [20] performed simulations aimed at predicting the right dependence of both the osmotic pressure and the elastic shear modulus on the droplet volume fraction. Their simulations have two particular features: (1) The energy of deformation per facet formed between two neighboring droplets has been taken as nonharmonic. (2) The real microscopic structure of concentrated emulsions has been taken into account. Instead of imposing a crystal-like lattice, the model deals with systems of disordered droplets. Lacasse et al. have "numerically" constructed disordered systems of soft spheres that interact through a two-body nonharmonic potential $U(\xi)$, represented by a power law (see earlier), with a form depending on the average coordination number of the droplets. Under pairwise repulsive potentials, the particles can be seen as soft compressible spheres, pushing one another and deforming when their center-to-center distance is smaller than their initial diameter. The total free energy F of the system is the sum of all the energies involved in the interacting pairs. A random distribution of $N > 1000$ monodisperse droplets was generated in a cubic box with periodic boundary conditions. The cubic box was then "numerically deformed" using isochoric uniaxial strains. For example, the z-axis was stretched by a factor $\lambda \geq 1$ and the perpendicular plane was compressed by a factor $\lambda^{-1/2}$. Numerical results for the osmotic pressure, Π and shear modulus G obtained from this model are in excellent agreement with the experimental results,as can be seen in Fig. 4.5. The simulations provide a physical insight into the origin of the shear modulus of emulsions. The nonharmonicity of the potential has profound consequences on the deformations: even under a

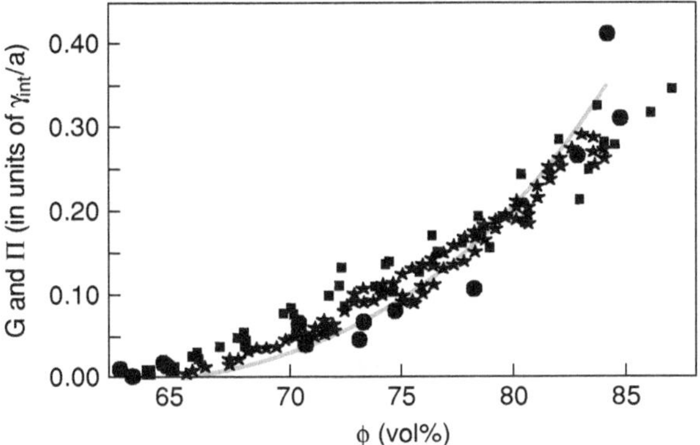

FIGURE 4.5. The computed shear modulus G' (stars) and osmotic pressure (line), compared with the experimental values for G'_p(squares) and Π (full circles). All data are normalized by γ_{int}/a. (Adapted from [21].)

uniform compression, it implies that there are nonaffine particle displacements. Lacasse et al. [20] have shown that the shear produces a positional relaxation (displacement) of the droplets that leads to a much smaller elasticity than in the absence of such relaxation. By subtracting the actual motion of the droplets from the affine motion, caused by the uniaxial strain, it appears that the droplet motion is clearly nonaffine, but random in direction. Such an effect is responsible for the "smooth" scaling of G, instead of a rapid jump in the vicinity of ϕ^*, as predicted from a mean-field picture [4,5,15–18].

4.5. Compressibility and Elasticity of Solid-Stabilized Emulsions

A growing interest is being devoted to the so-called Pickering or solid-stabilized emulsions [22–27], as they may advantageously replace conventional emulsions containing organic surfactants. To adsorb at an interface, particles need to be partly wetted by both phases. The contact angle of a particle adsorbed at an oil–water interface, is defined as the angle between the particle tangent at contact and the interface, through the water phase (see Fig. 4.6). For a preferentially water (respectively oil) wetted particle the contact angle is smaller (respectively larger) than 90°. Finkle et al. [28] empirically stated that the phase that preferentially wets the particle becomes the emulsion continuous phase. Hence O/W (respectively W/O) emulsions are obtained for contact angles smaller (respectively larger) than 90° [28,29]. The desorption energy of a particle from the interface can be simply estimated through the following relationship:

$$E = \gamma_{\mathrm{int}} \pi R^2 (1 - |\cos\theta|)^2 \tag{4.19}$$

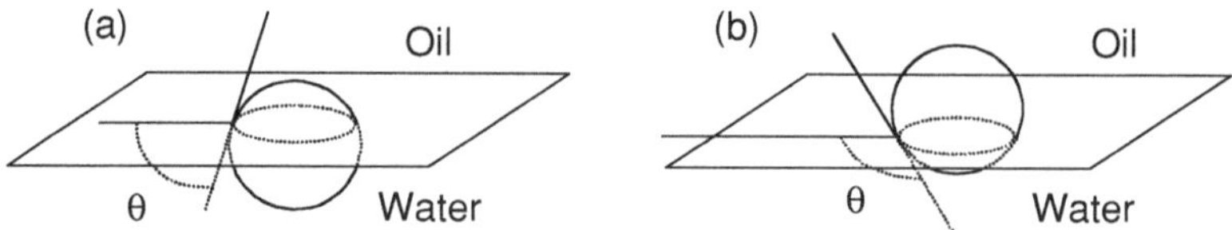

FIGURE 4.6. Definition of the contact angle θ for a particle adsorbed at the water–oil interface. Case where the particle is preferentially wetted by **(a)** water and by **(b)** oil.

where γ_{int} is the interfacial tension and R is the particle radius. The adsorption and desorption energy of a surfactant molecule at the interface is of the order of $k_B T$, while it can easily reach several thousands of $k_B T$ for nanometric particles. Consequently surfactant molecules are in dynamic equilibrium between interface and bulk whereas particles are irreversibly anchored at the interface. This difference is at the origin of peculiar properties of solid-stabilized emulsions [27,30–33]. In solid-stabilized emulsions, the stabilizing film in between the droplets is composed of very rigid layers that may provide outstanding resistance against coalescence. Arditty et al. [27,31] have measured the bulk properties of solid-stabilized concentrated emulsions and have deduced some characteristic properties of the interfacial layers covering the droplets. The osmotic pressure, Π, of the emulsions was measured for different oil volume fractions above the random close packing. The dimensionless osmotic pressure, $\Pi/(\gamma_{\mathrm{int}}/a)$, was always substantially higher than the corresponding values obtained for surfactant-stabilized emulsions. It can be concluded that droplet deformation in solid-stabilized emulsions is not controlled by the capillary pressure, γ_{int}/a, of the nondeformed droplets but rather by σ_0/a, σ_0 being a parameter characterizing the rigidity of the droplets surfaces. The data can be interpreted considering that the interfacial layers are elastic at small deformations and exhibit plasticity at intermediate deformations. σ_0 corresponds to the surface yield stress, that is, the transition between elastic and plastic regimes.

4.5.1. Osmotic Stress Resistance Measurements

Concentrated monodisperse polydimetyhylsiloxane (PDMS)-in-water emulsions stabilized by solid particles were fabricated [32–34]. Emulsions were stabilized by spherical, hydrophilic precipitated silica. The surface of such particles was made partially hydrophobic by chemically grafting n-octyltriethoxysilane in order to favor adsorption at the oil–water interface. The specific surface area, that is, the droplet surface covered per unit mass of silica was estimated to $s_f = 24\ \mathrm{m}^2/\mathrm{g}$. No residual free particles were present in the continuous phase after emulsification. The osmotic pressure, Π, of the emulsions was obtained by centrifugation for different oil volume fractions above the random close packing ϕ^*. Owing to droplet deformation, the function $\Pi(\phi)$ in the concentrated regime ($\phi > \phi^*$) reflects the mechanical properties of the layers formed by the solid particles at the oil–water interface. Right after centrifugation, the samples were stored at room temperature for several days. Within this period of time, there was no visible swelling of the

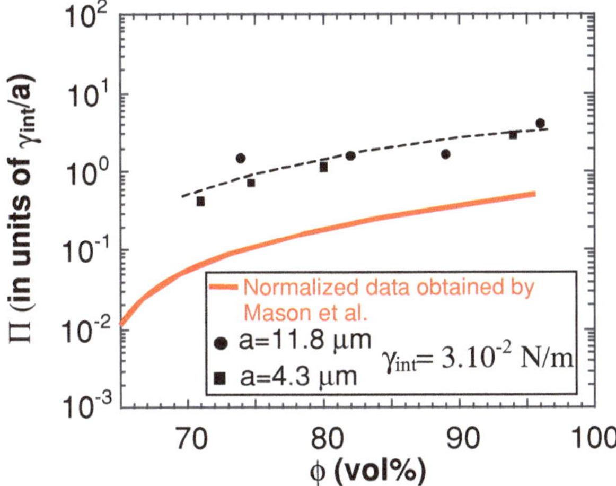

FIGURE 4.7. Evolution of the osmotic resistance Π normalized by γ_{int}/a for PDMS-in-water emulsions. The dashed line is a visual guide. The solid line corresponds to the best fit of the normalized data obtained by Mason et al. [7]. (Adapted from [31].)

cream, suggesting that either compaction of the droplets was irreversible or decompression was extremely slow. The osmotic pressure is a force per unit surface capable of causing the expansion of the system. This is the case for the osmotic pressure of molecular solutions and of nonaggregated colloidal dispersions. Because the emulsions did not expand after several days if the applied pressure was set to zero, Arditty et al. preferred to employ the term "osmotic stress resistance" rather than osmotic pressure in order to account for the eventual irreversibility of the compression. Hereafter, we summarize their main findings.

A set of experiments was performed at variable droplet sizes. The graph in Fig. 4.7 shows the dependence of the normalized (by γ_{int}/a) osmotic resistance as a function of the oil volume fraction. The normalized values fall onto a single curve within reasonable experimental uncertainty. The results were compared to the normalized data obtained by Mason et al. [7] in the presence of surfactants. These latter are represented as a solid line that corresponds to the best fit to the experimental points (Eq. (4.18)). It is worth noting that the normalized pressures in solid-stabilized emulsions are much larger than the ones obtained in the presence of surfactants.

4.5.2. *Surface Properties: Elasticity and Plasticity*

As explained in Section 4.2, the osmotic pressure can be expressed as a product of two independent terms, the derivative of the free energy with respect to the amount of interface $\sigma = \left(\frac{\partial F}{\partial S}\right)_N$ and a geometrical factor $\left(\frac{\partial S}{\partial V}\right)_N$ representing the effect of compression (see Eq. (4.1)). Since in solid-stabilized emulsions the interfaces contain particles with attractive interactions, they are expected to behave like 2-D solids whose properties should be revealed by σ. Assuming that $\left(\frac{\partial S}{\partial V}\right)_N$ is independent of the interface nature, from Eqs. (4.1) and (4.18), we obtain:

$$\left(\frac{\partial S}{\partial V}\right)_N \approx 1.7\frac{\phi^2(\phi^* - \phi)}{a} \tag{4.20}$$

σ can therefore be calculated and we get:

$$\sigma = \left(\frac{\partial F}{\partial S}\right)_N = \frac{\Pi a}{1.7\,\phi^2(\phi - \phi^*)} \tag{4.21}$$

Actually σ measures the bidimensional stress of the interface.

In Fig. 4.8, the evolution of σ as a function of ϕ for all the emulsions is reported. In the explored ϕ range, there is no visible variation of σ. This result reveals that the osmotic resistance can be described by a simple parameter, σ_0, independent of $\phi : \sigma = \sigma_0 \approx 0.2$ N/m. On the one hand, the order of magnitude of σ_0 is not compatible with the interfacial tension, γ_{int} measured between PDMS and water: $\gamma_{\text{int}} = 0.03$ N/m. On the other hand, this surface stress originates from the bidimensional network of the adsorbed solid particles that are known to exhibit strong lateral interactions. The invariance of σ suggests that this parameter corresponds to a bidimensional yield stress for plasticity as explained below. In Fig. 4.9, different possible mechanical behaviors of the interface are represented schematically. The abscissa represents the relative deformation of the interfaces, which is experimentally controlled by the droplet volume fraction ϕ following Eq. (4.20). The purely elastic behavior in Fig. 4.9a can be ruled out because it corresponds to a measured stress that strongly varies with ϕ, incompatible with the result of Fig. 4.8. In contrast, a plastic behavior should result in either a constant stress, σ_0, for perfect plastic behavior, as sketched in Fig. 4.9b, or a weak variation of σ for standard plastic behavior, as indicated in Fig. 4.9c. The precision of the measurements does not allow distinguishing pure and standard plasticity. Hence, at first order, σ can be considered constant.

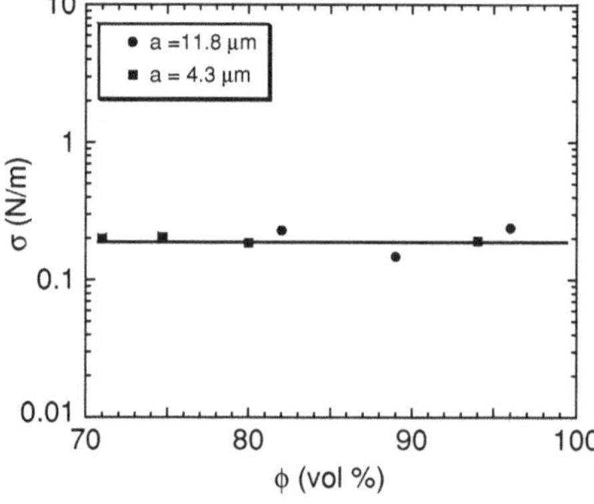

FIGURE 4.8. Evolution of the surface modulus σ as a function of ϕ for emulsions comprising PDMS droplets. (Adapted from [31].)

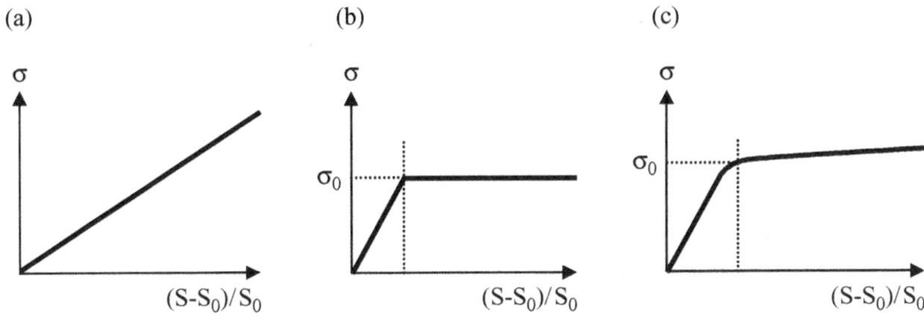

FIGURE 4.9. Schematic mechanical behavior of the interface **(a)** elastic behavior: the 2-D stress is proportional to the relative surface extension, **(b)** ideal plastic behavior: after a narrow elastic regime the stress becomes constant and equal to σ_0, and **(c)** standard plastic behavior: in the plastic regime the stress slowly increases with the relative surface extension. (Adapted from [31].)

The plastic behavior necessarily follows an elastic regime that could not be probed within the explored ϕ range ($\phi > 70\%$). By integrating Eq. (4.20), the relative surface excess $(S - S_0)/S_0$ can be calculated:

$$\int_{S_0}^{S} \frac{dS}{S_0} = 0.3\,(\phi - \phi^*)^2 \tag{4.22}$$

At $\phi = 70\%$, the relative surface excess is of the order of 0.1%. At this volume fraction, the surface stress has already reached its asymptotic value. Thus, the plastic strain of the surface is smaller than 0.1%.

Arditty et al. have proposed an independent route to measure the 2-D yield stress σ_0. It is based on the characteristic shear stress, τ_c, that produces droplet fragmentation. τ_c is expected to be of the order of the droplet deformability, that is, σ_0/a, which allows an independent determination of σ_0. Dilute monodisperse PDMS-in-water emulsions were submitted to shear stresses of variable intensity. After shearing cessation, the emulsions were observed under a microscope to determine whether fragmentation had occurred or not. Droplet fragmentation was revealed by a discrete jump of the droplets size before and after application of the stress. Droplet rupturing took place above a well-defined critical stress τ_c. Below τ_c, the droplet size remained invariant even if the stress was applied for several minutes while for $\tau > \tau_c$, droplet fragmentation was already accomplished after 10 s. Taking into account the viscosity ratio between the dispersed and continuous phases, the type and history of the shear flow (see Chapter 1, Section 1.7.5), Arditty et al. obtained $\sigma_0 = 0.19$ N/m, in very good agreement with the value deduced from osmotic resistance measurements.

4.5.2.1. Origin of σ_0. Influence of the Interparticle Interactions

In solid-stabilized emulsions, the droplet surface can be regarded as a compact 2-D network of solid particles with strong lateral attractive interactions. The surface stress tensor $\tilde{\sigma}$ is defined as

$$\tilde{\sigma} = \frac{\left\langle \vec{r} \otimes \vec{f} \right\rangle}{S} \tag{4.23}$$

where $\otimes$ is the tensorial product, $\vec{r}$ and $\vec{f}$ are the position and force vectors, respectively. At the onset of the plastic regime, surface clusters start to be dissociated and the yield stress σ_0 corresponds to the tangential component σ_{12}. If e_a is the pair interaction energy of the adsorbed particles, then the yield stress can be estimated:

$$\sigma_0 \simeq n_s \frac{z}{2} e_a \tag{4.24}$$

where z is the coordination number ($z = 6$ for a 2-D hexagonal packing), and n_s is the particle density at the interface. From $s_f = 24$ m^2/g, it can be deduced that $n_s = 2.4\ 10^{15}$ particles/m^2. The pair energy corresponding to the experimentally measured value $\sigma_0 = 0.2$ N/m is then: $e_a \approx 7\ 10^3 k_B T$. Arditty et al. argue that such high attractive energy mainly results from capillary forces [35,36] and from the interpenetration of the octyl layers in a bad solvent [37]. The magnitude of such interactions was evaluated and is of the same order than the total interaction energy deduced from σ_0.

The surface rigidity in solid-stabilized emulsions certainly determines the remarkable resistance of concentrated droplets against coalescence. A recent article reports that some oil-in-water solid-stabilized emulsions can even be dried without being destroyed until almost all the water is evaporated [30]. The original shape of the sample is preserved during the drying process, with absolutely no oil leakage, confirming the existence of a strong protective layer around the oil droplets. The resistance to coalescence is certainly determined by particle interactions, as can be inferred from the general phenomenology established in the field of solid-stabilized emulsions: the most stable emulsions are very often obtained with solid particles that are weakly or strongly aggregated in the bulk continuous phase [24]. However, this correlation has to be handled with care, as lateral interactions at the oil–water interface may be different from bulk interactions in the continuous phase. More experimental and theoretical work is needed to consolidate (or refute) the validity of this empirical link.

4.6. Conclusion

As reported in this chapter, the microscopic origin of both compressibility and elasticity of dense emulsions is rather well understood. Emulsions have elastic properties arising from either surface tension or surface elasticity and plasticity. Some protein-stabilized emulsions obey the same phenomenology as solid-stabilized emulsions: they exhibit substantially higher osmotic resistances and higher shear moduli than surfactant-stabilized emulsions [38–40]. Moreover, they are strongly resistant to water evaporation. Proteins possess the ability to form rigid interfacial aggregates (see Chapter 2) and it is likely that the same description of the interface in terms of elasticity and plasticity can be applied to these systems.

References

[1] J.W. Goodwin and W.B. Russel: "Rheology and Rheological Techniques." Curr. Opin. Colloidal Interface Sci. **2**, 409 and references therein (1997).

[2] E. Dickinson and M. Golding: "Influence of Alcohol on Stability of Oil-in-Water Emulsions Containing Sodium Caseinate." J. Colloid Interface Sci. **197**, 133 (1998).

[3] E. Dickinson and J. Chen: "Viscoelastic Properties of Protein-Stabilized Emulsions: Effect of Protein-Surfactant Interactions." J. Agric. Food Chem. **46**, 91 (1998).

[4] H.M. Princen and A.D. Kiss: "Osmotic Pressure of Foams and Highly Concentrated Emulsions. 2. Determination from the Variation in Volume Fraction with Height in an Equilibrated Column." Langmuir **3**, 36 (1987).

[5] H.M. Princen and A.D. Kiss: "Rheology of Foams and Highly Concentrated Emulsions: III. Static Shear Modulus." J. Colloid Interface Sci. **112**, 427 (1986).

[6] H.M. Princen: "Osmotic Pressure of Foams and Highly Concentrated Emulsions. I. Theoretical Considerations." Langmuir **2**, 519 (1986).

[7] T.M. Mason, J. Bibette, and D.A. Weitz: "Elasticity of Compressed Emulsions." Phys. Rev. Lett. **75**, 2051 (1995).

[8] T.G. Mason, A.H. Krall, H. Gang, J. Bibette, and D.A. Weitz: "Monodisperse Emulsions: Properties and uses." P. Becher (ed), *Encyclopaedia of Emulsion Technology*, Ch. 6, p. 299, Marcel Dekker, New York, Basel, Hong Kong (1996).

[9] P. Hébraud, F. Lequeux, J.-P. Munch, and D.J. Pine: "Yielding and Rearrangements in Disordered Emulsions." Phys. Rev. Lett. **78**, 4657 (1997).

[10] T. Mason: "Rheology of Monodisperse Emulsions." Ph.D Thesis, Princeton University (1995).

[11] H.M. Princen, M.P. Aronson, and J.C. Moser: "Highly Concentrated Emulsions. II Real Systems. The Effect of Film Thickness and Contact Angle on the Volume Fraction in Creamed Emulsions." J. Colloid Interface Sci. **75**, 246 (1980).

[12] F. Leal-Calderon, T. Stora, O. Mondain Monval, P. Poulin, and J. Bibette: "Direct Measurement of Colloidal Forces." Phys. Rev. Lett **72**, 2959 (1994).

[13] P.M. Kruglyakov, D. Exerowa, and K. Khristov: "New Possibilities for Foam Investigation: Creating a Pressure Difference in the Foam Liquid Phase." Adv. Colloid Interface Sci. **40**, 257 (1992).

[14] D.J. Stamenovic: "A Model of Foam Elasticity Based Upon the Laws of Plateau." J. Colloid Interface Sci. **145**, 255 (1991).

[15] H.M. Princen: "Highly Concentrated Emulsions. I Cylindrical Systems." J. Colloid Interface Sci. **71**, 55 (1979).

[16] H.M. Princen: "Rheology of Foams and Highly Concentrated Emulsions I. Elastic Properties and Yield Stress of a Cylindrical Model System." J. Colloid Interface Sci. **91**, 160 (1983).

[17] D.M.A. Buzza and M.E. Cates: "Uniaxial Elastic Modulus of Concentrated Emulsions." Langmuir **10**, 4503 (1994).

[18] D.M.A. Buzza, C.-Y.D. Lu, and M.E. Cates: "Linear Shear Rheology of Incompressible Foams." J. Phys. France II **5**, 37 (1995).

[19] D.C. Morse and T.A. Witten: "Droplet Elasticity in Weakly Compressed Emulsions." Europhys. Lett. **22**, 549 (1993).

[20] M.-D. Lacasse, G.S. Grest, D. Levine, T.G. Mason, and D.A. Weitz: "Model for the Elasticity of Compressed Emulsions." Phys. Rev. Lett **76**, 3448 (1996).

[21] T.G. Mason, M.-D. Lacasse, G.S. Grest, D. Levine, J. Bibette, and D.A. Weitz: "Osmotic Pressure and Viscoelastic Shear Moduli of Concentrated Emulsions." Phys. Rev. E **56**, 3150 (1997).

[22] W. Ramsden: "Separation of Solids in the Surface-Layers of Solutions and "Suspensions"—Preliminary Account." Proc. R. Soc. **72**, 156 (1903).

[23] S.U. Pickering: "Emulsions." J. Chem. Soc. **91**, 2001 (1907).

[24] R. Aveyard, B.P. Binks, and J.H. Clint: "Emulsions Stabilized Solely by Colloidal Particles." Adv. Colloid Interface Sci. **100–102**, 503 (2003).

[25] B. Binks and S.O. Lumsdon: "Stability of Oil-in-Water Emulsions Stabilised by Silica Particles." Phys. Chem. Chem. Phys. **1**, 3007 (1999).

[26] B.P. Binks: "Particles as Surfactants—Similarities and Differences." Curr. Opin. Colloid Interface Sci. **7**, 21 (2002).

[27] S. Arditty: "Fabrication, Stability and Rheological Properties of Solid-Stabilzed Emulsions." Ph.D thesis, Bordeaux I University (2004).

[28] P. Finkle, H.D. Draper, and J.H. Hildebrand: "The Theory of Emulsification." J. Am. Chem. Soc. **45**, 2780 (1923).

[29] B.P. Binks and S.O. Lumsdon: "Influence of Particle Wettability on the Type and Stability of Surfactant-Free Emulsions." Langmuir **16**, 8622 (2000).

[30] S. Arditty, J. Kahn-Giermanska, V. Schmitt, and F. Leal-Calderon: "Materials Based on Solid-Stabilized Emulsions." J. Colloid Interface Sci. **275**, 659 (2004).

[31] S. Arditty, V. Schmitt, F. Lequeux, and F. Leal-Calderon: "Interfacial Properties in Solid-Stabilized Emulsions." Eur. Phys. J. B **44**, 381 (2005).

[32] S. Arditty, C.P. Whitby, B.P. Binks, V. Schmitt, and F. Leal-Calderon: "Erratum—Some General Features of Limited Coalescence in Solid-Stabilized Emulsions." Eur. Phys. J. E **12**, 355 (2003).

[33] S. Arditty, C.P. Whitby, B.P. Binks, V. Schmitt, and F. Leal-Calderon: "Some General Features of Limited Coalescence in Solid-Stabilized Emulsions." Eur. Phys. J. E **11**, 273 (2003).

[34] T.H. Whitesides and D.S. Ross: "Experimental and Theoretical Analysis of the Limited Coalescence Process: Stepwise Limited Coalescence." J. Colloid Interface Sci. **169**, 48 (1995).

[35] P.A. Kralchevsky and K. Nagayama: "Capillary Interactions Between Particles Bound to Interfaces, Liquid Films and Biomembranes." Adv. Colloid Interface Sci. **85**, 145 (2000).

[36] P.A. Kralchevsky and N.D. Denkov: "Capillary Forces and Structuring in Layers of Colloid Particles." Curr. Opin. Colloid Interface Sci. **6**, 383 (2001).

[37] S.R. Raghavan, J. Hou, G.L. Baker, and S.A. Kahn: "Colloidal Interactions between Particles with Tethered Nonpolar Chains Dispersed in Polar Media: Direct Correlation Between Dynamic Rheology and Interaction Parameters." Langmuir **16**, 1066 (2000).

[38] T.D. Dimitrova and F. Leal Calderon: "Bulk Elasticity of Concentrated Protein-Stabilized Emulsions." Langmuir **17**, 3235 (2001).

[39] T.D. Dimitrova and F. Leal Calderon: "Rheological Properties of Highly Concentrated Protein-Stabilized Emulsions." Adv. Colloid Interface Sci. **108–109**, 49 (2004).

[40] L. Bressy, P. Hébraud, V. Schmitt, and J. Bibette: "Rheology of Emulsions Stabilized by Solid Interfaces." Langmuir **19**, 598 (2003).

5 Stability of Concentrated Emulsions

5.1. Introduction

The lifetime of emulsions may vary considerably from one system to another; it can change from minutes to many years, depending on the nature of the surfactants, the nature of both phases, and their volume ratio. Despite the large amount of work devoted to this issue, predicting the destruction scenario and the emulsion lifetime still raises challenging questions, especially with regard to concentrated emulsions. Irreversible coarsening of emulsions may proceed through two distinct mechanisms. The first mechanism, known as Ostwald ripening [1], is driven by the difference in Laplace pressure between droplets having different radii: the dispersed phase is transferred from the smaller to the larger droplets. The rate of droplet growth may be determined by the molecular diffusion across the continuous phase and/or by the permeation across the surfactant films. The second mechanism, known as coalescence, consists of the rupture of the thin film that forms between droplets, leading them to fuse into a single one. At a microscopic scale, a coalescence event proceeds through the nucleation of a thermally activated hole that reaches a critical size above which it becomes unstable and grows.

In principle, understanding the metastability of emulsions requires two types of information that account for two distinct phenomena. The first one concerns the microscopic mechanism of the instabilities. The second one concerns the scenario of destruction, that is, the time and space distributions of the coarsening events. Generally, the destruction scenario of emulsions results from the interplay between coalescence and Ostwald ripening. For the sake of simplicity, most studies have been performed under conditions such that one type of instability is dominating the other one, enabling one to monitor its progress quite precisely. In this limit, theoretical models as well as experiments have revealed that Ostwald ripening generates emulsions with narrow size distributions in the asymptotic regime [2–7]. In contrast, coalescence favors the diverging growth of large droplets that rapidly form a free layer at the top or the bottom of the sample [8,9]. Owing to their different consequences on the droplet evolution and size distribution, coalescence and Ostwald ripening can be easily identified: a system evolving under the effect of Ostwald ripening alone exhibits a narrow size distribution with a droplet growth rate that progressively vanishes; instead, coalescence events increase the polydispersity and accelerate the rate of coarsening as a result of the divergent growth of large nuclei.

In this chapter, we review some results in the field of emulsion metastability, emphasizing the destruction of concentrated emulsions (droplet volume fraction $\phi > 70\%$) through coalescence. The review concerning Oswald ripening (Section 5.2) is more concise, as this mechanism is fairly well understood and has been extensively documented in the literature. So far, the destruction of concentrated emulsions through coalescence is much less understood and has motivated many recent studies and developments that we summarize (Sections 5.3 to 5.6).

5.2. Ostwald Ripening

Ostwald ripening consists of a diffusive transfer of the dispersed phase from smaller to larger droplets. Ostwald ripening is characterized by either a constant volume rate Ω_3 [4,5] (diffusion-controlled ripening) or a constant surface rate Ω_2 [6] (surface-controlled ripening), depending on the origin of the transfer mechanism:

$$\frac{dD^{\alpha}}{dt} = \Omega_{\alpha} \tag{5.1}$$

where D is the average droplet diameter. Diffusion-controlled ripening ($\alpha = 3$) has been recognized in submicron diluted emulsions stabilized by ionic or nonionic surfactants [4,5]; so far, permeation-controlled ripening ($\alpha = 2$), has been proposed to account for the coarsening of concentrated air foams [6].

If the ripening is controlled by diffusion across the continuous phase, then the cube of the diameter increases linearly with time ($\alpha = 3$) and the ripening rate Ω_3 can be derived using the Lifshitz and Slyozov theory [2,3]:

$$\Omega_3 = 64\frac{\gamma_{\text{int}}\, \mathit{Diff}\, S V_m}{9RT} \tag{5.2}$$

where S is the molecular solubility, V_m is the molar volume, Diff is the molecular diffusion coefficient of the dispersed phase in the continuous one, γ_{int} is the interfacial tension of the oil–water interface, and R is the molar gas constant. The theory [2,3] also predicts that the size distribution becomes quite narrow and asymptotically self-similar, in agreement with experiments [4,5]. In principle, Eq. (5.2) is valid in the limit of very dilute emulsions. In general, it is expected that emulsions with higher volume fractions of disperse phase will have faster absolute growth rates than those predicted by the Lifshitz and Slyozov model. The theoretical rate of ripening must therefore be corrected by a factor $f(\phi)$ that reflects the dependence of the coarsening rate on the dispersed phase volume fraction ϕ [10–14].

Kabalnov [15] and Taylor [16] found that the presence of ionic micelles in the continuous phase had a surprisingly small effect on the rate of ripening, despite the fact that the solubility of the dispersed phase is largely enhanced. It was argued that owing to electrostatic repulsion, ionic micelles cannot absorb oil directly from emulsion drops. In the presence of nonionic surfactants, much larger increases in the rate of ripening are expected owing to larger solubilization capacities and to the absence of electrostatic repulsion between droplets and micelles. Weiss et al. [17] showed that a significant increase in average diameter can be achieved for tetradecane-in-water emulsions diluted with a fresh solution of nonionic surfactant. In their experiments, the variation of droplet diameter resulted from a complex interplay between oil solubilization by the micelles, which tended to reduce droplet diameter, and Ostwald ripening, which tended to increase it.

When the dispersed phase is composed of a binary mixture, growth may be arrested if one component is almost insoluble in the continuous phase, therefore retaining the soluble one, owing to the gradual loss of mixing entropy [18]. The osmotic pressure of the trapped species within the droplets can overcome the Laplace pressure differences that drive the coarsening and "osmotically stabilize" the emulsions. Webster and Cates [19] gave rigorous criteria (in terms of osmotic pressure within the droplets) for stabilization of monodisperse and polydisperse emulsions in the dilute regime. The same authors [20] also examined the concentrated regime in which the droplets are strongly deformed and therefore possess a high Laplace pressure. These authors conclude that osmotic stabilization of dense emulsions requires a pressure of trapped molecules in each droplet which is comparable to the Laplace pressure that the droplets would have if they were spherical, as opposed to the much larger pressures actually present in the system.

Mass transfers in emulsions may be driven not only by differences in droplet curvatures but also by differences in their compositions. This is observed when, for example, two chemically different oils are emulsified separately and the resulting emulsions are mixed. This phenomenon is called composition ripening [21,22]. Mass transfer from one emulsion to the other is controlled by the entropy of mixing and proceeds until the compositions of the droplets become identical. The most spectacular evidence of composition ripening comes from the so-called reverse recondensation that occurs when the two emulsions differ significantly both in their initial droplet size and in their rate of molecular diffusion. If the larger sized emulsion is composed of the faster diffusing oil, then molecular diffusion occurs in the "reverse" direction, that is, from large to small droplets [23].

Most experimental studies concerning Ostwald ripening have been performed in the limit of highly dilute emulsions ($\phi < 1\%$), in conditions such that the droplets are not in permanent contact. Ostwald ripening is then preferentially controlled by the molecular diffusion of the dispersed molecules across the continuous phase, as deduced from experimental measurements [4,5,17,24,25]. Taisne and Cabane [26] have examined the coarsening of concentrated alkane-in-water droplets ($\phi \approx$ 20–40%) stabilized by a nonionic poly(ethoxylated) surfactant, following a temperature quench. Because the rate of ripening does not depend on the alkane chain length, they conclude that the transfer of oil from the smaller drops to the larger ones does not occur by diffusion across the continuous phase but rather through the direct contact of the droplets (permeation) as observed in concentrated foams [6]. A similar conclusion was drawn by Schmitt et al. [27]. These authors produce different alkane-in-water concentrated emulsions ($\phi \approx 80\%$) stabilized by a nonionic surfactant and follow the kinetic evolution by granulometry using static light-scattering. The size distribution becomes remarkably narrow during the first stages of coarsening and progressively widens as time passes. They obtained evidence that the evolution is first determined by Ostwald ripening and then by coalescence. In the region dominated by Ostwald ripening, the experimental rates are not compatible with those predicted by the Lifshitz-Slyozov model assuming a volume-mediated transfer ($\alpha = 3$). The quadratic scaling ($\alpha = 2$) correctly accounts for the size evolution, suggesting that the oil transfer is surface-controlled. When a second hydrophobic species of large molecular size is dissolved in the droplets, both Ostwald ripening and coalescence are inhibited, even at very low concentration of the second component ($\sim$1 wt%). Schmitt et al. [27] argue that the presence of the second component modifies the properties of the surfactant monolayers (see Section 5.3.1.1). The fact that the two types of instability disappear simultaneously strongly suggests that they possess the same microscopic origin: hole nucleation in the thin liquid films. Owing to the activated nature of the process, only the largest holes grow spontaneously and produce coalescence events. Although the smaller holes are evanescent, they allow the transfer of matter between droplets, increasing and even dominating the total rate of transfer.

5.3. Coalescence

5.3.1. General Phenomenology and Microscopic Description

An emulsion that is, for instance, stable over many years at low droplet volume fraction may become unstable and coalesce when compressed above a critical osmotic pressure Π^*. As an example, when an oil-in-water emulsion stabilized with sodium dodecyl sulfate (SDS) is introduced in a dialysis bag and is stressed by the osmotic pressure imposed by an external polymer solution, coarsening occurs through the growth of a few randomly distributed large droplets [8]. A microscopic image of such a growth is shown in Fig. 5.1.

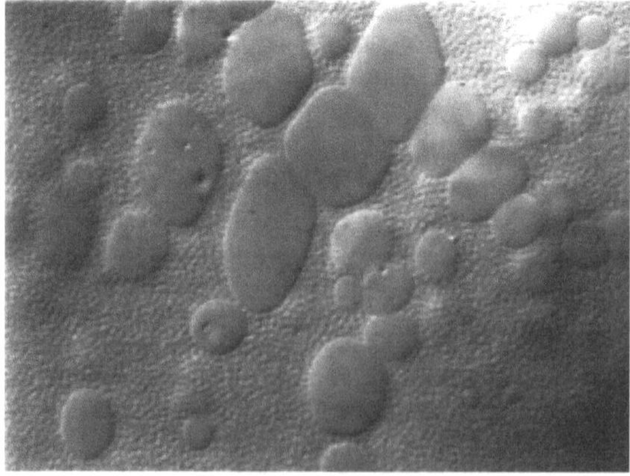

FIGURE 5.1. Microscopic image of an emulsion initially composed of monodisperse droplets having a diameter of about 1.5 μm, which has been submitted to an osmotic stress of 0.6 atm for 15 days at room temperature. (Reproduced from [8], with permission.)

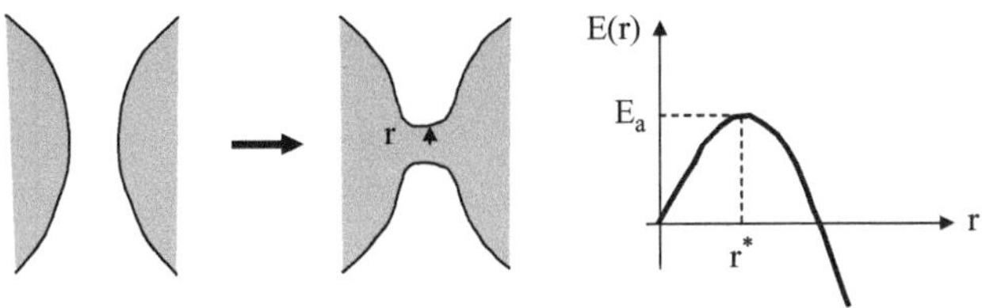

FIGURE 5.2. Scheme of the hole nucleation process and variation of the energy cost with hole radius r.

The same type of compression may be applied via centrifugation. In this case, the osmotic pressure results from the buoyant stress of the droplet layers below the top of the cream. The destabilization proceeds through the growth of a macroscopic domain that nucleates at one side of the sample and progresses as a front. Careful observation reveals that the front is a thin zone where the large drops grow before merging into a single macroscopic phase [28]. The progress of the coalescence front during centrifugation, $h(t)$, has been very instructive and reveals interesting issues. The function $h(t)$ is found linear at the beginning with a slope β and saturates at longer times at a value h^*, which depends on the applied acceleration. It can be deduced that below a certain height h^*, the rate of coalescence tends to zero, making it possible to define the threshold osmotic pressure Π^* to induce coalescence. β is found to depend on the thermodynamic properties of the surfactant layer. For instance, adding electrolytes in the presence of ionic surfactants or increasing the length of the hydrophobic tail of nonionic surfactants reduces the rate of coalescence β.

At a microscopic scale, a single coalescence event proceeds through the nucleation of a thermally activated hole that reaches a critical size, above which it becomes unstable and grows [29]. We shall term $E(r)$ the energy cost for reaching a hole of size r. A maximum of E occurs at a critical size r^*, $E(r^*) = E_a$ being the activation energy of the hole nucleation process (Fig. 5.2).

5.3.1.1. Surfactant-Stabilized Emulsions

The origin of the activation energy is still a matter of debate but it is now generally agreed that the so-called spontaneous curvature of the surfactant monolayer covering the emulsion drops is one of the most determining parameters [30]. Convincing evidence is given by a general experimental observation: it is well known that the micellar phase always tends to be the continuous phase of an emulsion. Shaking an O/W microemulsion coexisting with excess oil preferentially leads to the formation of an O/W emulsion, the continuous phase being the O/W microemulsion. Conversely, shaking a W/O microemulsion coexisting with excess water leads to the formation of a W/O emulsion with W/O swollen micelles in the continuous phase [31]. An interpretation for this correlation was proposed by Kabalnov and Wennerström [30]. Let us consider an oil film separating two water droplets (Fig. 5.3). If the surfactant covering the droplets is essentially oil-soluble,

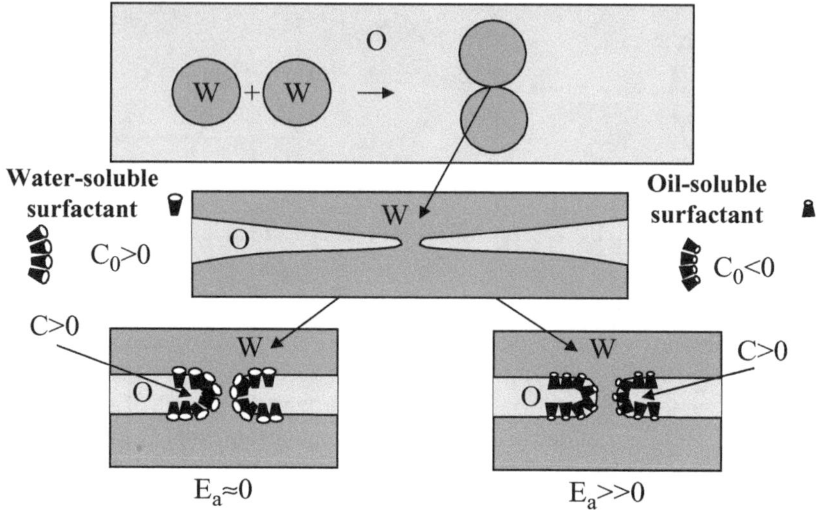

FIGURE 5.3. Scheme explaining the influence of the spontaneous curvature on the activation energy for coalescence in a W/O emulsion.

it possesses a large tail and a small polar head. The spontaneous curvature C_0 of the monolayer is negative, meaning that it tends to curve toward oil. If a tiny hole is formed in this film, the monolayer at the edge of the hole is curved against the direction favored by the spontaneous curvature. Because of this, for film rupture to occur, the system must pass through an energy barrier, after which the growth becomes spontaneous. This state can be reached only by a thermal fluctuation and has a low probability because of the unfavorable spontaneous curvature C_0. However, if the surfactant covering the droplets is essentially water-soluble, the spontaneous curvature C_0 of the monolayer is positive. Hole propagation occurs with a very moderate energy barrier because the local curvature C of the edge fits the surfactant monolayer spontaneous curvature. The consequence is that W/O emulsions persist in the presence of oil-soluble surfactants and are rapidly destroyed in the presence of water-soluble surfactants. Conversely, O/W emulsions persist in the presence of water-soluble surfactants. This model is in agreement with the phenomenology previously described and also with the well known empirical Bancroft rule [32].

The spontaneous curvature is not the only parameter controlling the activation energy. The experiments of Sonneville-Aubrun et al. [33] suggest that E_a is influenced by the short-range surface forces and by surfactant packing at the oil–water interface. These authors carefully examine the influence of centrifugation on O/W emulsions stabilized by ionic surfactants. The cream that forms after most of the water has been removed has the structure of a biliquid foam. Examination of this cream via electron microscopy shows polyhedral oil cells separated by thin films. The thickness of these films was measured through small-angle neutron-scattering. The results yield a disjoining pressure isotherm, where the film thickness is determined solely by the pressure applied to extract water during centrifugation. For hexadecane-in-water biliquid foams stabilized with SDS, this isotherm has two states, the common black film (CBF; water thickness beyond 25 Å) and the Newton black film (NBF; water thickness of 13 Å). The thickness of the NBF is stabilized by hydration forces, which resist the dehydration of counterions and head groups. The surface density of SDS molecules in these films has also been measured. As water is extracted, the concentration of counterions increases, and the head groups are more efficiently screened; as a result, the surface density of SDS in the monolayers rises. In the NBF state, the monolayers are tightly packed, with an orientational order that exceeds that of the lamellar phase. Consequently, some fluctuations that could cause the rupture of the films (vacancies in the monolayers and defects where opposing monolayers recombine) will be inhibited. This suggests a general route for improving the metastability of biliquid foams, which is simply to improve the packing of the surfactant molecules in the monolayers. This tighter packing of surfactant molecules may explain the surprisingly high metastability of biliquid foams when the films are in the NBF state.

Other experiments performed by Bergeron [34] on air foams stabilized with ionic surfactants reveal that the so-called Gibbs or dilatational elasticity ε may play an important role in the coalescence process. The Gibbs elasticity measures the variation of surface tension γ_{int} associated to the variation of the surfactant surface concentration Γ:

$$\varepsilon = -\Gamma d\gamma_{\text{int}}/d\Gamma \tag{5.3}$$

It is worth noting that ε is also linked to the packing of the surfactant molecules in the monolayers. Foam and emulsion films undergo both spatial and surfactant density (i.e., charge) fluctuations occurring at the interface (Fig. 5.4). The hydrodynamic influence associated with this surface moduli effect is often qualitatively expressed as a Gibbs–Marangoni stabilization mechanism. Bergeron argues that the surface elasticity plays a key role in dampening both spatial and density fluctuations in foam and emulsion films. When these fluctuations are dampened, the probability of overcoming the activation barrier that holds a film in a metastable state is lower and the film will be more stable. Whether or not disturbances are thermally or mechanically induced, a cohesive surfactant monolayer with a high surface elasticity will promote film stability. The influence of ε was revealed by

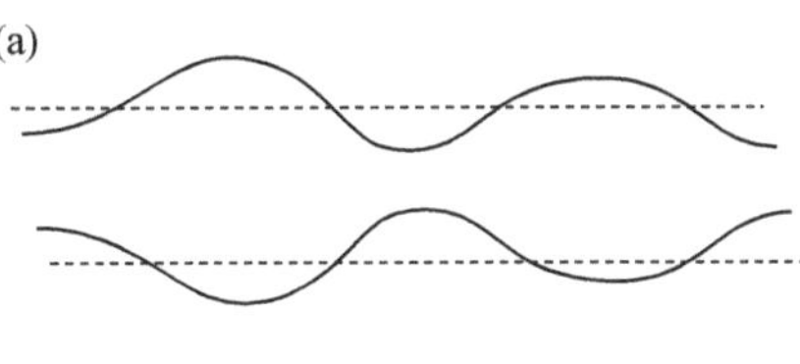

FIGURE 5.4. Scheme of **(a)** a spatial fluctuation and **(b)** a local depletion zone due to monolayer density fluctuations in thin liquid films.

comparing the relative coalescence rate of foams stabilized by ionic surfactants (alkyltrimethylammonium bromides) with the same polar head but different alkyl chain lengths: the resistance to coalescence was clearly correlated to ε, which was an increasing function of the alkyl chain length.

5.3.1.2. Surfactant-Free Emulsions

In the absence of surfactants, it is well known that hydrocarbon (alkane)-in-water emulsions coalesce and are totally destroyed after several hours of storage. However, Pashley et al. [35,36] were able to fabricate surfactant-free, alkane-in-water emulsions ($\phi < 6\%$) with remarkable kinetic stability after almost completely removing the dissolved gases. The reintroduction of dissolved gases did not destabilize already-formed emulsions. In addition, a systematic trend was found between the molecular chain length of the alkane and the lifetime of the degassed oil droplets in water: the larger the alkane chain length, the more stable the emulsion.

Considering the measured electrostatic surface potential of the bare hydrocarbon–water interface and the Hamaker constant for that system, the emulsions studied by Pashley et al. should resist coalescence for a long period of time (several weeks) owing to the high electrostatic barrier. The failure of double-layer (DLVO) theory to predict the behavior of nondegassed emulsions raises the issue of whether additional forces might be involved. To account for their observations, Pashley et al. discuss the potential role of a long-range attractive force, called the "hydrophobic interaction." Such interaction acts over relatively large distances (>10 nm) between hydrophobic surfaces in water [37] and is sensitive to dissolved gases [38]. The results of Pashley et al. [36,37] have confirmed the importance of dissolved gas, but a complete explanation for the effect of degassing on the hydrophobic dispersion still cannot be provided.

5.4. Measurements of the Coalescence Frequency

The evolution of emulsions through coalescence can be characterized by a kinetic parameter, ω, describing the number of coalescence events per unit time and per unit surface area of the drops. Following the mean field description of Arrhenius, ω can be expressed as:

$$\omega = \omega_0 \exp(-E_a/k_B T) \tag{5.4}$$

In this expression, $k_B T$ is the thermal energy and ω_0 is the attempt frequency of the hole nucleation process. Assuming the existence of a unique rupturing frequency ω per unit film area, numerical simulations performed on a 2-D cellular material [9], reveal that the scenario of destruction through coalescence is intrinsically inhomogeneous with a few giant cells growing much more rapidly than the average, in qualitative agreement with the experimental observations previously described. Owing to the intrinsic complexity of the destruction scenario, the measurements of ω_0 and E_a are scarce. However, any progress in understanding the metastability of emulsions is possible only through a quantitative determination of ω. Hereafter, we describe some attempts to measure ω for three different systems.

5.4.1. Simple Emulsions Stabilized by Surfactants

In the pioneering work of Deminière et al. [39], the emulsion is modeled as a stack of monodisperse cells with characteristic size D. The total number of drops per unit volume, n, is related to the volume fraction ϕ of the dispersed phase through the following relationship:

$$\phi = n\pi D^3/6 \tag{5.5}$$

Because coalescence is a completely random process, the total number of coalescence events per unit time is assumed to be proportional to the total surface area A of the droplets:

$$-\frac{dn}{dt} = \omega A = \omega n\pi D^2 \tag{5.6}$$

In Eq. (5.6), ω is defined as a coalescence frequency per unit surface area of the droplets. Considering Eqs. (5.5) and (5.6) and assuming that ω is constant (independent of D), it can be concluded that the mean size in the emulsion increases with time according the following law:

$$\frac{1}{D_0^2} - \frac{1}{D^2} = \frac{2\pi}{3}\omega t \tag{5.7}$$

where D_0 is the initial droplet diameter. This model predicts a divergence of the diameter after a finite time τ given by:

$$\tau = \frac{3}{2\,\pi\omega D_0^2} \tag{5.8}$$

The first kinetics measurements about coalescence were reported by Kabalnov and Weers in water-in-oil emulsions [40]. These authors measured the characteristic time at which the layer of free water formed at the bottom of the emulsions corresponded approximately to half of the volume of the dispersed phase. This time was assumed to be equal to τ. By measuring τ at different temperatures, the activation energy was deduced from an Arrhenius plot. Kabalnov and Weers were able to obtain the activation energy for a water-in-octane emulsion at $\phi \approx 50\%$, stabilized by the nonionic surfactant C12E5 (pentaethylene glycol mono n-dodecyl ether), above the phase inversion temperature (PIT), and found a value of 47 $k_B T_r$, T_r being the room temperature.

The above-mentioned method is based on a mean-field description that assumes spatially homogeneous monodisperse growth and constant coalescence frequency. The first assumption poorly agrees with experiments that reveal that the size distribution becomes not only extremely wide but also spatially inhomogeneous during the destruction process. At $\phi \approx 50\%$, the emulsion viscosity is quite low and local rearrangements take place: the larger drops can migrate and concentrate at the top (or bottom) of the samples under the effect of buoyancy (or gravity). Because the biggest drops become progressively predominant in the cream (or sediment), coalescence is locally accelerated. Indeed, neighboring drops are in permanent contact and larger drops will grow faster because they exhibit a larger surface contact area with their neighbors. Instead, in highly concentrated emulsions ($\phi >$ 75%), rearrangements due to creaming are quite slow and the size distribution remains spatially homogeneous for a long period of time. In this limit, it has been shown that the average droplet size does not exhibit the diverging growth predicted by Eq. (5.7) [41]. The mean-field model assuming a time-independent coalescence frequency provides a useful but oversimplified picture of the destruction process.

A different approach has been proposed by Schmitt et al. [42]. They produced O/W emulsions at $\phi = 78\%$, stabilized by ionic and nonionic surfactants. The emulsions were stored in that concentrated state at 20°C. The droplet size distribution was measured at regular time intervals via static light-scattering using Mie theory. Starting from the volume distribution, a little bit of algebra allowed a straightforward calculation of the average diameter $D[p, q]$ defined as:

$$D[p,q] = \left(\frac{\sum_i N_i D_i^p}{\sum_i N_i D_i^q} \right)^{\frac{1}{p-q}} \tag{5.9}$$

where N_i is the total number of droplets with diameter D_i. The emulsions were characterized in terms of their surface-averaged diameter $D_s = D[3, 2]$ and polydispersity:

$$P = \frac{1}{\overline{D}} \frac{\sum_i N_i D_i^3 \left| \overline{D} - D_i \right|}{\sum_i N_i D_i^3} \tag{5.10}$$

where $\overline{D}$ is the median diameter for which the cumulative undersized volume fraction is equal to 50%.

For all the emulsions under study, the size distribution follows the same qualitative evolution as the one reported in Fig. 5.5 for octane-in-water droplets stabilized with SDS. In Fig. 5.6, the evolutions of D_s, and P as a function of time for the same emulsion are reported. The time interval where the polydispersity is lower than 20% has been shaded: in terms of droplet diameter, the interval lies between lower and upper limits, denoted D_{s1} and D_{s2}, respectively (Fig. 5.6b). Initially, the polydispersity is of the order of 40–50%. As time passes, D_s increases until reaching D_{s1} and P decreases to around 13–20%, a low value indicating the formation of an emulsion with high degree of monodispersity. For $D_s < D_{s2}$, P stays at a low level and the diameter growth rate is continuously decreasing ($d^2 D_s / dt^2 \leq 0$). These two observations reveal that Ostwald ripening is the rate-determining mechanism. Indeed, theoretical calculations predict that the rate is self-saturating and that polydispersity is lower than 20% in the asymptotic

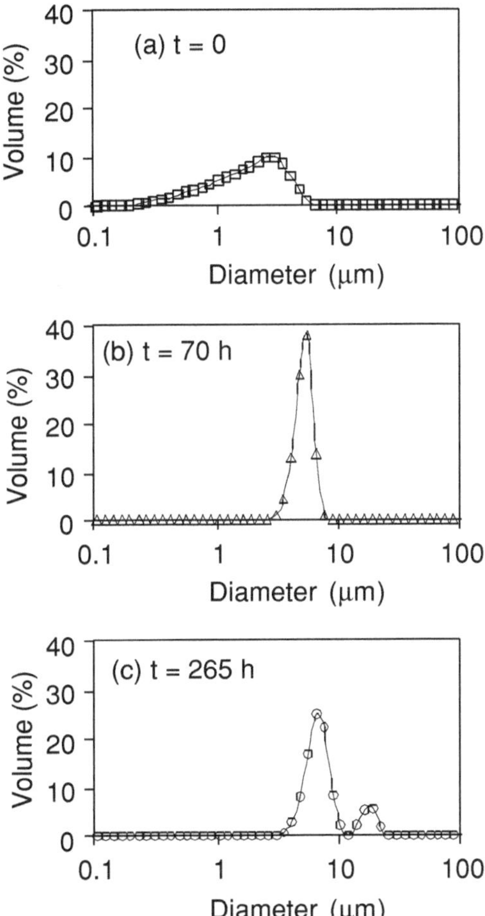

FIGURE 5.5. Size distribution of an octane-in-water emulsion stabilized with SDS. a) just after preparation, b) 70 hours and c) 265 hours after preparation. (Adapted from [42].)

regime: $P = 10.4\%$ for diffusion-controlled ripening and $P = 15.7\%$ for interface-controlled ripening. The experimental precision is insufficient to differentiate the two possible origins on the basis of P measurements. The same mechanism is operative when the average diameter is smaller than D_{s1} but the asymptotic regime is not yet achieved because of the large polydispersity of the initial emulsions. At longer times, for $D_s > D_{s2}$, the distribution becomes broad again ($P > 20\%$ and continuously increases) with the appearance of drops much larger than the average (Fig. 5.5). A general feature of this regime is the divergent diameter growth ($d^2 D_s/dt^2 \geq 0$) which allows one to conclude that this evolution is consistent with a coalescence-driven mechanism. Hence, for all the systems under study, the coarsening is determined first by Ostwald ripening followed by coalescence. It is only the time scale of these two mechanisms that varies depending on the surfactant and oil chemical natures.

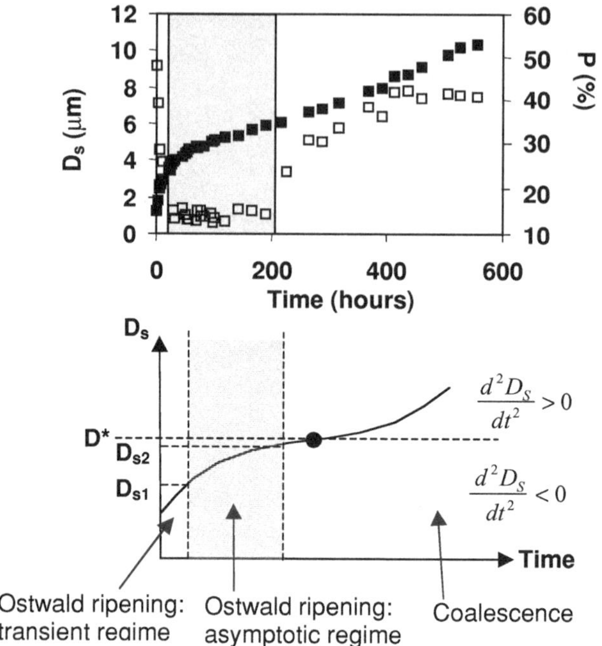

FIGURE 5.6. a) Experimental time evolution of the diameter D_s (■) and the polydispersity P (□, right scale) for an octane-in-water emulsion stabilized with SDS. b) Schematic representation for the evolution of D_s as a function of time. (Adapted from [42].)

The homogeneous growth observed in the shaded part of Fig. 5.6 ($D_{s1} < D_s < D_{s2}$) is governed by molecular diffusion through Eq. (5.1). The regime following the shaded zone is dominated by coalescence and the authors adopt the simple model previously described (Eq. (5.6)) to account for the droplet growth. However, they restrict the application of this mean field description to the very first steps of the coalescence regime. In this limit, the size distribution remains sufficiently narrow to describe the system in terms of a unique droplet diameter. For all the studied emulsions, a well-defined threshold diameter D^* (inflexion point) defines the transition between the two regimes. For $D_s < D^*$, Ostwald ripening is the rate-determining mechanism, while for $D_s > D^*$ coalescence becomes the dominating instability (Fig. 5.6b). At $D_s = D^*$, the two coarsening mechanisms occur at comparable rates:

$$\left(\frac{dD_s}{dt}\right)_{\text{Coalescence}} \approx \left(\frac{dD_s}{dt}\right)_{\text{Ostwald ripening}} \tag{5.11}$$

thus:

$$\frac{\Omega_\alpha}{\alpha(D^*)^{\alpha-1}} \approx \omega^* \frac{\pi}{3} \left(D^*\right)^3 \tag{5.12}$$

TABLE 5.1. Characteristic coalescence frequency for different alkane-in-water emulsions

Surfactant	Oil	Ω_3 (10^{-22} m^3 s^{-1})	Ω_2 (m^2 s^{-1})	D* (μm)	ω^*(m^{-2} s^{-1}) $D_s = D^*$ Eq. (5.12)	ω(m^{-2} s^{-1}) $D_s = D^* + 1$ μm Eq. (5.14)
Ifralan	Heptane	—	$(1.0 \pm 0.1)\ 10^{-16}$	3.1 ± 0.2	$(5.2 \pm 1.5)\ 10^5$	$(3.4 \pm 0.6)\ 10^5$
Ifralan	Octane	—	$(1.9 \pm 0.1)\ 10^{-17}$	3.0 ± 0.2	$(1.1 \pm 0.4)\ 10^5$	$(0.8 \pm 0.2)\ 10^5$
Ifralan	Nonane	—	$(5.9 \pm 0.3)\ 10^{-18}$	2.5 ± 0.1	$(7.3 \pm 2.1)\ 10^4$	$(6.5 \pm 1.3)\ 10^4$
Ifralan	Dodecane	—	$(2.5 \pm 0.1)\ 10^{-19}$	2.4 ± 0.1	$(3.6 \pm 1.2)\ 10^3$	$(4.2 \pm 0.8)\ 10^3$
SDS	heptane	(6.3 ± 0.3)	—	6.6 ± 0.3	$(1.6 \pm 0.5)\ 10^4$	$(2.0 \pm 0.4)\ 10^4$
SDS	Octane	(2.5 ± 0.1)	—	6.8 ± 0.4	$(5.4 \pm 1.6)\ 10^3$	$(7.2 \pm 1.4)\ 10^3$

Ifralan is a commercial nonionic surfactant essentially composed of a mixture of polyethylene glycols C12E5 and C10E5. In Eq. (5.12), $\alpha = 2$ was adopted for Ifralan surfactant and $\alpha = 3$ was adopted for SDS. (From [42].)

The right-hand side was derived from the variation of the drops number considering the volume conservation principle. From Eq. (5.12), the authors deduce an estimation of the frequency ω^* valid for $D_s \approx D^*$ and some values are reported in Table 5.1.

It is interesting to examine the impact of the alkane molecular weight on the coalescence rate, for a given surfactant: smaller alkanes coalesce more rapidly. The authors [27] argue that this dependence is related to the spontaneous curvature of the surfactant monolayers. The longer alkane chains, such as hexadecane, can hardly penetrate the hydrophobic surfactant brush covering the surfaces and therefore the natural spontaneous curvature is quite elevated, thus stabilizing the direct films against hole nucleation. Instead, shorter oil chains such as octane can more easily penetrate and swell the surfactant brush, providing a less positive average curvature, which allows rapid formation of holes in the O/W/O films.

To test the reliability of the previous method, the authors compared it to an independent measurement of ω. They thus propose an extended version of the previous mean-field model, valid at any stage of the coalescence regime, even in presence of broad droplet size distributions. It is obtained by considering that the variation of the total number of coalescence events is proportional to the total surface area per unit volume developed by the droplets of different sizes. The total number of drops and total surface are replaced by summations over all the granulometric size intervals:

$$\frac{d}{dt}\left(\sum_i N_i\right) = -\omega\left(\sum_i N_i \pi D_i^2\right) \tag{5.13}$$

Using the definition of the averaged diameters $D[3,0]$ and $D_s = \mathrm{D}[3,2]$ (Eq. (5.9)), the previous equation can be rewritten as:

$$\omega = -\frac{1}{\pi}\frac{d}{dt}\left(\frac{1}{D[3,0]^3}\right)D_s \tag{5.14}$$

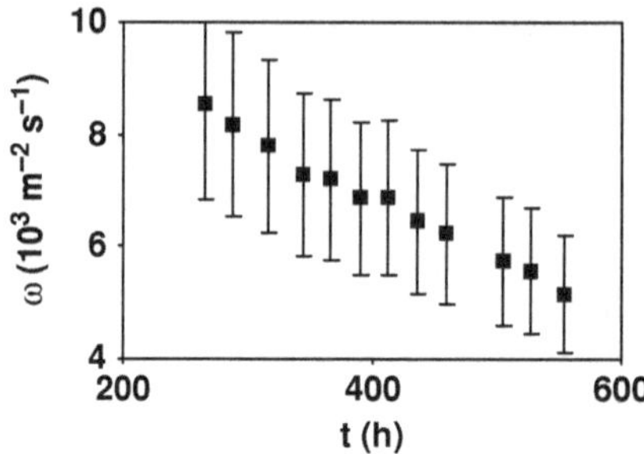

FIGURE 5.7. Evolution of the coalescence frequency with time deduced from Eq. (5.14) for an octane-in-water emulsion stabilized with SDS. (Adapted from [42].)

This equation reflects the possibility to measure ω from the experimental evolution of $D[3,0]$ and D_s, both diameters being directly deduced from the experimental droplet size distributions. Of course, this procedure is to be applied at long times, that is, in the regime governed by coalescence ($D_s > D^*$). In Fig. 5.7, it appears that ω exhibits a regular decrease with time.

Following the pioneering approach of Bibette et al. [8], it is argued that the decrease of ω essentially reflects the effect of the disjoining pressure in the films [42]. In concentrated emulsions, the droplet surfaces contain facets, along which droplets press against each other across thin films of water. The inward pressure exerted on the films, called the disjoining pressure Π_d, is opposed by repulsive interactions between the oil–water interfaces, and determines the film thickness. This pressure also determines the mean radius of curvature r_c on the remaining curved sections of the droplets surfaces by the Laplace equation: $\Pi_d = 2\gamma_{\text{int}}/r_c$. For $\phi = 78\%$ and polydisperse systems, the droplets are weakly deformed spheres, and press against each other across small facets. Hence, Π_d approaches the average Laplace pressure $\Pi_L = 4\gamma_{\text{int}}/D_s$ of the undeformed droplets: $\Pi_d \approx \Pi_L$. With an increase on the droplets diameter, the average Laplace pressure decreases and each plane parallel film is moving along the curve of the repulsive disjoining pressure isotherm so that to increase the average film thickness. Thus the coalescence frequency is expected to decrease as experimentally observed.

For the sake of comparison, Table 5.1. contains the numerical values of ω obtained following the two previous independent methods, at the same $\phi = 78\%$. For the second one (Eq. (5.14)), the reported data are those obtained in the vicinity of D^*(at $D_s \approx D^* + 1$ μm). The comparison was made at a D_s value larger than D^* to be sure that the measured frequency is negligibly perturbed by Ostwald ripening. The agreement between the two methods is rather satisfactory considering that the calculation procedures are completely different. Thus, using Eq. (5.12), it is possible to characterize the metastability of concentrated emulsions in terms of two independent parameters: the rate of Ostwald ripening Ω_α and the critical diameter D^* that defines the crossover between Ostwald ripening and coalescence. The characteristic coalescence frequency nearby D^* can then be deduced from Eq. (5.12). The phenomenology previously described is quite general, as it was reproduced in the presence of different alkanes stabilized by both ionic and non-ionic surfactants. Ionic surfactants generally provide a better stabilization against coalescence, revealed by larger D^* values.

5.4.2. Double Emulsions Stabilized by Surfactants

A different method for the measurement of the coalescence frequency has been proposed by Pays et al. [43–45]. It is based on the use of monodisperse (W/O/W) double emulsions. The so-called water-in-oil-in-water (W/O/W) emulsions are composed of oil globules dispersed in water, each globule containing smaller water droplets. Sodium chloride was initially dissolved in the internal droplets as a tracer to probe the coalescence kinetics. The amount of salt released through coalescence from the internal droplets to the external water phase was measured via an electrode selective to chloride ions immersed in the continuous phase of the double emulsion. The method of Pays et al. exploits the fact that the total number of internal droplets adsorbed on a globule surface governs the rate of release. An attractive interaction exists between the small internal droplets and between the droplets and the globule surface. However, because the globules are at least 10 times larger than the entrapped droplets, the attraction between the almost flat globule and a small droplet is nearly twice as large as that between inner droplets. This attraction is small enough for the small droplets to behave like a gas that adsorbs reversibly onto the globule surface. By varying the concentration of the hydrophilic surfactant within the external water phase, the authors found a regime in which the leakage is controlled by the droplet–globule coalescence only. Under such conditions, measuring the rate of release allows a direct determination of the average lifetime of the thin film that forms between a small internal droplet and the globule surface. A determination of both the activation energy and the natural frequency of the hole nucleation process is then possible, by exploring the temperature dependence of the rate of release.

The thin liquid film that forms between the internal droplets and the globule surface is composed of two mixed monolayers covered by both oil and water-soluble surfactant molecules. Because the water-soluble molecules migrate rapidly from the external to the internal water phase, the film can be regarded as close to thermodynamic equilibrium with respect to surfactant adsorption. As was previously explained, such inverse films possess a long-range stability when essentially covered by hydrophobic surfactant ($C_0 < 0$) but become very unstable when a large proportion of hydrophilic surfactant ($C_0 > 0$) is adsorbed. The transition from long-time to short-time stability may be achieved by varying the concentration of the hydrophilic surfactant in the external water phase.

In the experiments described by Pays et al. [44], the double globules were composed of dodecane and the surfactants used were sorbitan monooleate (SMO), which is oil soluble, and the water-soluble SDS. The concentrations of both surfactants were fixed and the initial internal droplet volume fraction was varied between 5% and 35%. The coalescence frequency was determined from a simple experiment in which the globule surface was totally saturated by the water droplets. For that purpose, a silicone polymer (polydimethylsiloxane), that added an additional depletion attraction to the inherent van der Waals interaction, was dissolved in the oil phase. However, this polymer did not adsorb at the water–oil interface and therefore, it did not perturb the film characteristics. The saturation of the globule surface by the droplets produced a spherical shell observable under a microscope when the oil globules were sufficiently large. n_i was defined as the total internal droplet concentration within the globules and n_a was the concentration of adsorbed droplets (per unit volume of the globules). The number of coalescence events per unit time was assumed to be proportional to the concentration of adsorbed droplets:

$$\frac{dn_i}{dt} = -\Lambda \cdot n_a \tag{5.15}$$

where Λ is the characteristic frequency of coalescence (note that in this definition, the coalescence frequency is not normalized by surface area). This rate of coalescence dn_i/dt remained constant as long as the globule surface was saturated, and this was confirmed by the appearance of a plateau in Fig. 5.8. Dividing the plateau value by the total number of available sites for adsorption (calculated from the diameters of the globules and internal droplets), the following Λ value was obtained: $\Lambda = 6\ 10^{-3}\ \text{min}^{-1}$. Measurements of the same type were performed with oils of different molecular weights: for octane globules Λ was very small and could not be measured, while for hexadecane $\Lambda = 2.5\ 10^{-2}\ \text{min}^{-1}$ [45]. Again, the spontaneous curvature may be evoked to explain the impact of the molecular chain length on Λ. Short alkanes such as octane easily penetrate the surfactant brush and provide a negative spontaneous curvature to the surfactant monolayer, which tends to stabilize W/O/W films (note that octane had the opposite destabilizing

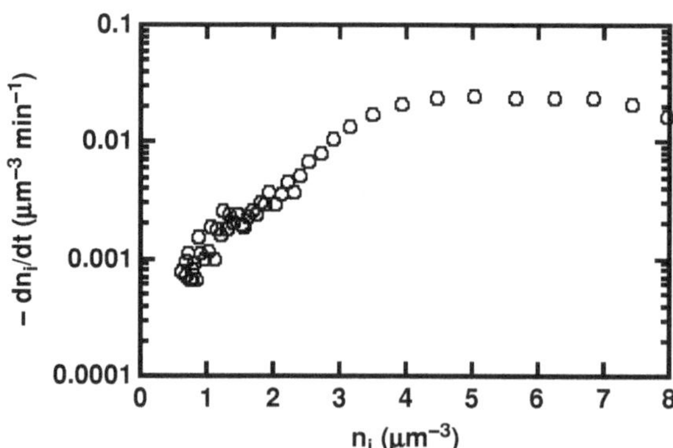

FIGURE 5.8. Rate of coalescence as a function of the number density of internal droplets in the globules. Globule diameter = 11.5 μm; droplet diameter = 0.36 μm; initial droplet volume fraction in the globules = 25%; globule volume fraction = 10%; SDS concentration = 2.4 10^{-2} mol/l; SMO concentration in the oil phase = 2 wt%; 0.1% of silicone oil with a gyration radius of 12 nm was added to dodecane. (Reproduced from [44], with permission.)

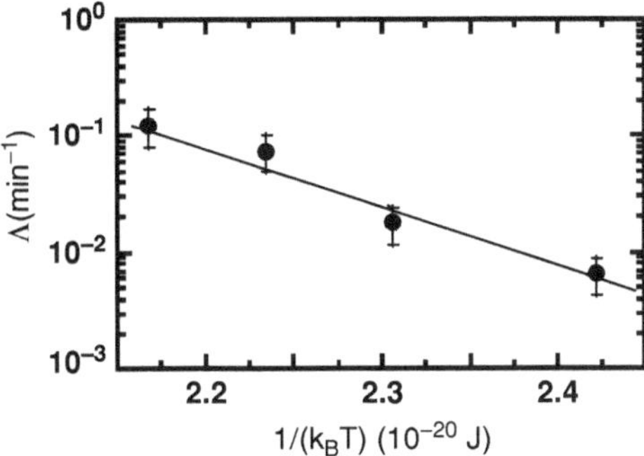

FIGURE 5.9. Frequency of coalescence as a function of $1/(k_B T)$. Globule diameter = 3.6 μm; droplet diameter = 0.36 μm; globule volume fraction = 10%; SDS concentration = 2.4 10^{-2} mol/l; SMO concentration in the oil phase = 2 wt%; initial internal droplet volume fraction = 20%. (Adapted from reference [44].)

effect in the case of O/W/O films, as described in §5.4.1). In contrast, it is likely that large alkanes (hexadecane) are excluded from the short surfactant brush and the spontaneous curvature becomes less negative.

The previous experiments were performed at room temperature. Within the same conditions the temperature was varied between 20°C and 60°C for double emulsions comprising dodecane globules. In Fig. 5.9, the evolution of Λ as a function of $1/(k_B T)$ in a semilog scale is plotted. The experimental points roughly follow an Arrhenius law and from the best fit to the data, the activation energy was obtained: $E_a = 30\, k_B T_r$. From the intercept, it was possible to obtain the attempt frequency which corresponds to the total number of holes generated per unit time in the film between the internal droplets and the globule surface: $\Lambda_0 = 4\;10^{10}\,\text{min}^{-1}$. Only a fraction of them will grow and ultimately produce a coalescence event while the other ones simply vanish. From the ratio between the measured values of Λ and Λ_0, it can be concluded that only one per 10^{13} holes generated in the film leads to a coalescence event at room temperature.

5.4.3. Simple Emulsions Stabilized by Solid Particles

Arditty et al. have described the scenario of coalescence in solid-stabilized emulsions and they have proposed a precise determination of the coalescence frequency [46,47]. Solid-stabilized emulsions can be obtained with a wide variety of solid organic or mineral powders. The surface of such particles is made partially hydrophobic to promote adsorption at the oil–water interface and generally the adsorption is totally irreversible. Emulsions can be prepared by manually shaking a mixture of oil and an aqueous dispersion of silica. When shaking is stopped, the oil drops coalesce to form drops of macroscopic size. The sequence in Fig. 5.10 reveals the typical evolution of a concentrated emulsion. The average droplet size increases at short times and approaches an asymptotic value at long times. It can be noticed that the emulsion remains remarkably monodisperse over the whole process. Because the solid content is initially insufficient to fully cover the oil–water interfaces, emulsion droplets coalesce to a limited extent. Under the effect of coalescence, the total interfacial area between oil and water is progressively reduced, thus increasing the degree of coverage by particles because they

FIGURE 5.10. Images of an O/W emulsion containing 90 wt% of oil and 24.4 mg of hydrophobically modified silica particles taken at different times since formation. The times are **(a)** 9 s, **(b)** 21 s, **(c)** 54 s, **(d)** 141 s and the scale bar = 7.5 mm. (Reproduced from [46], with permission.)

are irreversibly adsorbed. This results in the formation of dense solid interfacial films that ultimately inhibit coalescence and kinetically stabilize the emulsions.

Figure 5.11 shows the images of O/W emulsions obtained for different solid contents, m_p, expressed as a mass of particles in the emulsion. It can be clearly

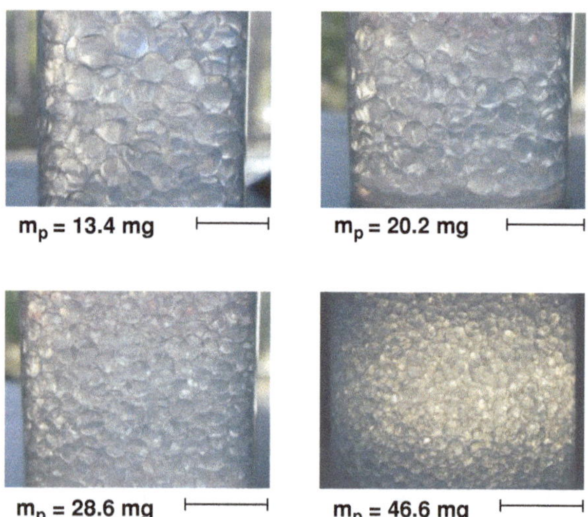

FIGURE 5.11. Images of O/W emulsions at long times containing 80 wt% oil obtained for different masses of hydrophobically modified silica particles, m_p (given). $V_d = 40$ ml. Scale bar = 1.2 cm. (Reproduced from [46], with permission.)

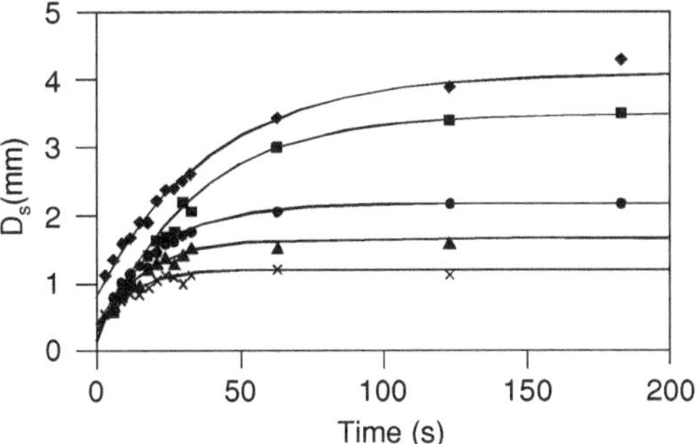

FIGURE 5.12. Evolution in time of the surface-averaged droplet diameter in O/W emulsions containing 90 wt% oil and different masses of silica particles: (♦) 18.3 mg, (■) 24.4 mg, (▲) 30.5 mg, (●) 36.6 mg and (×) 42.7 mg (Adapted from [46].)

seen that the final droplet size decreases when increasing the total amount of solid particles. The volume fraction of the dispersed phase is quite elevated, being larger than 80%. Such emulsions can therefore be considered as biliquid foams. Surprisingly, these foams can be stored at rest for months without any structural evolution.

The surface-averaged final diameter D_{sf} can be easily controlled by adjusting the amount of particles. Because the solid particles are irreversibly adsorbed at the oil–water interface, the inverse average droplet diameter varies linearly with the amount of particles according to the simple equation:

$$\frac{1}{D_{sf}} = \frac{s_f m_p}{6V_d} \tag{5.16}$$

where V_d is the volume of dispersed phase and s_f the specific surface area, that is, the droplet surface covered per unit mass of silica.

Figure 5.12 represents the evolution in time of the surface-averaged droplet diameter for different amounts of solid particles. The kinetic curves confirm the qualitative evolution previously described. The droplet growth is initially rapid but the coalescence rate progressively decreases until the average diameter reaches an asymptotic value. Figure 5.13 shows the change in the droplet size distribution

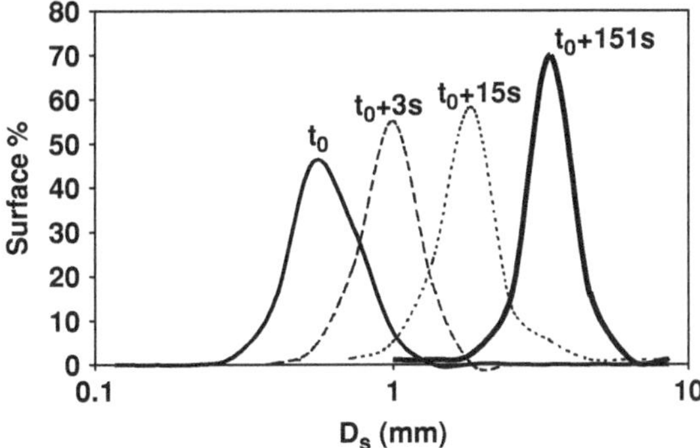

FIGURE 5.13. Evolution of the droplet size distribution with time for an O/W emulsion containing 90 wt% oil and 24.4 mg of silica particles. (Reproduced from [46], with permission.)

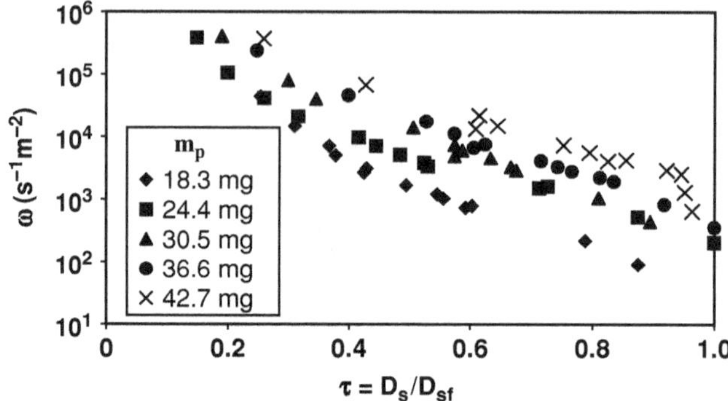

FIGURE 5.14. Variation of the coalescence frequency ω with τ for O/W emulsions. The emulsion contains 90 wt% oil. (Reproduced from [46], with permission.)

with time. For all the plots reported in this figure, the polydispersity (defined by Eq. (5.10)) is lower than 30%, a surprisingly low value considering that the distribution is achieved via coalescence. The change in the droplet concentration n as a function of time is deduced from Fig. 5.12 considering Eq. (5.5). The curves obtained were numerically differentiated with respect to time in order to obtain the characteristic frequency of coalescence through Eq. (5.6). Application of this equation is justified in this case owing to the existence of a well-defined droplet size over the whole growth process.

In Fig. 5.14 the evolution of ω is reported as a function of the degree of surface coverage. The degree of surface coverage τ is defined as the ratio of the particle density at any time to the final particle density. It can also be expressed as:

$$\tau = \frac{D_s}{D_{sf}} \tag{5.17}$$

As could be expected, the curves reveal a dramatic decrease of the coalescence frequency with τ. As τ approaches unity, the degree of surface coverage becomes sufficient to kinetically stabilize the droplets such that no further structural evolution is observed. In this limit, one expects that ω reaches very small, but nonzero values, the stability being only of kinetic order. In Fig. 5.15, the evolution of ω as a function of the average droplet diameter D_s at constant degree of surface coverage τ is reported. It is interesting to note that ω is a decreasing function of droplet diameter, at constant concentration of solid particles at the oil–water interface.

It is argued that the kinetics of the limited coalescence process is determined by the uncovered surface fraction $1 - \tau$ and by the rate of thinning (drainage) of the films formed between the deformable droplets [46,47]. The homogeneous and monodisperse growth generated by limited coalescence is intrinsically different from the polydisperse evolution observed for surfactant-stabilized emulsions. As noted by Whitesides and Ross [48], the mere fact that coalescence halts as a result of surface saturation does not provide an obvious explanation of the very

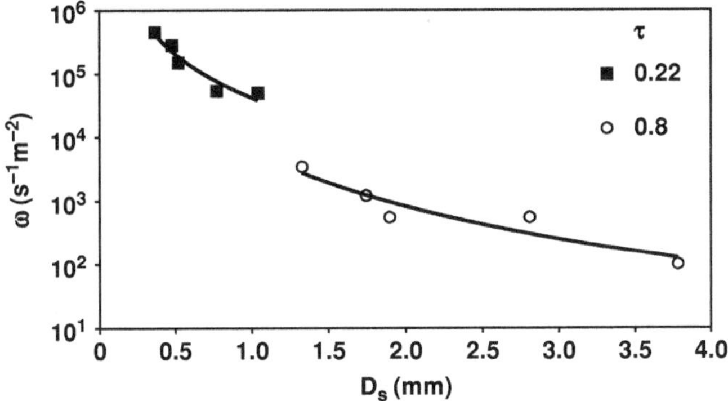

FIGURE 5.15. Variation of the coalescence frequency ω with D_s at constant τ for O/W emulsions stabilized by silica particles. The lines are only visual guides. (Reproduced from [46,47], with permission.)

narrow droplet size distributions that are frequently obtained in solid-stabilized emulsions. The same authors propose a theoretical analysis for either diffusional or turbulence-driven droplet collisions. Assuming that the coalescence probability between two drops is simply proportional to their individual uncovered surface fraction $1 - \tau$, Monte Carlo simulations predict droplet size distributions that are in close accord with experimental results, and that are much narrower than those resulting from unlimited coalescence. One interesting conclusion from the numerical calculations is that the final narrow size distribution is insensitive to the details of the initial droplet size distribution (uniformity, surface coverage) within fairly wide limits [48].

5.5. Gelation and Homothetic Contraction

The following section is devoted to a totally distinct scenario of destruction occurring in emulsions comprising highly viscous drops. Coalescence involves two different steps. The first step consists of the nucleation of a small channel between two neighboring droplets. Hereafter, we shall term τ_n the characteristic time separating two nucleation events. This first nucleation step is followed by a shape-relaxation driven by surface tension, which causes two droplets to fuse into a unique one. The characteristic time τ_r for shape relaxation is governed by the competition between surface tension and viscous dissipation and is given by:

$$\tau_r = \eta D / \gamma_{\text{int}} \tag{5.18}$$

where η is the viscosity of the droplets, D is their characteristic diameter, and γ_{int} is the interfacial tension between the two phases. When there is no energy barrier for coalescence, the droplets coalesce as soon as they collide under the effect of Brownian motion. In the limit where the characteristic shape relaxation time τ_r is short compared to the time τ_n separating two droplet collisions, it was found

FIGURE 5.16. Sequence showing the homothetic contraction of a bitumen-in-water emulsion. (Reproduced from [49], with permission.)

both theoretically and experimentally that the average droplet size scales with time t as $t^{1/3}$. A very different scenario is expected in the limit where τ_r is much larger than τ_n. The coarsening is now limited by shape relaxation, leading to very different structures and kinetics than in the previous case. This limit is frequently encountered in systems such as emulsions of highly viscous substances (bitumen) or phase separations in binary mixtures of polymers.

We now summarize the work of Philip et al. [49], who described the limit where $\tau_r \gg \tau_n$. They used model emulsions of highly viscous bitumen droplets that can be made suddenly unstable toward coalescence under addition of a suitable chemical. Once the emulsion is made unstable, the droplets form a macroscopic gel made of an array of fused droplets. Then the gel continuously contracts with time in order to reduce its surface area. To study the gelation and the contraction phenomena, the emulsion was introduced in a rectangular cuvette and a destabilizing agent was added in the continuous phase. Initially, the system remains fluid, but after some time, the emulsion does not flow any longer. At this stage, microscopic observation reveals that the droplets stick one to another and form a three-dimensional gel network. Once this network is formed, the gel starts to contract by reducing its surface area. In this process water is expelled from the space-filling network. The contraction remains remarkably homothetic, meaning that it preserves the geometry of the container (Fig. 5.16). Figure 5.17 shows the evolution of the

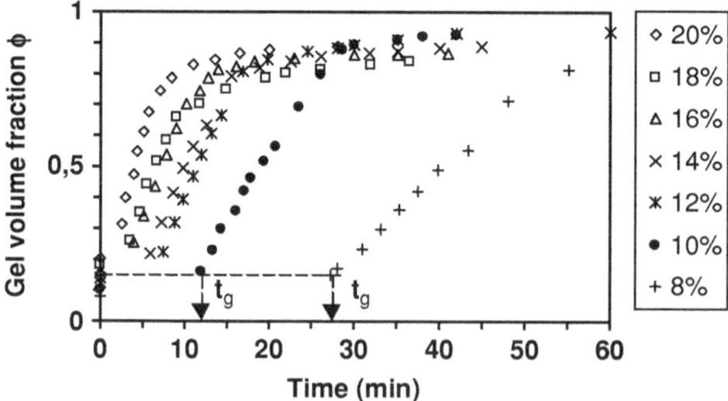

FIGURE 5.17. Evolution of the volume fraction in the bitumen contracting gel as a function of time for emulsions with different initial volume fractions. (Reproduced from [49], with permission.)

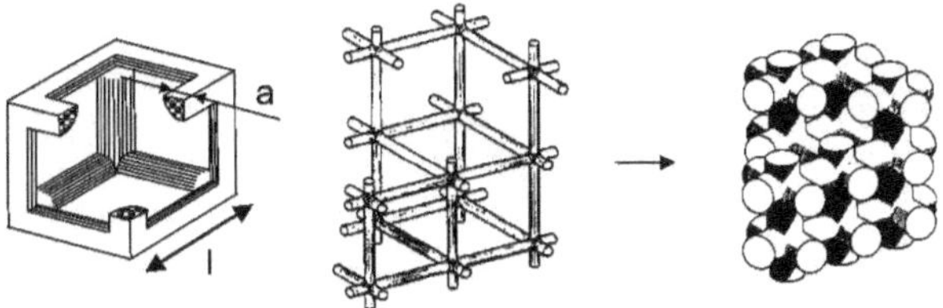

FIGURE 5.18. Cylindrical model for sintering. The gel is modeled as a cubic array made of intersecting cylinders with length l and radius a. (Reproduced from [52], with permission.)

bitumen volume fraction in the contracting gel as a function of time for emulsions with different initial volume fractions, at room temperature (25°C). When the destabilizing agent is introduced in the emulsion, a delay t_g has to be expected before the gel starts to shrink. t_g corresponds to the time required for the droplets to form a continuously interconnected network that fills the whole volume. Once the contraction starts, two different regimes can be clearly distinguished. The rate of contraction at the initial stage is quite rapid and becomes much slower toward the final stages of contraction. The first regime is roughly linear, the slope being an increasing function of the initial droplet volume fraction.

The contraction kinetics can be described using the so-called "cylindrical model for sintering" [50–53]. The model considers a cubic array formed by intersecting cylinders that are made up of strings of particles. The initial cylinder radius corresponds to the average radius of the particles in the material. Although the choice of cubic array is an idealized approximation compared to the complicated microstructures formed in real situations, studies have shown that the geometry of the unit cell chosen (e.g., octahedral, tetrahedron, inverse tetrahedron) has very little influence on the kinetics. To reduce their surface area, the cylinders tend to become shorter and thicker. In these calculations, it is assumed that the geometry of the cell is preserved. A brief description of the calculations follows.

If l and a are the length and radius of the cylinders, respectively (Fig. 5.18), the energy dissipated in viscous flow ($\dot{E}_f$) varies as [50–53]:

$$\dot{E}_f = \frac{3\pi\eta a^2}{l}\left(\frac{dl}{dt}\right)^2 \tag{5.19}$$

The superscript dot indicates a derivative with respect to time. The energy change due to the reduction of surface area ($\dot{E}_s$) is given by:

$$\dot{E}_s = \gamma_{\text{int}}\frac{dS_c}{dt} \tag{5.20}$$

where S_c is the surface area of a full cylinder. The energy balance requires the following condition:

$$\dot{E}_f + \dot{E}_s = 0 \tag{5.21}$$

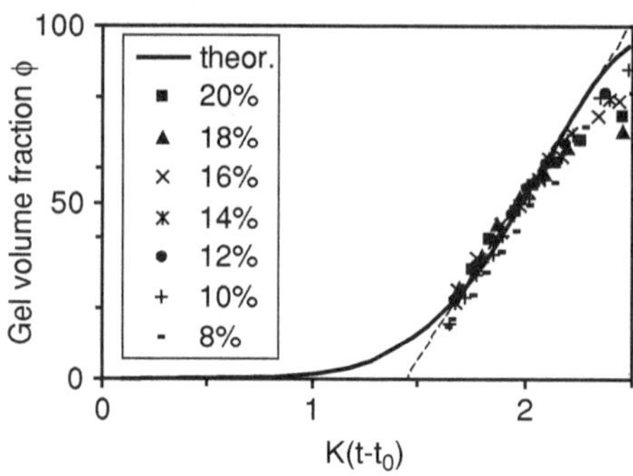

FIGURE 5.19. Evolution of the gel volume fraction ϕ as a function of the reduced time $K(t - t_0)$. The solid line represents the theoretical curve obtained using Eqs. (5.22 and 5.23). (Adapted from [49].)

From Eqs. (5.19), (5.20) and (5.21), the rate of densification is deduced:

$$\frac{\gamma_{\text{int}}}{\eta l_0}\left(\frac{1}{\phi_0}\right)^{1/3}(t - t_0) = \int_0^x \frac{2dx'}{\left(3\pi - 8\sqrt{2}x'\right)^{1/3} x'^{2/3}} \tag{5.22}$$

where $x = a/l$. For a cubic cell, x is related to the cylinder volume fraction as:

$$\phi = 3\pi(a/l)^2 - 8\sqrt{2}\,(a/l)^3 \tag{5.23}$$

ϕ corresponds to the measured volume fraction of bitumen inside the gel. t_0 is a fictitious time at which $x = 0$. In Eq. (5.22), $(\gamma_{\text{int}}/\eta l_0)(1/\phi_0)^{1/3} = K$ is a constant for a given initial volume fraction ϕ_0. Indeed ϕ_0 sets the initial cylinder length l_o. When the ratio of cylinder radius to its length is equal to 1/2, the neighboring cylinders touch and the cell contains only closed pores. The corresponding theoretical density (volume fraction) of the sample would be 94.2%. Therefore, the cylindrical model is not valid any longer for ϕ values larger than 94.2%.

Figure 5.19 shows the evolution of the gel volume fraction ϕ as a function of reduced time $K(t - t_0)$. The solid line represents the theoretical curve obtained using Eq. (5.22). For ϕ values between 20% and 80%, the theoretical curve is roughly linear with a slope of 1 (see dashed line). Equivalently, within the same ϕ range, the volume fraction should vary linearly with time with a slope equal to K. The experimental data (Fig. 5.17) were recalculated in order to be plotted in reduced coordinates. For each initial volume fraction, K is deduced from the initial slope of the curve $\phi = f(t)$ (for ϕ between roughly 20% and 60%). All the data lie within a unique curve that is in reasonable agreement with the theoretical one.

Philip et al. explored the influence of the dispersed phase viscosity on the rate of contraction by changing the temperature. According to viscous sintering theory, the rate of densification K should vary as the inverse viscosity. Figure 5.20 shows the evolution of K as a function of viscosity. It can be noted that lowering the viscosity has the effect of increasing the rate of contraction K. As expected, the observed contraction is controlled by the viscous flow of asphalt through the gel network. The experimentally observed slopes are in reasonable agreement with the expected value (−1).

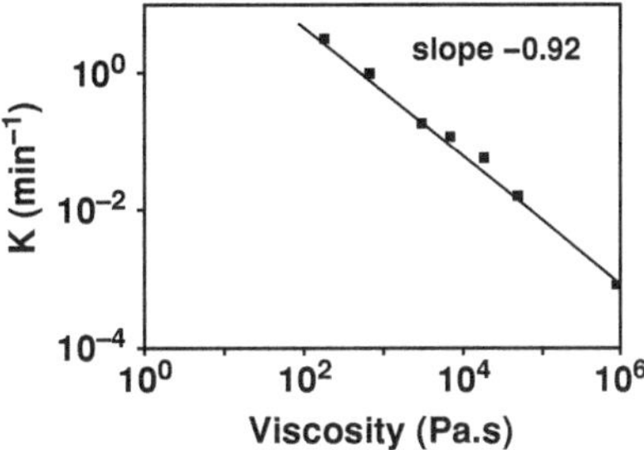

FIGURE 5.20. Evolution of K as a function of viscosity. (Adapted from [49].)

To see whether this contraction phenomenon is a general one, various O/W emulsions using highly viscous oils were prepared [54]. As in the case of bitumen, similar contraction mechanisms were observed when the emulsion was allowed to break by adding a suitable destabilizing agent. From all the experiments performed with different types of oils and destabilizing agents, it can be stated that the gel contracts in a homothetic way when drop viscosity exceeds around 100 Pa·s. The sintering process may be of great technological importance because it allows transforming an initially liquid emulsion, into a dense and highly viscous material within a short period of time at room temperature [55].

5.6. Partial Coalescence in Emulsions Comprising Partially Crystallized Droplets

We finally describe the behavior of oil-in-water emulsions in which the dispersed phase is composed of partially crystallized oils, that is, mixtures of hydrophobic compounds whose melting domain extends over one or several tens of degrees on both sides of ambient temperature. The emulsions are generally fabricated at sufficiently high temperature so that oil is totally melted. On cooling, the spherical shape of the warm dispersed droplets, which is controlled by surface tension, evolves into a rough and rippled surface owing to the formation of irregularly shaped/oriented crystals. Gelation of the emulsions can be induced by the so-called partial coalescence phenomenon [56–59]. Crystals located near the oil–water interface can protrude into the continuous phase depending on their orientation, the crystallization kinetics, the surface active species employed to stabilize the emulsion, and so forth [57,60,61]. When such crystals are present within the thin film separating two droplets, they may pierce the film and bridge the surfaces, causing the droplets to coalesce. If the crystallized fraction within the globules is sufficient, the intrinsic rigidity inhibits relaxation to the spherical shape after each coalescence event. The "pin" effect may occur between partially crystallized droplets or between crystallized and liquid (undercooled) droplets. In this latter case, coalescence induces fast crystallization of the undercooled droplets so that shape relaxation does not take place (Fig. 5.21). As time passes, large clusters appear and grow by the accretion of any other primary droplet or cluster until a rigid network made of

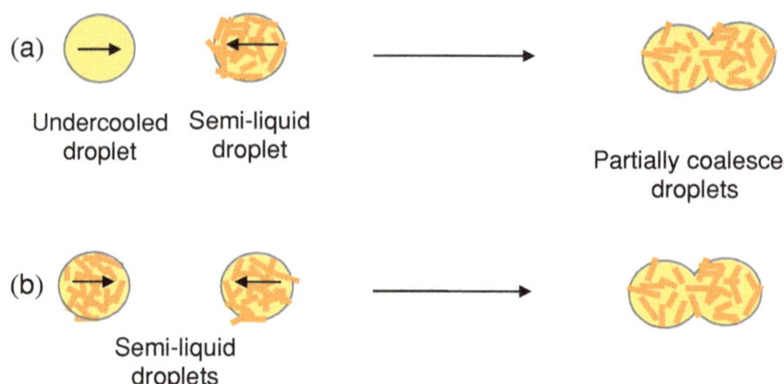

FIGURE 5.21. Possible scenarios for partial coalescence. (a) Crystallization induced by contact between solid particle and undercooled droplet. (b) Partial coalescence between two semiliquid droplets.

partially coalesced globules is formed, in which the original fat droplets remain recognizable. Partial coalescence may occur in quiescent storage conditions and is accelerated on application of a shear [62].

This "pin" effect of solid particles is well documented in the studies of the antifoam effect of oil–solid compounds used to destroy undesirable foam [63]. The fabrication of viscoelastic aerated food emulsions such as whipped creams or ice creams is also based on the unrelaxed coalescence between semiliquid milk fat globules. The colloidal structure of aerated food emulsions such as whipped creams or ice creams is complex, with the dispersed phase comprising gas bubbles, fat globules, and eventually ice crystals. Aerated materials are obtained through a whipping process during which the gas phase (air, nitrogen) is progressively incorporated in a primary oil-in-water emulsion under vigorous agitation [64]. An important feature of these materials is their rheological behavior: very often, they exhibit considerable yield stress and viscoelastic properties [65,66]. The agitation applied during the whipping process favors coalescence phenomena and many studies have demonstrated that the characteristic mechanical properties of whipped creams arise mainly from the formation of a network of partially coalesced globules [67–70].

5.7. Conclusion

The experimental studies described in this chapter certainly led to a better understanding of the coalescence phenomena in concentrated emulsions. Despite the complexity and variety of the destruction scenarios, different methods for measuring the coalescence frequency, ω, have been proposed. It should be within the reach of future work to measure ω for a large variety of systems in order to establish a comparative stability scale. This is a necessary step to determine the microscopic parameters that control the activation energy E_a and the attempt frequency ω_0.

It is probable that numerous interfacial parameters are involved (surface tension, spontaneous curvature, Gibbs elasticity, surface forces) and differ from one system to the other, according the nature of the surfactants and of the dispersed phase. Only systematic measurements of ω will allow going beyond empirics. Besides the numerous fundamental questions, it is also necessary to measure ω for an evident practical reason, which is predicting the emulsion lifetime. This remains a serious challenge for anyone working in the field of emulsions because of the polydisperse and complex evolution of the droplet size distribution. Finally, it is clear that the mean-field approaches adopted to measure ω are acceptable as long as the droplet polydispersity remains quite low ($P < 50\%$) and that more elaborate models are required for very polydisperse systems to account for the spatial fluctuations in the droplet distribution.

References

[1] W. Ostwald: In "Die Wissenschaptlichen grundlegen der analytischen Chemie." Analytisch Chemi, 221 pp. 3rd Ed. Wilhelm, Ingelmann, Leipzig (1901).

[2] I.M. Lifshitz and V.V. Slyozov: "Kinetics of Diffuse Decomposition of Supersaturated Solid Solutions." Soviet Physics JETP **35**, 331 (1959).

[3] I.M. Lifshitz and V.V. Slyozov: "The Kinetics of Precipitation from Supersaturated Solid Solutions." J. Phys. Chem. Solids **19**, 35 (1961).

[4] A.S. Kabalnov, A.V. Pertzov, and E.D. Shchukin: "Ostwald Ripening in Emulsions 1. Direct Observations of Ostwald Ripening in Emulsions." J. Colloid Interface Sci. **118**, 590 (1987).

[5] A.S. Kabalnov, K.N. Makarov, A.V. Pertzov, and E.D. Shchukin: "Ostwald Ripening in Emulsions 2. Ostwald Ripening in Hydrocarbon Emulsions: Experimental Verification of Equation for Absolute Rates." J. Colloid Interface Sci. **138**, 98 (1990).

[6] D.J. Durian, D.A. Weitz, and D.J. Pine: "Scaling Behavior in Shaving Cream." Phys. Rev. A **44**, R7902 (1991).

[7] C. Wagner: "Theory for the Coarsening of Solid Precipitates Caused by Ostwald Ripening." Z. Elektrochem. **65**, 581 (1961).

[8] J. Bibette, D.C. Morse, T.A. Witten, and D.A. Weitz: "Stability Criteria for Emulsions." Phys. Rev. Lett. **69**, 2439 (1992).

[9] A. Hasmy, R. Paredes, O. Sonnneville-Aubrun, B. Cabane, and R. Botet: "Dynamical Transition in a Model for Dry Foams." Phys. Rev. Lett. **82**, 3368 (1999).

[10] R. Lemlich: "Prediction of Changes in Bubble Size Distribution Due to Interbubble Gas Diffusion in Foam." Ind. Eng. Chem. Fund. **17**, 89 (1978).

[11] M. Tokuyama and K. Kawasaki: "Statistical-Mechanical Theory of Coarsening of Spherical Droplets." Physica A **123**, 386 (1984).

[12] Y. Enomoto, K. Kawasaki, and M. Tokuyama: "Computer Modelling of Ostwald Ripening." Acta Metall. **35**, 907 (1987).

[13] T.W. Patzek: "Self-Similar Collapse of Stationary Bulk Foams." AIChE J. **39**, 1697 (1993).

[14] S.P. Marsh and M.E. Glicksman: "Kinetics of Phase Coarsening in Dense Systems." Acta Mater. **44**, 3761 (1996).

[15] A.S. Kabalnov: "Can Micelles Mediate a Mass Transfer Between Oil Droplets?" Langmuir **10**, 680 (1994).

[16] P. Taylor: "Ostwald Ripening in Emulsions." Colloids Surfaces **99**, 175 (1995).
[17] J. Weiss, N. Herrmann, and D.J. McClements: "Ostwald Ripening of Hydrocarbon Emulsion Droplets in Surfactant Solutions." Langmuir **15**, 6652 (1999).
[18] W.I. Higuchi and J. Misra: "Physical Degradation of Emulsions via the Molecular Diffusion Route and Its Possible Prevention." J. Pharm. Sci. **51**, 459 (1962).
[19] A.J. Webster and M.E. Cates: "Stabilization of Emulsions by Trapped Species." Langmuir **14**, 2068 (1998).
[20] A.J. Webster and M.E. Cates: "Osmotic Stabilization of Concentrated Emulsions and Foams." Langmuir **17**, 595 (2001).
[21] A.S. Kabalnov, A.V. Pertsov and E.D. Shchukin: "Ostwald Ripening in Two-Component Disperse Phase Systems: Application to Emulsion Stability." Colloid Surfaces **24**, 19 (1987).
[22] L. Taisne, P. Walstra, and B. Cabane: "Transfer of Oil Between Emulsion Droplets." J. Colloid Interface Sci. **184**, 378 (1996).
[23] R.A. Arlauskas and J.G. Weers: "Sedimentation Field Flow Fractionation Studies of Composition Ripening in Emulsions." Langmuir **12**, 1923 (1996).
[24] J. Weiss, J.N. Coupland, D. Brathwaite, and D.J. McClements: "Influence of Molecular Structure of Hydrocarbon Emulsion Droplets on Their Solubilization in Nonionic Surfactant Micelles." Colloids Surfaces A **121**, 53 (1997).
[25] N. Hedin and I. Furo: "Ostwald Ripening of an Emulsion Monitored by PGSE NMR." Langmuir **17**, 4746 (2001).
[26] L. Taisne and B. Cabane: "Emulsification and Ripening Following a Temperature Quench." Langmuir **14**, 4744 (1998).
[27] V. Schmitt, C. Cattelet, and F. Leal-Calderon: "Coarsening of Alkane-in-Water Emulsions Stabilized by Nonionic Poly(Oxyethylene) Surfactants: The Role of Molecular Permeation and Coalescence." Langmuir **20**, 46 (2004).
[28] O. Sonneville: "Biliquid Foams." Ph.D thesis, Paris VI University (1997).
[29] A.J. de Vries: "Foam Stability. IV. Kinetics and Activation Energy of Film Rupture." Rec. Trav. Chim. Pays-Bas Belgique **77**, 383 (1958).
[30] A.S. Kabalnov and H. Wennerström: "Macroemulsion Stability: The Oriented Wedge Theory Revisited." Langmuir **12**, 276 (1996).
[31] K. Shinoda and H. Saito: "The Effect of Temperature on the Phase Equilibrium and the Types of Dispersions of the Ternary System Composed of Water, Cyclohexane and Nonionic Surfactant." J. Colloid Interface Sci. **26**, 70 (1968).
[32] W.D. Bancroft: "The Theory of Emulsification, V." J. Phys. Chem. **17**, 501 (1913).
[33] O. Sonneville-Aubrun, V. Bergeron, T. Gulik-Krzywicki, B. Jönsson, H. Wennerström, P. Lindner, and B. Cabane: "Surfactant Films in Biliquid Foams." Langmuir **16**, 1566 (2000).
[34] V. Bergeron: "Disjoining Pressures and Film Stability of Alkyltrimethylammonium Bromide Foam Films." Langmuir **13**, 3474 (1997).
[35] R.M. Pashley: "Effect of Degassing on the Formation and Stability of Surfactant-Free Emulsions and Fine Teflon Dispersions." J. Phys. Chem. B **107**, 1714 (2003).
[36] N. Maeda, K.J. Rosenberg, J.N. Israelachvili, and R.M. Pashley: "Further Studies on the Effect of Degassing on the Dispersion and Stability of Surfactant Free Emulsions." Langmuir **20**, 3129 (2004).
[37] J.N. Israelachvili and R.M. Pashley: "The Hydrophobic Interaction is Long Range, Decaying Exponentially with Distance." Nature **300**, 341 (1982).

[38] M.E. Karaman, B.W. Ninham, and R.M. Pashley: "Effects of Dissolved Gas on Emulsions, Emulsions Polymerization, and Surfactant Aggregation." J. Phys. Chem. **100**, 15503 (1996).

[39] B. Deminière, A. Colin, F. Leal-Calderon, J.F. Muzy, and J. Bibette: "Cell Growth in a 3d Cellular System Undergoing Coalescence." Phys. Rev. Lett. **82**, 229 (1999).

[40] A.S. Kabalnov and J. Weers: "Macroemulsion Stability Within the Winsor III Region: Theory Versus Experiment." Langmuir **12**, 1931 (1996).

[41] V. Carrier: "Compared Stability of Aqueous Foams and Emulsions." Ph.D thesis, Bordeaux I University (2001).

[42] V. Schmitt, C. Cattelet, and F. Leal-Calderon: "Measurement of the Coalescence Frequency in Concentrated Emulsions." Europhys. Lett. **67**, 662 (2004).

[43] K. Pays: "Double Emulsions: Coalescence and Compositional Ripening." Ph.D thesis, Bordeaux I University (2000).

[44] K. Pays, J. Kahn, P. Pouligny, J. Bibette, and F. Leal-Calderon: "Double Emulsions: A Tool for Probing Thin Film Metastability." Phys. Rev. Lett. **87**, 178304 (2001).

[45] K. Pays, J. Kahn, B. Pouligny, J. Bibette, and F. Leal-Calderon: "Coalescence in Surfactant-Stabilized Double Emulsions." Langmuir **17**, 7758 (2001).

[46] S. Arditty, C.P. Whitby, B.P. Binks, V. Schmitt, and F. Leal-Calderon: "Some General Features of Limited Coalescence in Solid-Stabilized Emulsions." Eur. Phys. J. E **11**, 273 (2003).

[47] S. Arditty, C.P. Whitby, B.P. Binks, V. Schmitt, and F. Leal-Calderon: "Erratum—Some General Features of Limited Coalescence in Solid-Stabilized Emulsions." Eur. Phys. J. E **12**, 355 (2003).

[48] T.H. Whitesides and D.S. Ross: "Experimental and Theoretical Analysis of the Limited Coalescence Process: Stepwise Limited Coalescence." J. Colloid Interface Sci. **169**, 48 (1995).

[49] P. Philip, L. Bonakdar, P. Poulin, J. Bibette, and F. Leal-Calderon: "Viscous Sintering Phenomena in Liquid-Liquid Dispersions." Phys. Rev. Lett. **84**, 2018 (2000).

[50] G.W. Scherer: "Sintering of Low-Density Glasses: I. Theory." J. Am. Ceram. Soc. **60**, 236 (1977).

[51] G.W. Scherer and D.L. Bachman: "Sintering of Low-Density Glasses: II. Experimental Study." J. Am. Ceram. Soc. **60**, 239 (1977).

[52] G.W. Scherer: "Sintering of Low-Density Glasses: III. Effect of a Distribution of Pore Sizes." J. Am. Ceram. Soc. **60**, 243 (1977).

[53] G.W. Scherer: "Viscous Sintering in Inorganic Gels." Colloid Surface Sci. **14**, 265 (1987).

[54] J. Philip, J.E. Poirier, J. Bibette, and F. Leal-Calderon: "Gelation and Coarsening in Dispersions of Highly Viscous Droplets." Langmuir **17**, 3545 (2001).

[55] F. Placin, M. Feder, and F. Leal-Calderon: "Viscous Sintering Phenomena in Liquid-Liquid Dispersions: Application to the Preparation of Silicone Macroporous Aerogels." J. Phys. Chem. B **107**, 9179 (2003).

[56] K. Boode and P. Walstra: "Partial Coalescence in Oil-in-Water Emulsions 1: Nature of the Aggregation." Colloids and Surfaces A : Physicochem. Eng. Aspects **81**, 121 (1993).

[57] K. Boode, P. Walstra, and A.E.A. de Groot-Mostert: "Partial Coalescence in Oil-in-Water Emulsions 2. Influence of the Properties of the Fat." Colloids Surfaces A Physicochem. Eng. Aspects **81**, 139 (1993).

[58] D. Rousseau: "Fat Crystals and Emulsion Stability." Food Res. Int. **33**, 3 (2000).

[59] M.A.J.S. van Boekel and P. Walstra: "Stability of Oil-in-Water Emulsions with Crystals in the Dispersed Phase." Colloids Surfaces **3**, 109 (1981).

[60] B.E. Brooker: "The Adsoprtion of Crystalline Fat to Air-Water Interface of Whipped Cream." Food Struct. **9**, 223 (1990).

[61] K. Golemanov, S. Tcholakova, N.D. Denkov, and T.D. Gurkov "Selection of Surfactants for Stable Paraffin-in-Water Dispersions, Undergoing Solid-Liquid Transition of the Dispersed Particles." Langmuir **22**, 3560 (2006).

[62] E. Davies, E. Dickinson, and R. Bee: "Shear Stability of Sodium Caseinate Emulsions Containing Monoglyceride and Triglyceride Crystals." Food Hydrocolloids **14**, 145 (2000).

[63] N.D. Denkov: "Oil Based Antifoams." Langmuir **20**, 9463 (2004).

[64] G.A. van Aken: "Aeration of Emulsions by Whipping." Colloids Surfaces A Physicochem. Eng. Aspects **190**, 333 (2001).

[65] A.K. Smith, H.D. Goff, and Y. Kakuda: "Microstructure and Rheological Properties of Whipped Cream as Affected by Heat Treatment and Addition of Stabilizer." Int. Dairy J. **10**, 295 (2000).

[66] D.W. Stanley, H.D. Goff, and A.K. Smith: "Texture-Structure Relationships in Foamed Dairy Emulsions." Food Res. Int. **29**, 1 (1996).

[67] B.E. Brooker: "The Stabilization of Air in Foods Containing Fat. A Review." Food Struct. **12**, 115 (1993).

[68] B.E. Brooker, M. Anderson, and A.T. Andrews: "The Development of Structure in Whipped Cream." Food Microstruct. **5**, 277 (1986).

[69] H.D. Goff: "Instability and Partial Coalescence in Whippable Dairy Emulsions." J. Dairy Sci. **80**, 2620 (1997).

[70] H.D. Goff: "Formation and Stabilization of Structure in Ice-Cream and Related Products." Curr. Opin. Colloid Interface Sci. **7**, 432 (2002).

6 Double Emulsions

6.1. Introduction

Double emulsions may be either of the water-in-oil-in-water type (W/O/W) (with dispersed oil globules containing smaller aqueous droplets) or of the oil-in-water-in-oil type (O/W/O) (with dispersed aqueous globules containing smaller oily dispersed droplets). Increasing attention has been devoted to these systems with the aim of taking advantage of their double (or multiple) compartment structure. Indeed, double emulsions present many interesting possibilities for the controlled release of chemical substances initially entrapped in the internal droplets. Various industries, including foods and cosmetics, are showing clear interest in the technological development of such complex systems. The field of human pharmaceuticals is the major area of application; W/O/W emulsions have been investigated mostly as potential vehicles for various hydrophilic drugs (vaccines, vitamins, enzymes, hormones) that would be then slowly released. Active substances may also migrate from the outer to the inner phase of multiple emulsions, providing in that case a kind of reservoir particularly suitable for detoxification (overdose treatment) or, in a different domain, for the removal of toxic materials from waste water. In any case, the impact of double emulsions designed as drug delivery systems would be of significant importance in the controlled release field, provided that the stability and release mechanisms were more clearly understood and monitored.

The problems associated with the stability of multiple emulsions have been extensively studied and one can find a rather large literature about the subject [1–3 and references therein]. Double emulsions are generally prepared with two emulsifiers of opposite solubility (water-soluble and oil-soluble). Both emulsifiers mix at the interfaces and the stability of the films with respect to coalescence is governed by the composition of the binary mixture. Coalescence in multiple emulsions can occur at several levels: (1) between small inner droplets, (2) between large globules, and (3) between the globule and the small droplets dispersed within it. The globules are also permeable to many different chemical species that can migrate from the internal phase to the external one or vice versa, without film rupturing. Diffusion processes have been investigated as a possible mechanism for transport of active substances [4,5]: here, the concentration gradient exerted by the whole set of various molecules (surfactant, electrolytes, and active species) is involved. The resulting breakdown of multiple emulsions, as a result of one or a combination of all of these mechanisms, leads, at various rates, to the release of the active ingredient(s) entrapped in the inner phase to the outer phase in an uncontrolled way. This is why the use of multiple emulsions as commercial products is so restricted, though much attention has been paid to their many potential practical applications [3,6,7].

In this chapter, we provide an overview of the most recent advances in the field of double emulsions, emphasizing the differences between three types of emulsifiers: short surfactants, polymers and solid colloidal particles.

6.2. W/O/W Surfactant-Stabilized Emulsions

A better understanding of the stability conditions and release properties in double emulsions requires the use of model systems with a well-defined droplet size. Goubault et al. [8] and Pays et al. [9] have proposed two distinct techniques to produce size-controlled double emulsions. The same authors then investigated the two mechanisms that are responsible for the release of a chemical substance in the presence of short surfactants [10–12]. (1) One is due to the coalescence of the thin liquid film separating the internal droplets and the globule surfaces. (2) The other mechanism, termed "compositional ripening," occurs without film rupturing; instead, it occurs by diffusion and/or permeation of the chemical substance across the oil phase. By varying the proportions and/or the chemical nature of the surface active species, Pays et al. show that it is possible to shift from one type of mechanism to the other one. They study separately both mechanisms and they establish some basic rules that govern the behavior of W/O/W double emulsions. In this section, we summarize and discuss the main results obtained by Pays et al. [10–12] and then complete the chapter with a consideration of several other major contributions.

6.2.1. *Emulsion Preparation*

W/O/W quasi-monodisperse emulsions were fabricated following a two-step procedure [8–12]. Size-controlled W/O emulsions, stabilized by an oil-soluble surfactant, were first prepared. Salt (NaCl) was added to the dispersed phase to avoid any rapid coarsening phenomenon [13] (see Chapter 5, Section 5.2.) and as a tracer to probe the different release mechanisms. Double emulsions were fabricated by dispersing the inverse one within an aqueous continuous phase containing a hydrophilic surfactant. Large oil globules were produced, each one containing smaller inverse water droplets. Salt or glucose was added in the external water phase to prevent any osmotic pressure mismatch between the internal and the external water compartments. Figure 6.1 shows a microscopic image of a quasi-monodisperse double emulsion immediately after its fabrication. Large oil globules uniform in size ($\sim$ 10 μm) are visible, each one containing smaller water droplets ($\sim$ 0.4 μm).

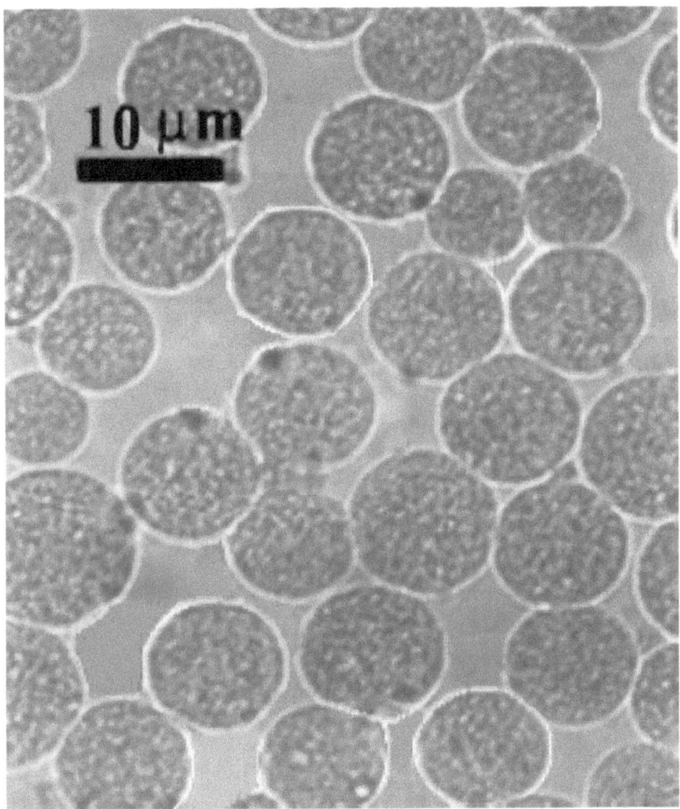

FIGURE 6.1. Microscopic image of a size-controlled W/O/W emulsion obtained following the two-step procedure described in [9] and [10]. $d_g = 9$ μm, $d_i = 0.36$ μm, $\phi_i = 20\%$, $\phi_g = 10\%$, $C_h = 0.1$ CMC, $C_l = 2$ wt% with respect to oil phase = dodecane. (Reproduced with permission from [10].)

In comparison with previous studies, the systems studied by Pays et al. possess the advantage of being size-controlled and reproducible. This is an important property because, as demonstrated by the authors, the rate of release of double emulsions is strongly dependent on the colloidal size of the dispersed objects. The following notations are used to characterize the composition and properties of double emulsions:

C_l = concentration of the lipophilic surfactant in the oil globules
C_h = initial concentration of the hydrophilic surfactant in the external water phase
ϕ_i = volume fraction of water droplets in the oil globules
ϕ_g = volume fraction of oil globules in the external water phase
d_i = diameter of the water droplets
d_g = diameter of the oil globules
CMC = critical micellar concentration

Two different techniques were used to study the kinetic evolution of the double emulsions. The concentration of salt (NaCl) present in the aqueous external phase was measured by means of specific electrodes (potentiometry). This technique was combined with direct microscopic observations as well as repeated single-globule creaming experiments using optical manipulation [14,15]. In the set-up, a unique non-Brownian globule (more than 10 μm in diameter) was illuminated by one or two focused laser beams. The radiation pressure exerted by the lasers allowed one to capture and to displace a globule at any position in a transparent cell. When the lasers were switched off, the globule moved upward because of buoyancy. The globule diameter, d_g, as well as its creaming velocity V_{creaming} in the stationary regime, were measured under a microscope. From the Stokes equation, the average density of the globule, ρ_g, is deduced:

$$V_{\text{creaming}} = \frac{1}{18} \frac{(\rho_w - \rho_g) g d_g^2}{\eta_c} \tag{6.1}$$

In Eq. (6.1), ρ_w is the density of the external water phase, g is the acceleration due to gravity (9.8 m/s^2), and η_c is the viscosity of the continuous phase. Then, it becomes possible to determine the internal droplet volume fraction ϕ_i inside the globules according to the relation:

$$\rho_g = \phi_i \rho_{wi} + (1 - \phi_i) \rho_o \tag{6.2}$$

where ρ_o is the oil density and ρ_{wi} is the density of the internal water phase.

6.2.2. General Phenomenology

Figure 6.2 represents the behavior of quasi-monodisperse double emulsions in (C_h, ϕ_i^0) coordinates, where ϕ_i^0 is the initial volume fraction of droplets in the globules. Sorbitan monooleate (SMO) was used for the stabilization of the primary W/O emulsion and sodium dodecyl sulfate (SDS, CMC = 8 10^{-3} mol/l) was used for the stabilization of the oil globules in the external aqueous phase. Three different compositions zones, referred to as A, B, and C can be defined; they differ by their qualitative behavior observed via microscopy. Moderate internal droplet volume fractions are considered first ($\phi_i^\circ \leq 20\%$). If C_h = CMC/10, the system does not exhibit any structural evolution after a few days of storage (zone A). If C_h is

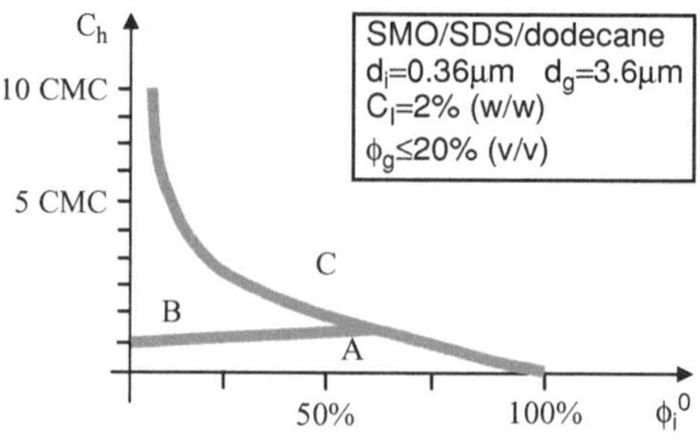

FIGURE 6.2. Behavior of a quasi-monodisperse double emulsion as a function of the two composition parameters C_h and ϕ_i^0. Observations were performed via microscopy for 24 h after preparation. (A) No coalescence. (B) Droplet—globule coalescence. (C) Droplet–droplet and droplet-globule coalescence. (Adapted from [11].)

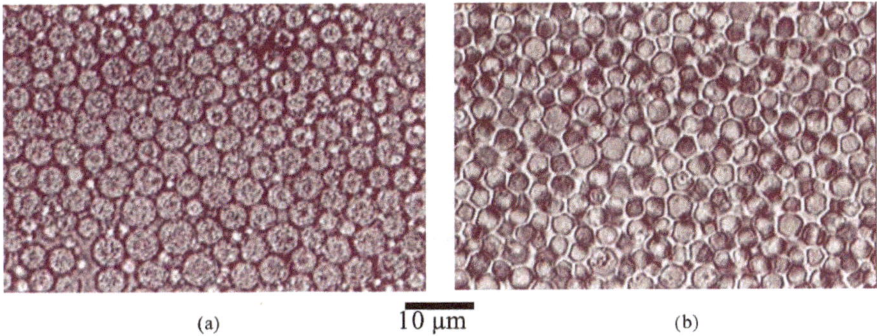

FIGURE 6.3. Transformation of an emulsion from double (W/O/W) to simple (O/W). $d_g = 3.6$ μm, $d_i = 0.36$ μm, $\phi_i^0 = 5\%$, $\phi_g = 10\%$, $C_h = 10$ CMC, $C_l = 2$ wt% with respect to oil phase = dodecane. **(a)** $t = 0$; **(b)** $t = 400$ min. (Reproduced with permission from [11].)

equal to or larger than approximately 1 CMC, the double W/O/W emulsion rapidly transforms into a simple O/W emulsion (see Fig. 6.3). The characteristic time scale for the transformation becomes shorter when C_h increases, in perfect agreement with the pioneering experiments of Ficheux et al. [16]. For globules with diameter $d_g = 3.6$ μm, $\phi_i^0 = 5\%$ and $C_h = 10$ CMC, it takes around 5 h for the transformation to occur. Repeated microscopy observations revealed that the globules become progressively empty and that there is apparently no coarsening of the internal droplets (zone B). To elucidate the origin of this evolution, polydisperse double emulsions were produced with large internal droplets that could be perfectly identified under the microscope. Under the same conditions, the internal droplets may be in contact with the globule surface some of the time without exhibiting any structural change, and then suddenly disappear, as shown in Fig. 6.4. From all these observations, it can be concluded that the mechanism responsible for the transformation from a double to a simple emulsion is the coalescence of the internal droplets on the globule surface.

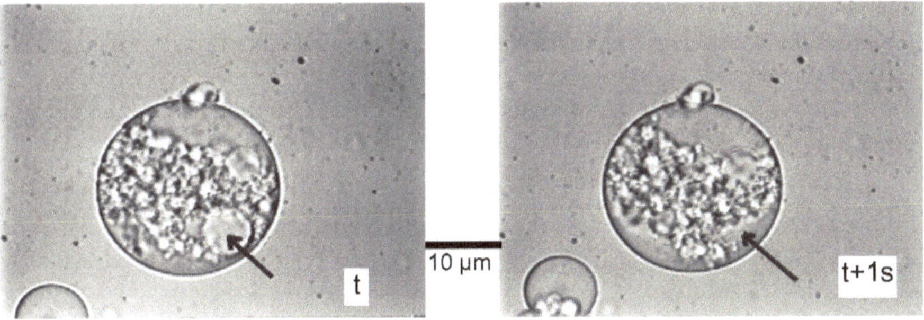

FIGURE 6.4. The internal droplet indicated by the arrow coalesces on the globule surface. The two images were taken at times 1 s apart. $C_h = 10$ CMC, $C_l = 2$ wt%. (Reproduced with permission from [11].)

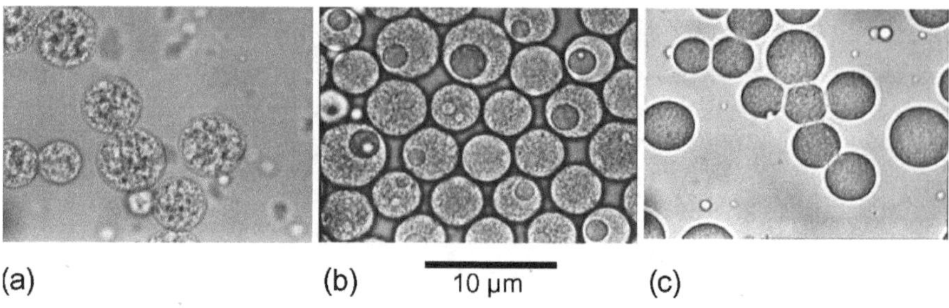

FIGURE 6.5. Destruction of a double emulsion with a high initial droplet volume fraction. $\phi_i^0 = 50\%$, $C_h = 3$ CMC, $C_l = 2$ wt%. **(a)** $t = 60$ min; **(b)** $t = 250$ min; **(c)** $t = 850$ min. (Reproduced with permission from [11].)

A slightly distinct scenario of destruction occurs when the initial droplet volume fraction ϕ_i^0 exceeds some critical value that depends on C_h. The transition from a double to a simple emulsion is still observed, but in this case there is some coarsening of the internal droplets during the process of destruction (zone C), as visible in the images in Fig. 6.5, taken at regular time intervals. Some large nuclei resulting from droplet—droplet coalescence are clearly distinguished microscopically. Once they reach the surface, they coalesce rapidly and disappear. In the last stages of the transformation, when the droplet concentration becomes small enough, inter droplet coalescence is no longer observed and the scenario becomes identical to the one observed for the systems with low initial droplet concentration (Fig. 6.5c).

In the limit of an extremely high concentration of droplets ($\phi_i^0 > 90\%$), the destruction involves droplet—droplet and droplet–globule coalescence throughout the process. Figure 6.6 is a sequence showing the destruction of a single globule as a function of time. To facilitate the observation, a very large globule with initial diameter of approximately 30 μm was considered. The optical contrast between the globule initially containing 95% (v/v) water and the aqueous external phase is quite low, which explains the difficulty in discerning the globule. The globule diameter decreases rapidly as a consequence of droplet—globule coalescence. On the other hand, the large nuclei observed inside the globule result from droplet—droplet coalescence. The total destruction of the globule at $C_h = 3$ CMC occurs over a time scale of the order of 10 min, a much shorter lifetime than for an initially diluted globule.

6.2.3. Role of the Hydrophilic Surfactant in Inducing Coalescence

The phenomenology described in the preceding section is general since it has been reproduced using different ionic surfactants, such as alkyl sulfonates or alkyl quaternary ammonia [11,12,16]. Villa et al. [17] have also investigated coalescence phenomena in W/O/W globules. They developed a capillary microscopy technique

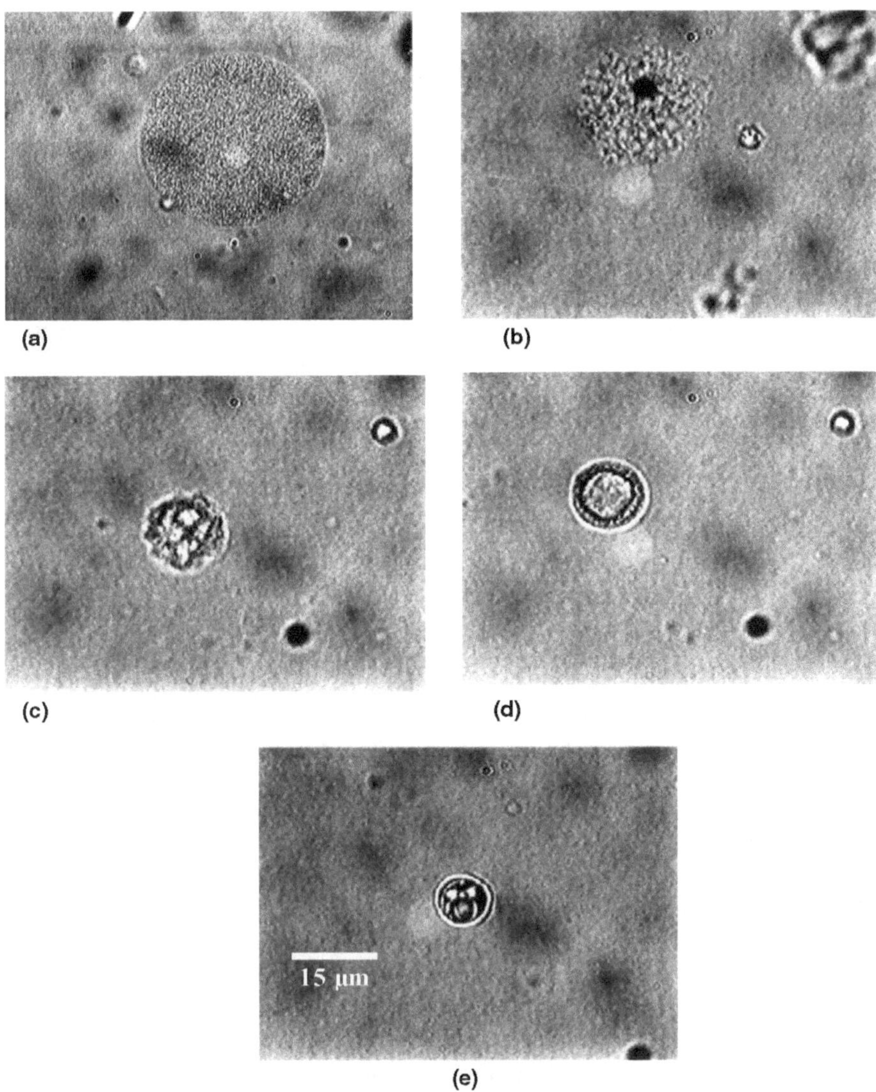

FIGURE 6.6. Destruction of a double globule with very high initial droplet volume fraction. $\phi_i^0 = 95\%$, $C_h = 3$ CMC, $C_l = 2$ wt%. **(a)** $t = 0$; **(b)** $t = 26$ s; **(c)** $t = 2$ min; **(d)** $t = 7$ min; **(e)** $t = 10$ min. (Reproduced with permission from [11].)

by which a single double-emulsion globule can be prepared directly within a circular capillary and the phenomena can be quantified using image analysis. The use of micropipets allows injection of oil and water droplets for double-emulsion formation within the microcapillary. Globules of similar size and with a desirable number of inner droplets can be constructed in this manner. This procedure enables both direct visualization of individual double emulsion globules for their

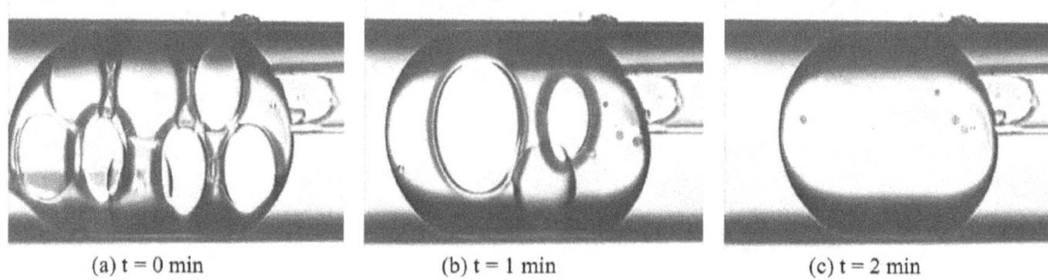

FIGURE 6.7. Internal coalescence followed by external coalescence in a W/O/W double-emulsion globule. Internal droplets = PSMO solution at 50 CMC; oil phase = hexadecane with 2 wt% SMO; external water phase = pure water. (Reproduced with permission from [17].)

entire life span, and micromanipulation of internal water droplets. By means of this technique, globules were prepared using either SDS (ionic) or polyoxyethylene (20) sorbitan monooleate PSMO (nonionic) as the water-soluble surfactants and SMO as the oil-soluble surfactant. The results confirmed that SMO has a stabilizing effect, while both ionic and nonionic hydrophilic surfactants have a destabilizing effect. By varying the surfactant concentrations in both the internal and external aqueous phases, droplet—droplet and droplet globule coalescence (see Fig. 6.7) were found to occur under a limited range of conditions. In general, these conditions require that a high overall concentration of hydrophilic surfactant (either ionic or nonionic) must be destined for the internal and external aqueous phases. These phenomena were shown to also depend on the concentration of the SMO in the oil phase; the stabilizing ability increases with SMO amount. González-Ochoa et al. [18] performed direct observations of single W/O/W globules of colloidal size, stabilized with SMO and SDS. They identical visualized droplet—droplet and droplet—globule coalescence at high SDS concentrations. In addition, they reported two-stage coalescence phenomena of large water droplets on the globule surface. The process was captured step by step via fast digital video microscopy. In the first stage, a transient structure is formed: the oil enveloping the water droplet peels off, leaving the droplet immersed in the continuous water phase supported by a film of oil and surfactants. The retraction of the oil occurs in a time span of 1 ms. In the second stage, the film covering the water droplet wears and the drop breaks, releasing its contents to the continuous water phase. The second stage occurs in a time span of a few tens of milliseconds (Fig. 6.8). It is noteworthy that two-stage coalescence was observed at rather large SDS concentrations (30 CMC) and for large internal droplets (~10 μm).

The previous experimental observations reported in the preceding text are, at least to a certain extent, in agreement with the well-known Bancroft rule. Indeed, a double W/O/W emulsion turns into a simple direct one when a sufficient quantity of the water-soluble surfactant is added. Similarly, by shaking a 1:1 mixture of water and oil, each phase containing one of the two types of surfactants, a direct emulsion is obtained if the aqueous phase contains a large amount of water-soluble

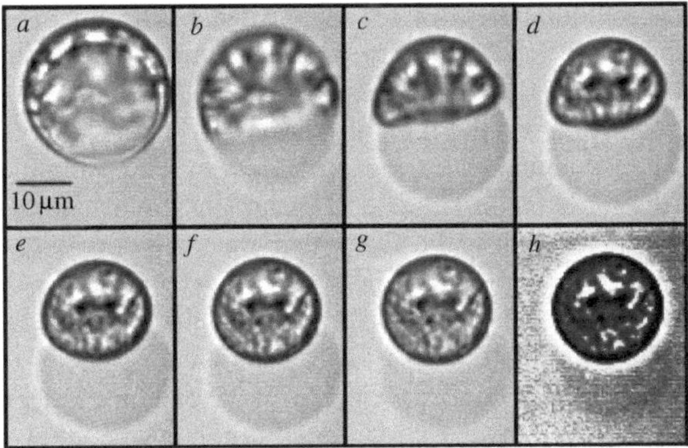

FIGURE 6.8. Two-stage coalescence in double emulsions. Sequence of side view pictures, taken at a frequency of 5000 frames/s, of a water droplet coalescing on the globule surface. $C_h = 30$ CMC of SDS; $C_l = 2$ wt% of SMO in dodecane. (Reproduced with permission from [18].)

surfactant. By contrast, when less of this surfactant is present, it is possible to obtain a double emulsion. From the above-described set of experiments, we end up with the same type of conclusion: the double emulsion persists as long as the water-soluble surfactant concentration is not too high. More generally, it is known that the micellar phase always tends to be the continuous phase of an emulsion. A mechanistic interpretation for this correlation was proposed, according to which the surfactant packing type (spontaneous curvature) affects both the phase behavior of microemulsions and the coalescence energy barrier of an emulsion (see Chapter 5, Section 5.3.1.1). Kabalnov and Wennerström[19] argue that the effect of the spontaneous curvature on emulsion stability comes from the kinetics of the hole nucleation in emulsion films. Consider an oil film separating an internal water droplet from the external water phase (Fig. 6.9). If C_h is low, the monolayer covering the droplet and globule surfaces is essentially composed of lipophilic molecules. The spontaneous curvature of oil-soluble surfactants is negative (W/O shape). Propagation of a hole is damped, because the monolayer at the edge of the

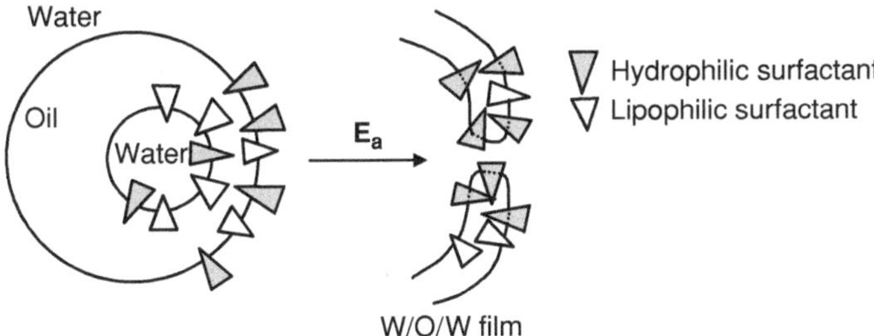

FIGURE 6.9. Scheme of the hole nucleation process in a W/O/W film. (Adapted from [11].)

nucleation hole is inhibited because it is curved against the direction favored by the spontaneous curvature. As a result, for the film rupture to occur, the system must pass through an energy barrier, after which growth becomes spontaneous. This state can be reached only via thermal fluctuation and has a low probability because of the unfavorable spontaneous curvature. However, increasing the concentration of the hydrophilic surfactant in the external water phase concentrates this molecule at the interfaces, thus rendering the average spontaneous curvature less negative or even positive. Hole propagation in the W/O/W film now occurs with a very moderate energy barrier because the curvature of the edge fits the spontaneous curvature. This could explain why coalescence of the internal droplets on the globule surface occurs above a critical concentration C_h in the external water phase. Because coalescence also occurs between the internal droplets when they are sufficiently concentrated, it can be deduced that the hydrophilic surfactant initially introduced in the external water phase is being transferred toward the internal droplets as a result of the entropy of mixing.

6.2.4. *Kinetics of Release*

6.2.4.1. Limit of Low SDS Concentration and Low Internal Droplet Volume Fraction

Pays et al. [10–12] investigated the influence of the hydrophilic surfactant concentration C_h on the kinetics of release. Because in their experiments the internal water volume was about 100 times smaller than the external volume, C_h fixed the chemical potential of the hydrophilic surfactant molecules. The thin liquid film formed between the internal droplets and the globule surface was composed of monolayers covered by SMO and SDS molecules, separated by oil. Because SDS molecules migrate from the external to the internal water phase within a very short period of time (1 min [12]), the film could be regarded as close to thermodynamic equilibrium with respect to surfactant adsorption a few minutes after preparation. Following the well-known Bancroft rule, such inverted films possess a long-term stability when essentially covered by a hydrophobic surfactant (such as SMO) but become very unstable when a strong proportion of hydrophilic surfactant (such as SDS) is adsorbed. As discussed above, the transition from long-range to short-range stability may be achieved by varying the concentration of the hydrophilic surfactant in the external water phase.

Several emulsions were prepared with SMO as the emulsifier of the primary emulsion, at constant concentration C_1 in the oil phase, and SDS at various concentrations C_h in the external aqueous phase. Figure 6.10 shows the quantity of salt released (expressed in relative percentage) as a function of time; for all the plots, the globule diameter, the initial droplet volume fraction, and the globule volume fraction were the same. For $C_h \leq$ CMC, the release is quite slow, occurring over a characteristic time scale of several days. The rate decreases when C_h increases, being minimal around 1 CMC. When the process is achieved (nearly 100% has been released), it appears via microscopy that the water droplet

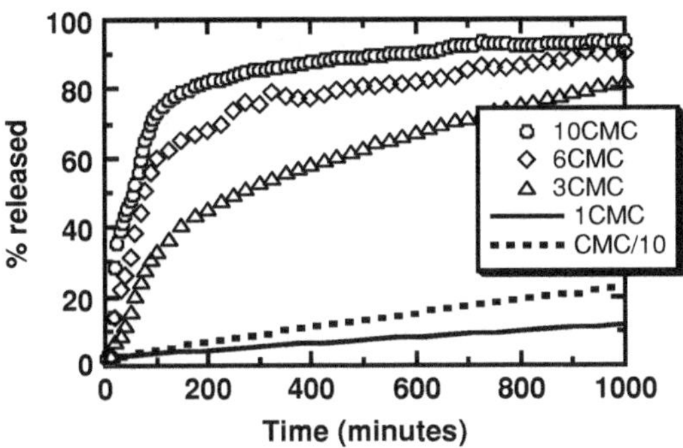

FIGURE 6.10. Influence of SDS concentration in the external water phase on the kinetics of release. $d_g = 4$ μm, $d_i = 0.36$ μm, $\phi_i^0 = 20\%$, $\phi_g = 10\%$, $C_l = 2$ wt%. (Adapted from [10].)

concentration ϕ_i in the globules has apparently not varied. This is confirmed by the creaming technique described in Section 6.2.1, since a constant rate of creaming of the globules was measured after 3 days of storage. The ϕ_i value deduced from Eqs. (6.1) and (6.2) corresponds to the initial value. It can therefore be concluded than in this regime, the salt release occurs without film rupture; instead it is produced by an entropically driven diffusion and/or permeation of the salt across the oil globule. For $C_h >$ CMC, the release is quite fast and the rate increases with the hydrophilic surfactant concentration. Repeated microscopic observations revealed a gradual decrease of the inner droplet concentration, which is almost zero when 100% of release is attained. When $C_h < 5$ CMC, no evolution of the water-in-oil droplets was observed. This definitely confirms that for 1 CMC $< C_h <$ 5 CMC, salt release is controlled by the coalescence of the internal droplets on the globule surface.

The regime governed by coalescence was examined in more detail. The process of film rupture is initiated by the spontaneous formation of a small hole. The nucleation frequency, Λ, of a hole that reaches a critical size, above which it becomes unstable and grows, determines the lifetime of the films with respect to coalescence. A mean field description [19] predicts that Λ varies with temperature T according to an Arrhenius law:

$$\Lambda = \Lambda_0 \exp(-E_a/k_B T) \tag{6.3}$$

where k_B is the Boltzmann constant, E_a is the activation energy for coalescence, and Λ_0 is the attempt frequency of the hole nucleation process. Pays et al. [10–12] have proposed an unambiguous method for the measurement of the microscopic parameters Λ_0 and E_a based on the use of monodisperse W/O/W double emulsions. The hydrophilic surfactant concentration, the globule diameter, and the globule volume fraction were fixed, while the initial internal droplet volume fraction was varied. Within the time scales that were explored (less than 1000 min), the contribution of diffusion/permeation across the oil globule was negligible. The data are represented in Fig. 6.11. Everything else being constant, the characteristic time of release decreases as the initial internal droplet volume fraction increases. Interestingly, microscopy revealed that a fraction of the internal droplets was adsorbed on the globule surface. This is a natural consequence of the van der Waals

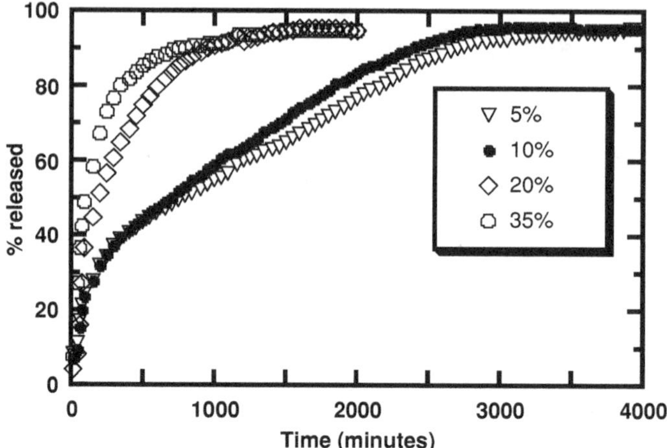

FIGURE 6.11. Influence of the initial internal droplet volume fraction on the kinetics of release. $d_g = 3.6$ μm, $d_i = 0.36$ μm, $\phi_g = 10\%$, $C_h = 3$ CMC, $C_l = 2$ wt%. (Adapted from [10].)

attraction that exists between the internal droplets and the external water phase [20]. In the following, n_i is defined as the total internal droplet concentration within the globules, and n_a is defined as the concentration of adsorbed droplets (per unit volume of the globules). It is reasonable to assume that the number of coalescence events per unit time is simply proportional to the concentration of adsorbed droplets:

$$dn_i/dt = -\Lambda n_a \tag{6.4}$$

where Λ is the characteristic frequency of coalescence between an adsorbed droplet and the globule surface. At any time t, n_i can be calculated from the ordinate of the curves (Fig. 6.11) and the number of coalescence events dn_i/dt may be deduced from the derivative of the curves. The experimental points were transformed and plotted again in (n_i, dn_i/dt) coordinates in Fig. 6.12. All the data lie on a single

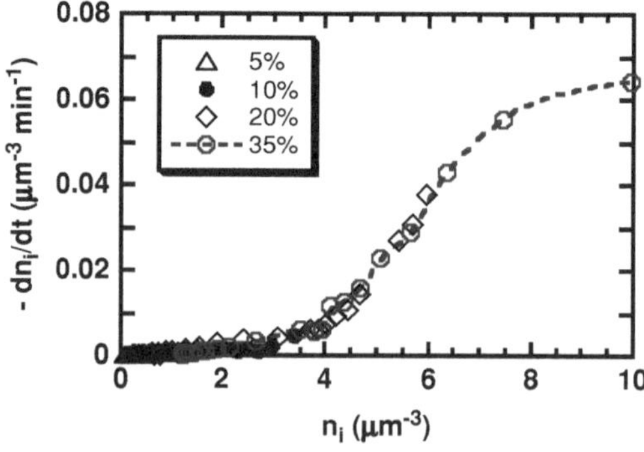

FIGURE 6.12. Rate of coalescence as a function of the number density of internal droplets in the globules. $d_g = 3.6$ μm, $d_i = 0.36$ μm, $\phi_g = 10\%$, $C_h = 3$ CMC, $C_l = 2$ wt%. The dashed line is a visual guide. (Adapted from [10].)

curve, meaning that the rate of coalescence dn_i/dt depends only on n_i [21]. Following Eq. (6.4), this function reflects the adsorption isotherm of the water droplets on the globule surface $n_a = f(n_i)$.

The adsorption isotherm was modeled in order to deduce a numerical value for Λ. Following the model of Frumkin and Fowler reported in [21], n_a is given by the following set of equations:

$$\frac{\Theta}{1-\Theta} = K(n_i - n_a)\exp\left(-\frac{u_a + 4u_l\Theta}{k_B T}\right) \tag{6.5}$$

$$\Theta = \frac{n_a}{n_0} \tag{6.6}$$

where n_0 is the total concentration (per unit volume) of available sites for adsorption easily deduced from geometrical considerations, u_a is the adsorption energy, u_l is the lateral energy of interaction between the droplets, and finally K is a constant calculated from the model [21].

Assuming that the internal droplets experience a hard-sphere-like repulsion when surfactant layers come in contact, an estimation of the van der Waals interactions can be obtained from the average length of the surfactant tails ($l \approx 3$ nm) [20]. The coalescence frequency is therefore the unique free parameter in the model and is determined from the best fit to the experimental curves. In Fig. 6.13, the kinetic evolution of n_i at constant C_h, d_i, and ϕ_g, is plotted. Using one and the same value of Λ, the theoretical points can be fitted correctly to the experimental data, whatever the globule diameter. This value is $\Lambda = 6\ 10^{-3}\ \text{min}^{-1}$ for $C_h = 3$ CMC, $d_i = 0.36\ \mu$m. The obtained value indicates that a droplet of diameter 0.36 μm spends on average 3 h on the globule surface before a coalescence event occurs. Measurements of the same type were performed in the same conditions, but with different alkanes composing the oil phase (octane, dodecane, hexadecane). As can be seen in Fig. 6.14, the coalescence frequency raises from almost zero to $2.5\ 10^{-2}\ \text{min}^{-1}$ as the oil chain length varies from C8 to C16 (see Chapter 5, Section 5.4.2 for a more detailed explanation).

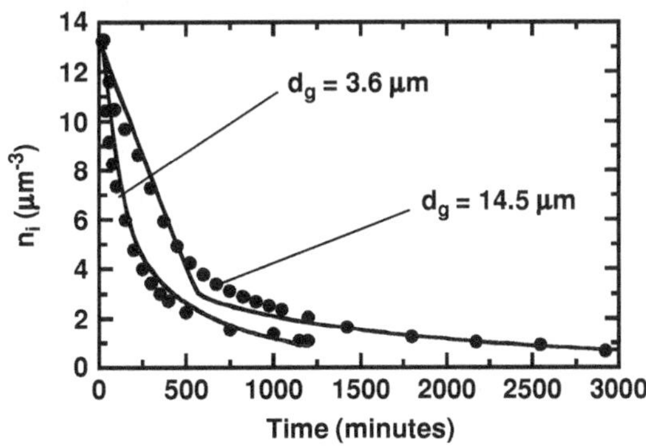

FIGURE 6.13. Number density of internal droplets for two globule diameters. $d_i = 0.36\ \mu$m, $\phi_g = 10\%$, $C_h = 3$ CMC, $C_l = 2$ wt%. The continuous lines are theoretical predictions. (Adapted from [10].)

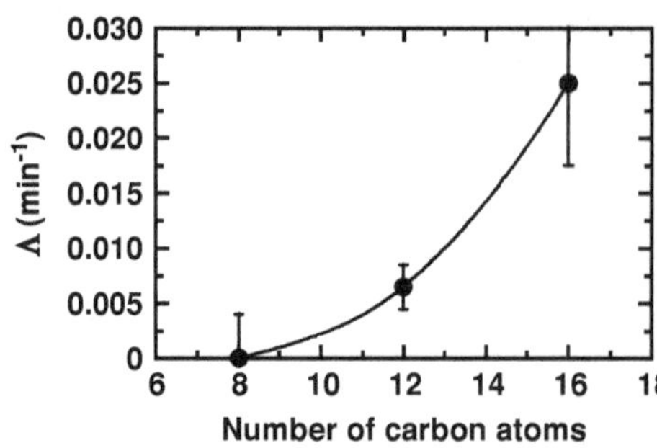

FIGURE 6.14. Evolution of the coalescence frequency, Λ, as a function of the oil hydrocarbon chain length. $d_g = 3.6\ \mu\text{m}$, $d_i = 0.36\ \mu\text{m}$, $\phi_i^0 = 20\%$, $\phi_g = 10\%$, $C_h = 3$ CMC, $C_l = 2$ wt%. (Adapted from reference [11].)

6.2.4.2. Limit of High SDS Concentration or High Internal Droplet Volume Fraction

In Fig. 6.15b,c, the effect of varying the hydrophilic surfactant concentration from 3 CMC to 10 CMC is illustrated. It is clear from the graph that increasing the hydrophilic surfactant concentration has the effect of accelerating the salt release.

Pays et al. [11] tried to fit the experimental curves with the adsorption—coalescence model, using Λ as the unique free parameter. As can be observed in Fig. 6.15a, the agreement between theory and experiments is fairly good at $C_h = 3$ CMC but large deviations appear at higher concentrations. In Fig. 6.15c, two different Λ values have been tried in an attempt to fit the experimental data, but neither of them is satisfactory. From this, it can be stated that the model valid at low hydrophilic surfactant concentration does not correctly describe the experimental results at high concentrations. In other words, the rate of release cannot be described any longer in terms of a unique coalescence frequency when C_h is higher than about 5 CMC. At such a high concentration, internal droplet–droplet coalescence accelerates the rate of release. Indeed, coalescence of internal droplets produces large nuclei which are preferentially adsorbed on the globule surface owing to their larger van der Waals attraction. Moreover, when a nucleus coalesces at the globule surface, a larger amount of matter is released at one time. The inherent polydispersity resulting from droplet–droplet coalescence can therefore explain the complex behavior of double emulsions with high hydrophilic surfactant concentrations and the fact that the process cannot be described via a single coalescence frequency. The same type of conclusion can be drawn when the initial-internal droplet concentration is large, even at low surfactant concentration.

The general behavior for the release of double emulsions stabilized by short surfactants has been described in the preceding text. By appropriately choosing the surfactant concentrations and their chemical nature, it is possible to dissociate the two mechanisms that are responsible for the release of chemical substances. From a practical point of view, it can be concluded that long-term encapsulation of small neutral or charged molecules is not really possible using short surfactants as stabilizing agents. When the hydrophilic surfactant concentration is lower than a critical value C_h*, the release occurs preferentially by diffusion and/or permeation across the oil phase, while above C_h* it occurs preferentially by coalescence. In both cases, the characteristic period of release does not exceed several days, which is

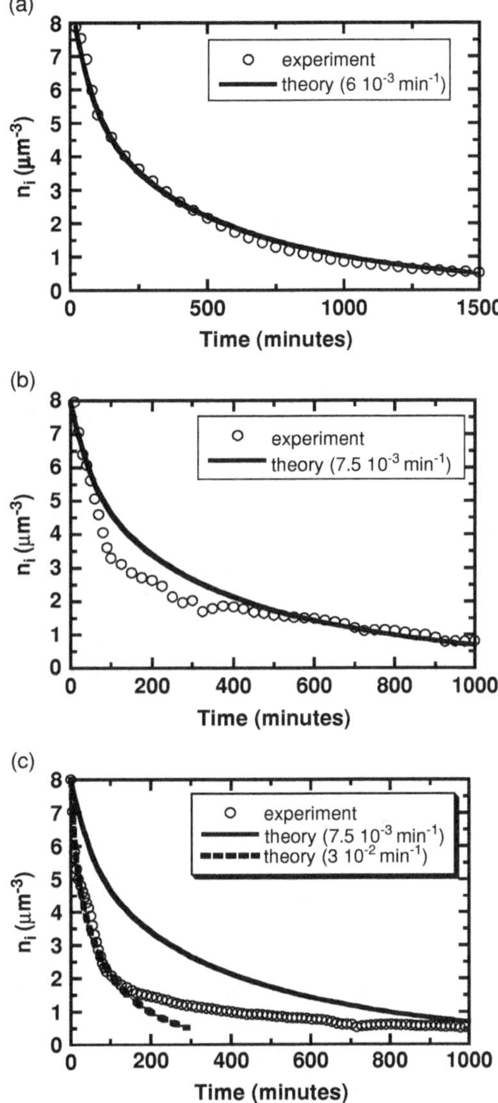

FIGURE 6.15. Number density of internal droplets in the globules as a function of time. $d_g = 3.6$ μm, $d_i = 0.36$ μm, $\phi_i^0 = 20\%$, $\phi_g = 12\%$, $C_l = 2$ wt% **(a)** $C_h = 3$ CMC; **(b)** $C_h = 6$ CMC; **(c)** $C_h = 10$ CMC. In the insets, the numbers correspond to the Λ values tested to fit the experimental data. (Adapted from [11].)

not sufficient for most practical applications. C_h* is of the order of 1 CMC for ionic surfactants and of the order of 100 CMC for nonionic surfactants [10–12, 16–18].

6.2.5. Water Transport Under Osmotic Pressure Mismatch

Under osmotic pressure gradients between the two aqueous phases of W/O/W emulsions, water may migrate either from the internal to the external phase or vice versa, depending on the direction of the osmotic pressure gradient. This process is entropically driven and is another manifestation of "compositional ripening." Such water transport may be critical in the various applications of multiple emulsions. For instance, swelling of the internal droplets may increase the effective globule fraction and produce significant changes in the rheological properties of double emulsions. In some cases, swelling is used as a way to press the internal droplets one against each other at a point that droplet—droplet and droplet—globule coalescence occurs, thus facilitating the release of a water-soluble encapsulated species. Water transport rates are affected by the magnitude of the osmotic pressure gradients between the water phases, the nature and concentration of the surfactants used for the preparation of the emulsions, the nature and viscosity of the oil phase, and so forth [22–29]. In their work on the water permeability of the oil phase in W/O/W emulsions under osmotic pressure gradients, Kita et al. [30] proposed two possible mechanisms for the permeation of water and water-soluble materials: water molecules (1) pass through the thin liquid film formed by the internal droplets in contact with the globule surface or (2) diffuse across the oil phase by being incorporated in "reverse micelles," while Colinart et al. [31] had earlier suggested reverse-micellar water transport as well as via hydrated surfactants as the two possible means of water migration. Garti et al. [27] and Jager-Lezer et al. [29] have established that oil-soluble surfactant is a major factor for water migration in multiple emulsions and that water transport rates increase with increasing oil-soluble surfactant concentration.

Wen and Papadopoulos [32] reported a visualization study of the water transport in W/O/W single globules under osmotic pressure mismatch. By using micropipettes, they could prepare a small oil compartment confined between two large water drops (W2) in a microcapillary tube. A unique aqueous drop (W1) was inserted into the oil compartment at a different position each time an experiment was conducted (Fig. 6.16). Two limiting conditions were examined: (1) the internal water droplet is in contact with the oil boundary and (2) there exists a visible minimum distance of separation between W1 and W2. Using video microscopy, the radius $R = d_g/2$ of W1 was measured at regular time intervals, making it possible to calculate the rate of radius change —dR/dt. Rates were found to rise linearly with increasing oil-soluble surfactant concentration in the oil phase, though the effects were more pronounced when W1 and W2 were in visual contact (Fig. 6.17). In the latter case, it was suggested that the variation of radius was controlled mainly by water diffusion or permeation across the thin liquid film separating W1 and W2. In the case of visually non contacting W1 and W2, spontaneous emulsification occurred and was revealed by the formation of small satellites around the large mother droplet (see Fig. 6.17b). In this limit, dR/dt was most probably controlled by spontaneous emulsification and water solubilization in reverse micelles.

6.3. W/O/W Polymer-Stabilized Emulsions

The lifetime of double emulsions is considerably shortened by the rapid diffusion of the water-soluble small-molecule surfactants toward the droplet interface.

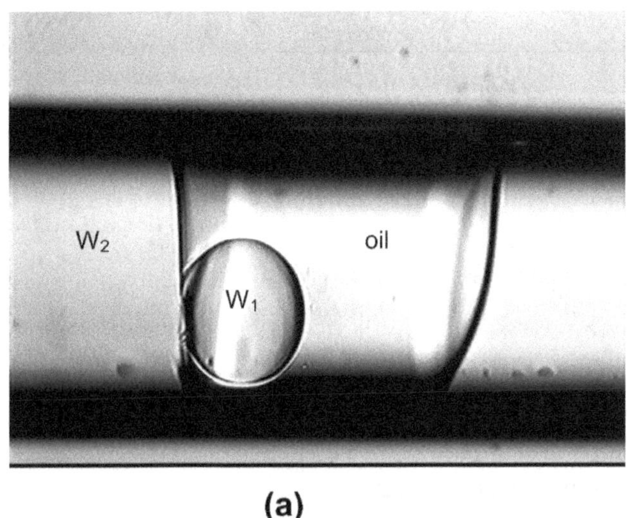

(a)

(b)

FIGURE 6.16. W1/O and O/W2 interfaces in (**a**) visual contact and (**b**) no visual contact. (Reproduced with permission from [32].)

Among other mechanisms, the formation of inverse micelles allows the diffusion of molecules through the oil phase. Because of the inefficiency of short surfactants to ensure long-term stability, much effort is being spent on finding out formulations incorporating polymeric stabilizers [33,34] or associating short surfactants and polymers or proteins [35]. Double emulsions with enhanced stability have been obtained, but more work needs to be done to perfectly control and understand the release properties of these new materials, as well as to ensure their reliability as commercial products. In principle, large hydrophilic polymers cannot cross the oil

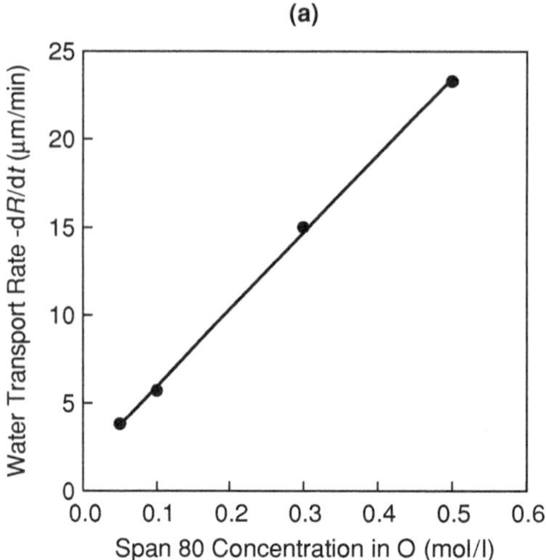

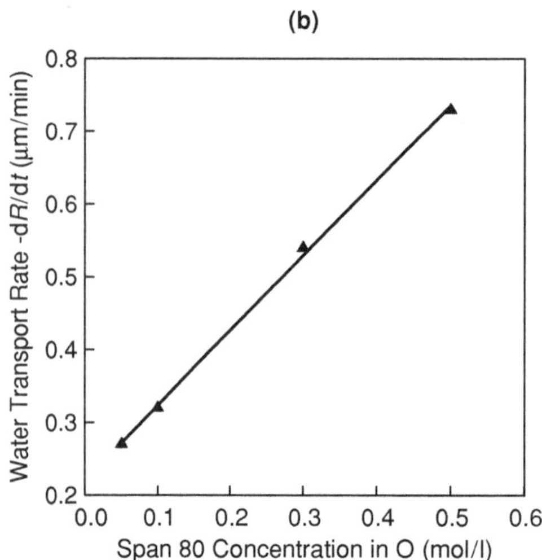

FIGURE 6.17. Effects of SMO (span 80) concentration in the rate of radius variation in W/OW, when W1/O and O/W2 interfaces are at **(a)** visual contact and **(b)** no visual contact. W1 = pure water; O = n-hexadecane + SMO, W2 = 5 10^{-3} mol/l NaCl solution. (Reproduced with permission from [32].)

phase and adsorb on the droplet surfaces. The elasticity of the interfaces and the steric repulsions between droplets and the inner surfaces of globules are the most probable reasons to explain the stability improvement.

Michaut et al. [36] have investigated the dynamic rheological properties of concentrated multiple W/O/W emulsions to characterize their amphiphile composition at interfaces. A small surfactant molecule (SMO) and an amphiphilic polymer (hydrophobically modified poly(sodium acrylate)) were used to stabilize the inverse emulsion and the oil globules, respectively. Rheological and interfacial tension measurements show that the interfaces remain asymmetric, that is, the polymeric surfactant adsorbed at the globule interface does not migrate to the droplet interfaces through the oil phase. The stability was tested through release kinetics of a marker (NaCl) initially encapsulated in the aqueous droplets, and by rheology

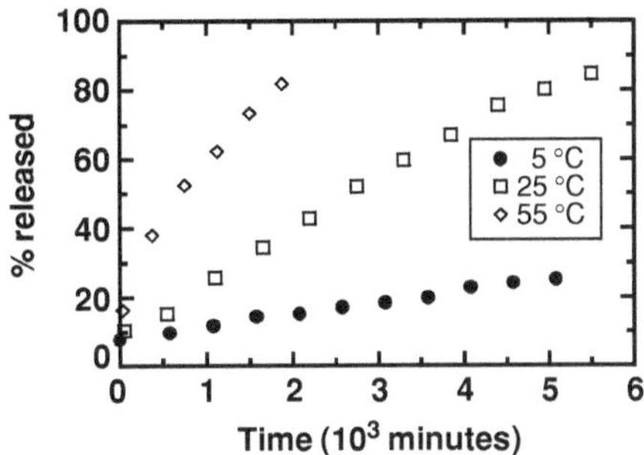

FIGURE 6.18. Kinetics of leakage of an emulsion stabilized with Arlacel P 135 and Synperonic PE/F68 as a function of temperature. $[\mathrm{NaCl}]^0 = 0.4$ mol/l, $\phi_i^0 = 20\%$, $\phi_g = 18\%$, $d_i = 0.36\mu\mathrm{m}$, $d_g = 4\mu\mathrm{m}$, $C_h = 5\%$, $C_l = 2$ wt%; oil: dodecane. (Adapted from [38].)

[37]. Slow release rates and remarkable long shelf-life (months) were obtained compared to typical multiple emulsions stabilized by two short surfactants (SMO and polyoxyethylene (20) sorbitan monolaurate). Finally, the long lifetime of the emulsions allowed study via diffusing wave spectroscopy (DWS) of the interactions between the droplets and the globule surface [37].

Pays et al. [38] explored the behavior of double emulsions stabilized by two polymeric surfactants. They showed that the release of the encapsulated species is controlled by compositional ripening solely and a simple model based on Fick's theory was proposed. In Fig. 6.18 the relative percentage of salt (NaCl) released during the first 5 days following the preparation of double emulsions stabilized by a couple of amphiphilic polymers is plotted (lipophilic polymer = polyethylene-30 dipolyhydroxystearate, Arlacel P135, $M_W \approx 5000$ g/mol; hydrophilic polymer = triblock copolymer of ethylene oxide (EO) and propylene oxide (PO) with average formula 75EO/30PO/75EO, Synperonic PE/F68, $M_W \approx 8400$ g/mol). When the release process had been completed (nearly 100% released), it was microscopy showed that the water droplet concentration ϕ_i in the globules had apparently not varied. It can therefore be concluded that in this regime, the release occurs without film rupturing.

6.3.1. Phenomenological Model for Compositional Ripening

In the absence of coalescence, the global molecular flow J of a species across the oil globules is entropic in origin and is described by Fick's law:

$$J = -\frac{dN_i}{dt} = PS(C_i - C_e) \tag{6.7}$$

where S is the total surface area involved in the transfer process, N_i is the number of encapsulated moles in the internal droplets, C_i and C_e are the concentrations in the internal and external water phases respectively. P, termed the permeability coefficient, characterizes the rate of release across the oil membrane. P can be seen as a phenomenological constant that reflects the influence of all the microscopic parameters involved in the transfer. Of course, P is expected to depend on the chemical nature of the encapsulated substance, the chemical nature of the oil, and the monolayer composition. By integrating the previous differential equation, we obtain:

$$N_i = \frac{N_i^0}{1+\alpha}\left[\alpha + \exp\left(-PS\frac{1+\alpha}{V_i}t\right)\right] \tag{6.8}$$

where N_i^0 is the initial number of encapsulated moles and $\alpha = V_i/V_e$, V_i and V_e being the internal and external water volumes, respectively, which are assumed to remain constant in time.

Combining Eq. (6.8) and the conservation condition, the following expression for the relative percentage of release is derived:

$$\%\text{released}(t) = 100\frac{C_e(t)}{C_\infty} = 100\left[1 - \exp\left(-PS\frac{1+\alpha}{V_i}t\right)\right] \tag{6.9}$$

where

$$C_\infty = \frac{N_i^0}{V_i + V_e} = \frac{\phi_g \phi_i C_i^0}{(1-\phi_g) + \phi_i \phi_g} \tag{6.10}$$

Here C_i^0 is the initial concentration of the encapsulated species in the internal droplets. The model predicts an exponential leakage as a function of time. If S is assumed to be equal to the total globule surface area, P can be numerically deduced from the initial slope p_0 of the experimental curves:

$$P = \frac{p_0}{100}\frac{d_g}{6}\frac{\phi_i(1-\phi_g)}{(1-\phi_g) + \phi_i \phi_g} \tag{6.11}$$

6.3.2. Microscopic Approaches of the Permeability: State of the Art

Compositional ripening in double emulsions is reminiscent of the so-called "passive" leakage measured across phospholipid bilayers and is partially responsible for chemical exchange across biological membranes. At a microscopic level, several models to explain the permeation phenomenon have been developed [39], all of them in agreement with Fick's law. Some models propose that permeation across the membranes results from a solubilization process followed by diffusion across the hydrophobic part of the phospholipid membrane. For hydrophilic substances, the rate-determining parameter is the so-called Born energy, which represents the energy cost for the transfer of a hydrophilic species from a medium of high dielectric constant to one of low dielectric. Other models describe the permeation as a passage through reversible and thermally activated holes that are permanently formed in the bilayer. The transient holes may be large enough to allow the passage of small hydrophilic substances. The rate of transfer is controlled by the energy cost for hole formation, which includes surface tension and curvature effects (see Section 6.2.3). Although they involve different microscopic mechanisms, these limiting models lead to quite similar expressions for the permeation coefficient. In all cases, P is predicted to follow an effective Arrhenius law:

$$P = P_0 \exp(-\varepsilon_a/k_B T) \tag{6.12}$$

where ε_a is the activation energy and P_0 a pre-factor that is not temperature-dependent or is only weakly so.

These models have been established for simple, well-defined phospholipid membranes. It is clear that the composition of the thin liquid film separating the internal and external water compartments in double emulsions is substantially more complex, with two mixed surfactant monolayers separated by an oil membrane, possibly containing micelles. The experiments of Wen and Papadopoulos [32] described in Section 6.2.5 suggest that the transport of water or of any hydrophilic substance does not occur at the same rate depending on the position of the internal droplets: the rate of exchange is faster for a droplet adsorbed on the globule surface than for a nonadsorbed droplet. In double emulsions, a complete description of the permeability should take into account the contribution of different modes of transport (thin film permeation, inverse micellar transport), as well as the transfer between the internal droplets within the same oil globule. For the sake of simplicity, the permeation process is most often described using a mean-field description with a unique effective permeation parameter [25,38] as in the set of equations from (6.7) to (6.11).

6.3.3. *Influence of Temperature on Compositional Ripening*

Pays et al. have measured the rate of leakage of two distinct emulsions at different temperatures. One was stabilized by a couple of short surfactants (SMO/SDS) and the other one by a couple of amphiphilic polymers (Arlacel P135/Synperonic PE/F68) [38]. In both cases, the concentrations were chosen such that the leakage occurred by compositional ripening only. This was confirmed by microscopic observations that revealed that both the internal droplet size and the droplet concentration inside the globules did not vary in time, even when 100% of leakage was attained. The exact composition of the emulsions and the colloidal diameters are given in the captions of Figs. 6.18 and 6.19, which represent the percentage released as a function of time at different temperatures. The profiles can be fitted reasonably by a mono-exponential function. By combining the initial slope of the experimental curves p_0 and Eq. (6.11), the permeation coefficient P is deduced at a given temperature. In Fig. 6.20 the evolution of log (P) as a function of the inverse thermal energy is plotted. Within experimental uncertainty, the variation is linear for both types of stabilizing agents, in agreement with the Arrhenius law. From the semilog plot, the following set of kinetic parameters is obtained:

Stabilizing system	SMO/SDS	Arlacel P135/Synperonic PE/F68
ε_a	20 k_BTr	20 k_BTr
P_0	2.8 10^{-3} m/s	1.8 10^{-5} m/s

Of course, the values cannot by themselves provide any insight on the mechanism of transport since P is a mean field parameter that only describes the global transfer.

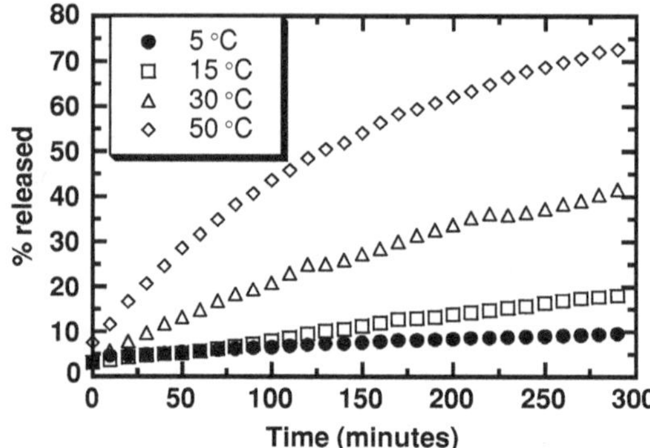

FIGURE 6.19. Kinetics of leakage of an emulsion stabilized by SMO and SDS as a function of temperature; $[NaCl]^0 = 0.4$ mol/l, $\phi_i^0 = 20\%$, $\phi_g = 12\%$, $d_i = 0.36$ μm, $d_g = 3.6$ μm, $C_l = 2$ wt%, $C_h = 0.3$ CMC, oil phase = dodecane. (Adapted from [38].)

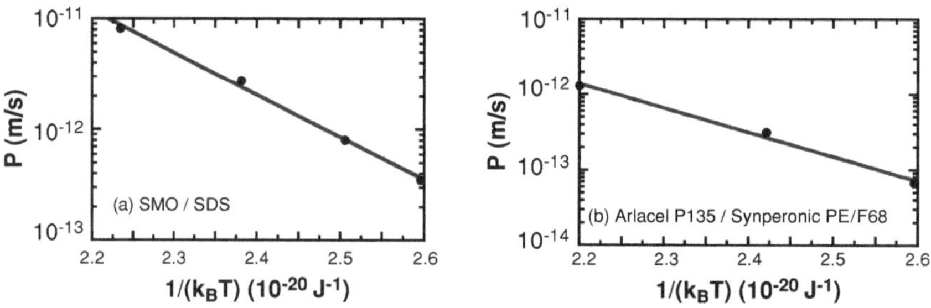

FIGURE 6.20. Arrhenius plots deduced from Figs. 6.18 and 6.19. (Adapted from [38].)

However, it appears that the pre-exponential factor P_0 of the polymer-stabilized double emulsion is two decades smaller, resulting in a slower rate of leakage (see Figs. 6.18 and 6.19). Therefore, polymers are more suitable surface-active species to ensure long-term encapsulation in double emulsions than are short surfactant systems.

6.4. Solid-Stabilized Double Emulsions

Some attempts have been made to improve the shelf-life of multiple emulsions by incorporating small colloidal particles. Binks et al. [40] have successfully prepared multiple emulsions using solely particles as emulsifiers. W/O/W emulsions with either toluene or triglyceride oil were prepared in a two-stage process. The first step involved the formation of a W/O emulsion by homogenizing water into a dispersion of hydrophobic silica particles in oil with high shear. In the second step, the W/O emulsion just made was re-emulsified into an aqueous dispersion of hydrophilic silica particles using low shear. The effects of particle concentration, oil type, and oil/water ratio have been investigated. Emulsions stable toward both coalescence and sedimentation can be prepared at high enough concentrations of

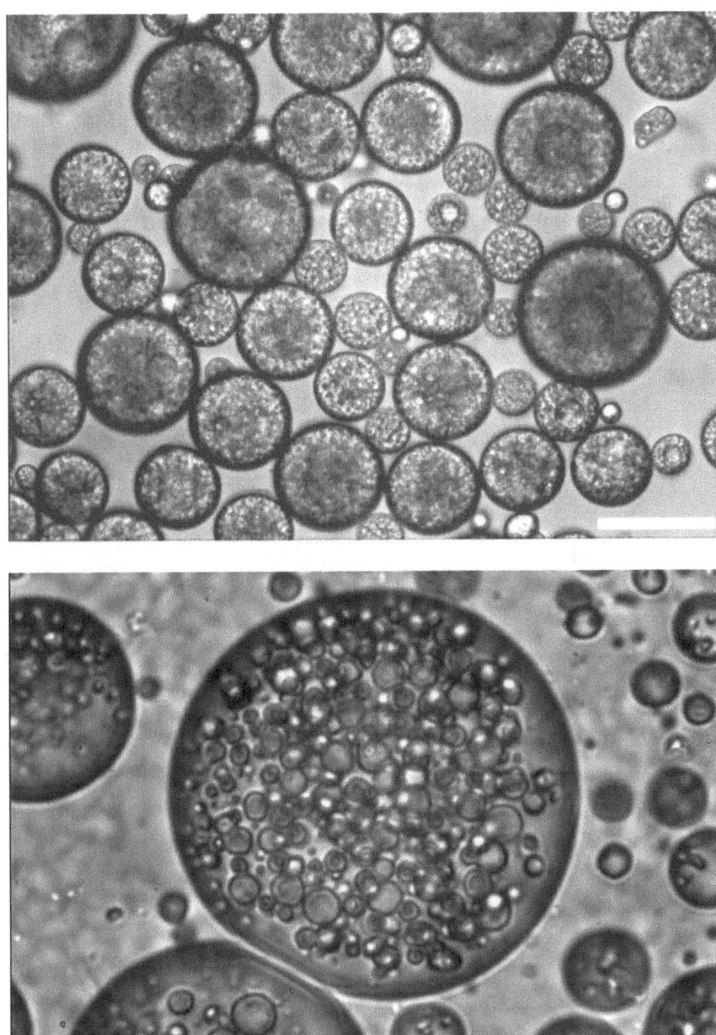

FIGURE 6.21. Microscopic images of double emulsions stabilized by two types of silica particles of different hydrophobicity. **(Top)** W/O/W with triglyceride oil (scale bar = 50 μm). **(Bottom)** O/W/O with toluene (scale bar = 20 μm). (Reproduced with permission from [40] and [41].)

the two particle types. An example of such an emulsion is shown in Fig. 6.21 in which oil globules are approximately 40 μm in diameter and water droplets are 2 μm. Both droplets and globules are stable to coalescence for more than 1 year in all emulsions. In solid-stabilized emulsions, the solid particles are irreversibly anchored at the oil–water interface and develop strong lateral interactions

[41], forming a kind of rigid membrane that mechanically protects the droplets against coalescence. Binks et al. have also studied the kinetics of release of electrolyte from inner drops to the outer aqueous phase under isotonic conditions in W/O/W emulsions of toluene using conductivity [40]. A model in which the diffusion of salt through the oil membrane including its partitioning between water and oil has been proposed [40]. The calculated rate constant for release, however, is much larger than the experimental one, suggesting that either an energy barrier to the interfacial transport of salt is present or that counter transport of glucose (required to maintain isotonicity) from outer to inner aqueous phases is significant.

In a similar manner, O/W/O emulsions were prepared by first emulsifying oil into an aqueous dispersion of hydrophilic silica particles, and then gently re-emulsifying the O/W emulsion so formed into an oil dispersion of hydrophobic silica particles. Figure 6.21 shows a typical microscopic image of a double emulsion with toluene as oil.

6.5. Conclusion

In this chapter, we have examined separately the two mechanisms that are responsible for the leakage of encapsulated substances in double emulsions: coalescence and compositional ripening. This study was possible after finding experimental conditions in which the two mechanisms occur over timescales that are sufficiently distinct to be decoupled. So far, the mechanism of compositional ripening is still controversial and requires further investigations. From a practical aspect, technological applications were first envisaged in the presence of short surfactants. The intrinsic instability (fast coalescence and compositional ripening) of such materials discouraged technologists and the promising formulation of double emulsions was almost abandoned. Studies using amphiphilic polymers, proteins, natural polyelectrolytes, and solid colloidal particles [36–38,40,42] reveal that coalescence can be inhibited and that compositional ripening may be extremely slow. In addition, several methods are now available for the preparation of size-controlled (monodisperse) double emulsions. Among the most efficient technologies are the use of specific porous membranes [43] or highly viscous media [44]. All these advances are quite encouraging for the prospects of fabricating materials with reproducible properties (size-controlled emulsions) and performing long-term encapsulation of active species.

References

[1] A.T. Florence and D. Whitehill: "The Formulation and Stability of Multiple Emulsions." Int. J. Pharm. **11,** 277 (1982).

[2] A.T. Florence, T.K. Law, and T.L. Wateley: "Nonaqueous Foam Structures from Osmotically Swollen W/O/W Emulsion Droplets." J. Colloid Interface Sci. **107,** 584 (1982).

[3] N. Garti: "Double Emulsions: Progress and Applications." Curr. Opin. Colloid Interface Sci. **3,** 657 (1998).

[4] C. Chiang, G.C. Fuller, J.W. Frankenfeld, and C.T. Rhodes: "Potential of Liquid Membranes for Drug Overdose Treatment: In Vitro Studies." J. Pharm. Sci. **67,** 63 (1978).

[5] S. Matsumoto and W.W. Kang: "Formation and Applications of Multiple Emulsions." J. Dispersion Sci. Technol. **10,** 455 (1989).

[6] N. Garti, M. Frenkel, and R. Schwartz: "Multiple Emulsions II. Proposed Technique to Overcome Unpleasant Taste of Drugs." J. Dispersion Sci. Technol. **4,** 237 (1983).

[7] P. Becher, S.S. Davis, J.D. Hadgrapt, K.J. Palin: "Medical and Pharmaceutical Applications of Emulsions." *Encyclopedia of Emulsion Technology*, vol. 2, Marcel Dekker New York (1985).

[8] C. Goubault, K. Pays, D. Olea, J. Bibette, V. Schmitt, and F. Leal-Calderon: "Shear Rupturing of Complex Fluids: Application to the Preparation of Quasi-Monodisperse W/O/W Double Emulsions." Langmuir **17,** 5184 (2001).

[9] K. Pays, F. Leal Calderon, and J. Bibette: "Method to Produce a Monodisperse Double Emulsion." French Patent 00 05880 (2000).

[10] K. Pays, J. Giermanska-Kahn, P. Pouligny, J. Bibette, and F. Leal-Calderon: "Double Emulsions: A Tool for Probing Thin Film Metastability." Phys. Rev. Lett. **87,** 178304 (2001).

[11] K. Pays, J. Kahn, B. Pouligny, J. Bibette, and F. Leal-Calderon: "Coalescence in Surfactant-Stabilized Double Emulsions." Langmuir **17,** 7758 (2001).

[12] K. Pays: "Double Emulsions: Coalescence and Compositional Ripening." Ph.D thesis, Bordeaux I University (2000).

[13] M.P. Aronson and M.F. Petko: "Highly Concentrated Water-in-Oil Emulsions: Influence of Electrolyte on Their Properties and Stability." J. Colloid Interface Sci. **159,** 134 (1993).

[14] A. Ashkin: "Acceleration and Trapping of Particles by Radiation Pressure." Phys. Rev. Lett. **24,** 156 (1970).

[15] M.I. Angelova and B. Pouligny: "Trapping and Levitation of a Dielectric Sphere with off-Centered Gaussian Beams: I Experimental." Pure Appl. Optics **2,** 261 (1993).

[16] M.F. Ficheux, F. Leal-Calderon, L. Bonakdar, and J. Bibette: "Some Stability Criteria for Double Emulsions." Langmuir **14,** 2702 (1998).

[17] C.H. Villa, L.B. Lawson, Y. Li, and K.D. Papadopoulos: "Internal Coalescence as a Mechanism of Instability in Water-in-Oil-in-Water Double-Emulsion Globules." Langmuir **19,** 244 (2003).

[18] H. González-Ochoa, L. Ibarra-Bracamontes, and J.L. Arauz-Lara: "Two-Stage Coalescence in Double Emulsions." Langmuir **19,** 7837 (2003).

[19] A.S. Kabalnov and H. Wennerström: "Macroemulsion Stability : The Oriented Wedge Theory Revisited." Langmuir **12,** 276 (1996).

[20] J.N. Israelachvili: *Intermolecular and Surface Forces.* Academic Press, New York (1992).

[21] R.H. Fowler and A. Guggenheim: *Statistical Thermodynamics.* Cambridge University Press, Cambridge, UK (1939).

[22] S. Matsumoto and M. Kohda: "The Viscosity of W/O/W Emulsions: An Attempt to Estimate the Water Permeation Coefficient of the Oil Layer from the Viscosity Changes in Diluted Systems on Aging under Osmotic Pressure Gradients." J. Colloid Interface Sci. **73,** 13 (1980).

[23] S. Matsumoto, T. Inoue, Masanori Kohda, and K. Ikura: "Water Permeability of Oil Layers in W/O/W Emulsions under Osmotic Pressure Gradients." J. Colloid Interface Sci. **77,** 555 (1980).

[24] S. Magdassi and N. Garti: "Release of Electrolytes in Multiple Emulsions: Coalescence and Breakdown or Diffusion Through Oil Phase?" Collois Surf. **12,** 367 (1984).

[25] S. Magdassi and N. Garti: "A Kinetic Model for Release of Electrolytes from W/O/W Multiple Emulsions." J. Controlled Release **3,** 273 (1986).

[26] J.A. Omotosho, T.L. Whateley, and A.T. Florence: "Methotrexate Transport from the Internal Phase of Multiple W/O/W Emulsions." J. Microencapsul. **6,** 183 (1989).

[27] N. Garti, S. Magdassi, and D. Whitehill: "Transfer Phenomena Across the Oil Phase in Water-Oil-Water Multiple Emulsions Evaluated by Coulter Counter : 1. Effect of Emulsifier 1 on Water Permeability." J. Colloid Interface Sci. **104,** 587 (1985).

[28] N. Garti, A. Romano-Pariente, and A. Aserin: "The Effect of Additives on Release from W/O/W Emulsions." Colloid Surf. **24,** 83 (1987).

[29] N. Jager-Lezer, I. Terrisse, F. Bruneau, S. Tokgoz, L. Ferreira, D. Clausse, M. Seiller,d and J.L. Grossiord: "Influence of Lipophilic Surfactant on the Release Kinetics of Water-Soluble Molecules Entrapped in a W/O/W Multiple Emulsion." J. Controlled Release **45,** 1 (1997).

[30] Y. Kita, S. Matsumoto, and D. Yonezawa: "Permeation of Water Through the Oil Layer in W/O/W-Type Multiple-Phase Emulsions." Nippon Kagaku Kaishi **1,** 11 (1978).

[31] P. Colinart, S. Delepine, G. Trouve, and H. Renon: "Water Transfer in Emulsified Liquid Membrane Processes." J. Membr. Sci. **20,** 167 (1984).

[32] L. Wen and K.D. Papadopoulos: "Effects of Surfactants on Water Transport in W1/O/W2 Emulsions." Langmuir **16,** 7612 (2000).

[33] Y. Sela, Y. Magdassi, and N. Garti: "Polymeric Surfactants Based on Polysiloxanes-Graft-Poly(Oxyethylene) for Stabilization of Multiple Emulsions." Colloids Surfaces **83,** 143 (1993).

[34] Y. Sela, Y. Magdassi, and N. Garti: "Release of Markers from the Inner Water Phase of W/O/W Emulsions Stabilized by Silicone Based Polymeric Surfactants." J. Controlled Release **33,** 1 (1995).

[35] Y. Sela, S. Magdassi, and N. Garti: "Newly Designed Polysiloxane-Graft-Poly (Oxyethylene) Copolymeric Surfactants: Preparation, Surface Activity and Emulsification Properties." Colloid Polym. Sci. **272,** 684 (1994).

[36] F. Michaut, P. Perrin, and P. Hébraud: "Interface Composition of Multiple Emulsions: Rheology as a Probe." Langmuir **20,** 8576 (2004).

[37] F. Michaut, P. Hébraud, and P. Perrin: "Amphiphilic Polyelectrolyte for Stabilization of Multiple Emulsions." Polym. Int. **52,** 594 (2003).

[38] K. Pays, J. Giermanska-Kahn, P. Pouligny, J. Bibette, and F. Leal-Calderon: "Double Emulsions: How Does Release Occur?" J. Controlled Release **79,** 193 (2002).

[39] R.T. Hamilton and E.W. Kaler: "Alkali Metal Ion Transport Through Thin Bilayers." J. Phys. Chem. **94,** 2560 (1990).

[40] B.P. Binks, A.K.F. Dyab, and P.D.I. Fletcher: "Multiple Emulsions Stabilised Solely by Nanoparticles." In Proceedings of the 3rd World Congress on Emulsions, Lyon, France (2002).

[41] R. Aveyard, B.P. Binks, and J.H. Clint: "Emulsions Stabilized Solely by Colloidal Particles." Adv. Colloid Interface Sci. **100–102,** 503 (2003).

[42] N. Garti and R. Lutz: "Recent Progress in Double Emulsions." In: D.N. Petsev (ed), *Emulsions: Structure, Stability and Interactions.* Elsevier, Amsterdam, Boston, Heidelberg (2004).

[43] S.M. Joscelyne and G. Trägardh: "Membrane Emulsification—a Literature Review." J. Membr. Sci. **169,** (2000).

[44] J.M.J. Bibette, F. Leal-Calderon, and P. Gorria: "Polydisperse Double Emulsion, Corresponding Monodisperse Double Emulsion and Method for Preparing the Monodisperse Emulsion." Patent WO 2001021297 (2001).

7
New Challenges for Emulsions: Biosensors, Nano-reactors, and Templates

7.1. Introduction

Emulsions have a widespread range of applications; the most important ones so far include cosmetics, food, detergents, adhesives, coatings, paints, surface treatments, road surfacing, and pharmaceuticals. Scientific information about emulsions can certainly direct their use in potentially new fields that are linked to new markets or can simply provide new tools for basic research. Nano- or microtechnologies are rapidly expanding today and in many cases they markedly overlap with colloidal science. Indeed, colloids have typical length scales that are intermediate between molecules and macroscopic objects; they can thus act either as minute substrates or as reservoirs. They can also be manipulated by applying external forces to selectively sort desirable products, or to create local stress-controlled conditions. Liquid droplets can compartmentalize minute amounts of defined reactants, either to screen a large compound library or parallelize a directed process imposed by confinement. We present in this chapter some particularly promising examples of new applications of emulsions in nano- or microtechnologies, related to biotechnologies, biophysics, and processing of high-tech materials for micro-optics.

This chapter contains three sections. The first introduces the possibility of employing emulsions as intermedia for making new biosensors. Two examples are presented: homogeneous bioassays and single immunocomplex micromechanical measurements. The second section introduces the possibility of using emulsions as nano-reactors for screening large libraries, such as, for instance, in the case of in vitro directed enzyme evolution. The third section introduces the possibility of using emulsion droplets as microtemplates to prepare controlled-shape colloidal clusters.

7.2. Emulsions as Biosensors

A biosensor is used to determine various parameters associated with biological species such as single biomolecules, molecular complexes, viruses, cells, and tissues. Understanding mechanisms, from the molecular to the cellular level, is certainly one of the most important issues in molecular biology. It requires information simultaneously dealing with concentrations, structures, conformations, stresses, and dynamics, both at different length and time scales. The possibility of measuring such parameters requires the development of sensitive, selective, real-time, and multiscale biosensors. A crucial aspect of biomolecules concerns their ability to build reversible complexes through specific molecular recognition mechanisms. Indeed, a major property of biomolecules is their ability to bind a variety of ligands to fulfill a specific function such as mediating cell adhesion, triggering receptor-mediated cell activation, or regulating intracellular networks and immuno-responses. This section describes the recent contribution of emulsion-based magnetic colloids to assay specific recognition phenomena and to probe the mechanical properties of the resulting complexes.

7.2.1. Emulsions for Homogeneous Assays

One important biotechnological application deriving from these recognition phenomena are immunoassays, which are essentially based on the formation of specific complexes and involve the detection and quantification of a large number of proteins present in serum or plasma, from hormones to viral infectious agents. The strategy employed for the detection of antigens is based on the formation of an immunocomplex, obtained from the specific molecular recognition of the antigen by two distinct antibodies. The first antibody is aimed at selectively and specifically identifying the antigen while the second one is used to quantify its concentration. The most convenient way of performing each of these steps consists of using colloids on which antibodies are grafted. Indeed, the colloidal particles do not substantially modify the antibody affinity for the antigens, while at the same time the colloidal particles add one or several essential properties for purification, sorting, and detection of antigens. For example, magnetic colloids have revolutionized methods for sorting and washing biomolecules to be detected. When the assay requires one step only, which means that no washing is necessary, the assay is termed homogeneous, in contrast to heterogeneous. Researchers working with emulsions are bringing new perspectives into this field that are discussed in the text that follows.

Magnetic colloidal particles are broadly employed in various biotechnological applications. They are used as carriers for sorting and probing proteins, nucleic acids, and cells. These materials can indeed be designed such that their surfaces can specifically recognize targets within many component biological mixtures. By grafting antibodies onto magnetic colloids that can specifically recognize antigens or cell receptors, it is possible to separate these targets from a mixture and quantify them. In terms of process, the synthesis of magnetic or fluorescent colloidal particles requires additional steps to incorporate these new features. Very different approaches exist today to prepare these commercially available materials. Indeed, many types of particles can be purchased with significantly different characteristics in terms of size, monodispersity, magnetic content and susceptibility, and surface chemistry. Among others, one route consists of magnetizing porous monodisperse latex particles from freely diffusing small magnetic nanoparticles into the particle pores. These materials are widely used for cell, nucleic acid, and protein sorting. In terms of processing, emulsions may be valuable intermediates because many different compounds can a priori be either solubilized or dispersed. This can lead to novel colloids having various new features [1]. However, it implies that both emulsification and surface chemistry are sufficiently elaborated so that these initially liquid droplets can be ultimately transformed into particles with controlled properties.

Emulsifying an oil-based ferrofluid, whose oil can be further evaporated, allows preparation of particles with an extremely high magnetic oxide content, with a size typically within the submicron range [1,2]. Figure 7.1 shows a transmission electron microscope (TEM) image of a single magnetic droplet. It is obtained from a shear-controlled emulsification (see Chapter 1, Section 1.7.6.4) of an octane-based ferrofluid, followed by slow evaporation under a nitrogen atmosphere. This evaporation step aims to eliminate most of the remaining oil (octane) originating from the oil-based ferrofluid, which ensures a maximum content of iron oxide nanoparticles within the droplet core. The TEM image reveals the texture of these particles: a core made out of closely packed small nanoparticles of iron oxide surrounded by a polymer shell. These monodisperse magnetic cores can indeed be encapsulated in a polymer shell that allows them to be grafted with various probes of interest for biotech applications such as immunodiagnosis and RNA or DNA sorting. Owing to this high magnetic content associated with a droplet diameter that remains in the submicron range, these colloids are both highly responsive to an external field and yet very Brownian. The two characteristics ensure a very rapid chaining of particles in the presence of an applied homogeneous field, within independent one-dimensional lines; this phenomenon was previously discussed in Chapter 2 in the context of direct measurement of colloidal forces [3]. So far, magnetic field gradient-induced forces were the only ones to be exploited in bioanalysis. Indeed, a field gradient exerts, on a polarized particle having a magnetic induced moment $\vec{m}$, a driving force proportional to $\nabla\vec{B}\,.\,\vec{m}$. This force can be externally actuated to sort the magnetic particles after they have captured the expected targets within a complex mixture. The technique is indeed currently employed for separating biomolecules or cells and quantifying them [4,5].

Baudry et al. [6] have reported that colloidal dipole–dipole interactions can have a direct impact on the rate of molecular recognition. In this case, the magnetic interaction that is considered is the colloid–colloid attraction that arises from the polarization of each particle. The authors reported the possibility of tuning the recognition rate between grafted ligands and receptors with magnetic dipolar forces. This may be of interest since when ligands and receptors are both attached on surfaces, their recognition kinetics may be substantially reduced. In nature, this effect has influenced the evolution of adhesive ligand–receptor couples depending on the conditions at which they are supposed to operate. For instance, the capture and adhesion of circulating cells such as leucocytes at the surface of blood vessels during the inflammatory process must be very rapid because the blood stream reduces the contact duration between ligands and receptors [7]. This

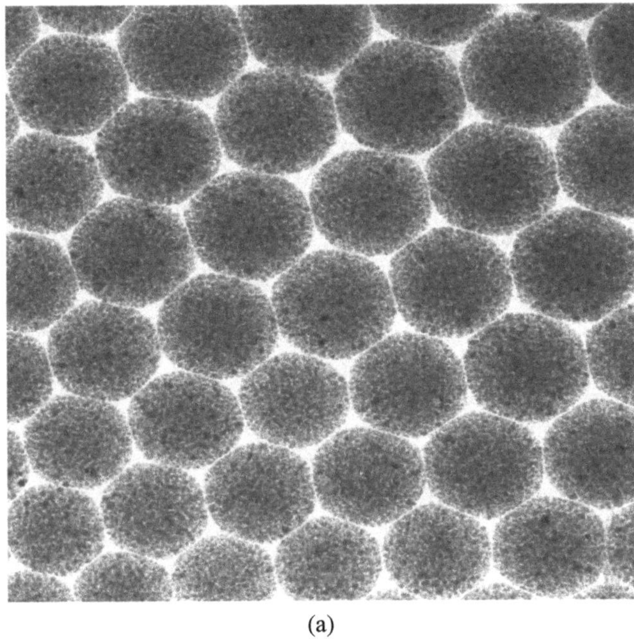

(a)

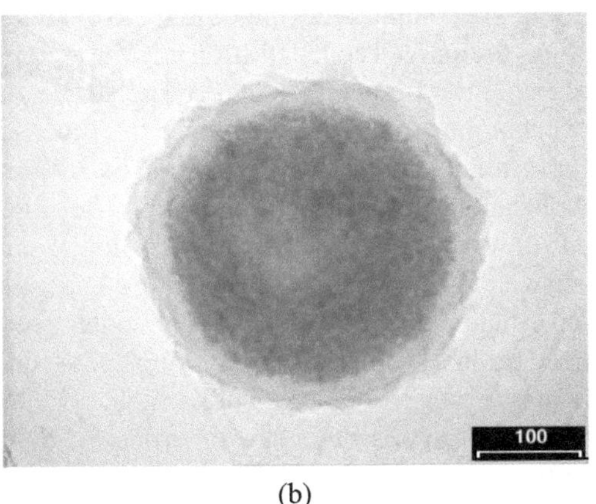

(b)

FIGURE 7.1. **(a)** Transmission electron microscopy image of a collection of 200-nm magnetic emulsion droplets obtained from emulsifying an octane-based ferrofluid. **(b)** One droplet is shown after polymerization. A polymer shell is visible that encapsulates the iron oxide nanoparticles. (With permission of Ademtech).

is achieved thanks to integrins that link to selection through an extremely rapid process though both ligands and receptors remain anchored on surfaces. In vitro, the same effects are therefore expected for colloids grafted with biomolecules that are used as probes for cell sorting and diagnosis. Baudry et al. [6] proposed using Brownian magnetic colloids that self-assemble into linear chains to improve considerably the rate of recognition between grafted receptors and their ligands, essentially via the one-dimensional restriction that augments the colliding frequency and in turn increases the recognition rate.

Diagnosis techniques are generally based on building a specific immunocomplex structure in which the antigen to be detected is recognized by two antibodies [4,8,9]. Early immunochemistry was based on precipitation of large complexes composed of antibodies and antigens. Following the same track, the use of Brownian particles, which quickly find their target, had significantly improved the detection sensitivity, because of the increase in scattered light when aggregation between grafted colloids takes place. As the most simple but very generic example, let us consider an antigen (Ag) having two epitopes for two antibodies A and B. To reveal the presence of such an antigen, particles grafted with A and B antibodies and the sample are mixed. The formation of small clusters is then expected, the rate of which depends on many factors. Change in light-scattering due to the presence of these small clusters will reveal the existence of sandwich-like structures: A–Ag–B (latex agglutination immunoassay). These homogeneous assays, as opposed to heterogeneous assays in which washing steps are necessary before detection, are today by far the most simple and straightforward: they were introduced in 1956 [8,10] and today several hundreds of different tests based on this principle can be found on the market, mainly for detection of infectious diseases and protein quantification [11].

The rate for building the complex sets the time scale of an assay. Meanwhile, in many diagnostic assays the antigen concentration is in the picomolar range. At these concentrations, the encounter frequency between species becomes a critical issue to consider, as low encounter frequencies increase the time required to complete the assay. More rapid homogeneous assays would be of great benefit both for increasing throughput and in medical emergencies where nearly instantaneous tests are required. It is worth noting here that microfluidic devices can certainly be one potential answer to this problem; indeed the controlled flow of minute volumes provides an efficient means to augment encounter events between antigen and their immobilized probes [12]. However, it brings new constraints for reliably manipulating and sampling such small volumes.

Because the emulsion-based particles are superparamagnetic with a high susceptibility, the resulting magnetic colloidal forces induce a fast chaining process [13]: the time scale for bringing two colloidal particles into contact in the presence of a magnetic field H, and at an initial volume fraction ϕ, is given by

$$\tau = \frac{6\eta}{\phi\,\mu_0 \chi^2 H^2} \tag{7.1}$$

where η is the viscosity of the surrounding fluid, μ_0 is the vacuum magnetic permeability, ϕ is the particle volume fraction, and χ is the magnetic susceptibility of the particles [3,14]. For $B = \mu_0 H = 20$ mT, $\chi = 0.1$, $\phi = 0.03\%$, and $\eta = 0.001$ Pa·s, the time τ to nucleate chains is less than 1 s. These chains persist as long as the field is maintained, which allows for a rapid formation of ligand–receptor–ligand links between pairs of particles within the chain. To quantify the influence of

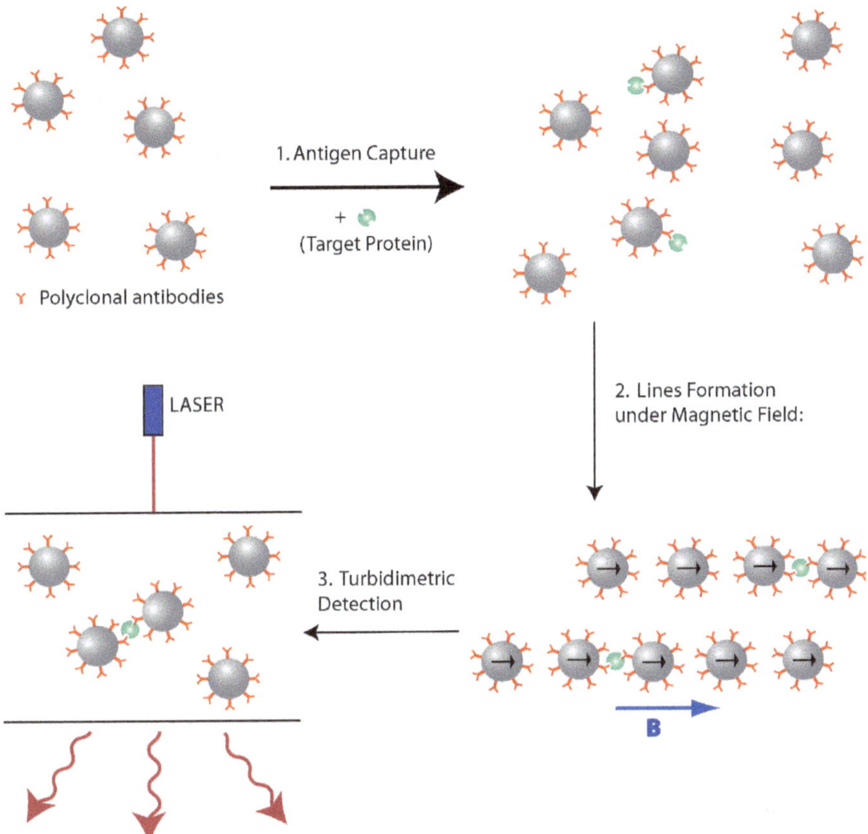

FIGURE 7.2. 1, Magnetic particles grafted with polyclonal antibodies (or two different monoclonal antibodies) are mixed with the sample which contains the antigens. Antigens are first captured by grafted beads. 2, Application of a magnetic field induces a magnetic dipole in each bead, allowing an almost instantaneous formation of chains. Essentially all particles that have captured an antigen will form links with their neighbors. After typically a few minutes only, the magnetic field is switched off, and as a result of Brownian motion only doublets that are linked remain: 3, Doublets are detected via a turbidimetric measurement.

this one-dimensional restriction on the recognition rate, the authors have measured the optical density change due to the amount of colloidal doublets that remain after the field is switched off, as illustrated in Fig. 7.2. In this experiment, ovalbumin is used as a model ligand and colloidal magnetic particles of 200 nm in diameter obtained from controlled emulsification are grafted with polyclonal IgG rabbit anti-ovalbumin antibodies as a model for receptors. In this experiment there are approximately 30 antibodies per particle and the colloid volume fraction ϕ is 0.03%, which corresponds to a particle concentration C_p of 120 pmol/l. A final concentration C_{ova} of ovalbumin is adjusted and each sample is first incubated for 1 min at 25°C and then a homogeneous field of 20 mT is applied for 5 min.

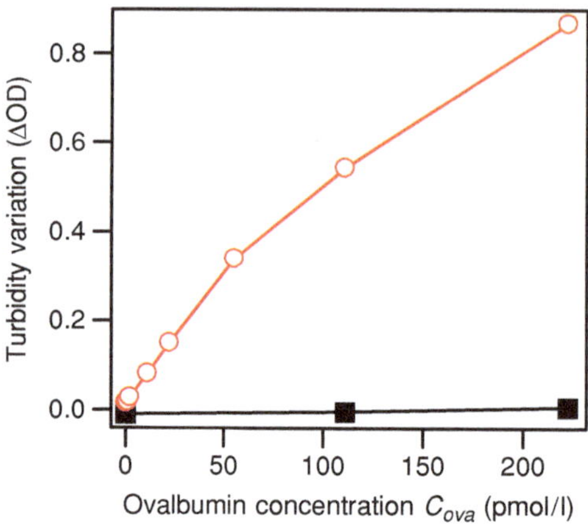

FIGURE 7.3. Optical density difference measured with an UV/VIS spectrophotometer, before and after the application of the magnetic field B, versus ovalbumin concentration. Open circle: $B = 20$ mT, square: $B = 0$ mT. The sample is first incubated for 1 min at 25°C and then a magnetic field is applied for 5 min.

The same experiment is also performed with a zero field. Optical density, defined as $OD = -\log I/I_0$, where I and I_0 are respectively the transmitted and incident intensity, is measured before and after the field is applied. The difference is plotted in Fig. 7.3 as a function of C_{ova}. In the absence of field, the signal is low, reflecting the very limited number of persisting doublets and therefore the inefficiency of free Brownian collisions on that time scale. By contrast, in the presence of a 20 mT field a turbidity difference is measured down to picomolar concentrations. The control of the force applied to each particle dramatically increases the rate of doublet formation. As a matter of fact, these results provide the basis of a more rapid and sensitive homogeneous assay than the classical latex agglutination immunoassay previously mentioned. Indeed, these findings are directly applicable to immunodiagnosis: since the optical density measurement is directly related to the Ag concentration, a fast titration down to picomolar antigen concentration can be routinely performed.

7.2.2. Emulsions for Single Immunocomplex Micromechanics

Forces and stresses that molecules exert on each other have become directly measurable via various single molecule techniques. These techniques include optical and magnetic tweezers, the atomic force microscopy, and the biomembrane force probe [15,16]. The mechanical properties of biomolecules and molecular bonds have been investigated, resulting in a better comprehension of the functional aspects of these molecules and in the measurement of binding-potential landscapes [17,18]. It has become clear that piconewton forces control many key processes in biology, involving receptor–ligand pairs, protein and nucleic acid structures, or identical self-assembled proteins. For instance, single-molecule measurements have provided significant understanding of biopolymer mechanics longer than their persistence length [19–21]. The entropic origin of the elasticity of long DNA molecules has been demonstrated in the low force regime as well as the enthalpic contribution governing structural transitions at very large extension forces.

By contrast, for small complexes with typical size of a few nanometers, or for biopolymers smaller than their persistence length, there are only few data reporting on their mechanical properties. Indeed, most of the single-molecule techniques do not apply, except the atomic force microscope (AFM), which to some extent has provided the first data at that scale [22–25]. Nevertheless, such measurements are expected to offer a better comprehension of the functional aspects of these complexes, as well as a better understanding of their binding modes under loading forces. Following this goal, Koenig et al. [26] have introduced a method to deduce the force-extension law of nanoscale biocomplexes, with a force resolution of 0.1 pN and a distance resolution of 1–2 nm. The method introduced by Koenig and co-workers uses emulsion-based magnetic colloids and exploits the principle of the magnetic chaining technique (MCT) [3], described in Chapter 2. More precisely, this method makes use of long-range forces between colloids, both attractive and repulsive, which are exploited as a force probing tool. As detailed in Chapter 2, Section 2.2.1.3, these emulsion-based magnetic colloids are indeed good candidates for this purpose because they spontaneously self-organize into linear chains when a magnetic field is applied. Within a chain, the monodisperse particles are regularly spaced. This separation results from the equilibrium between the attractive magnetic forces and the repulsive ones. The determination of the colloidal separation is obtained by measuring the Bragg diffraction pattern of that chain. From this distance and the magnetic field strength, the repulsive force–distance profile can be constructed.

To measure the force–extension law of a small biomolecule, these authors employed a two-step strategy. First, the background repulsive force–distance profile, in the absence of biomolecules, $F_{bg}(h)$, is measured, h being the interparticle spacing. Then, once the biocomplexes have been properly attached within each interval between colloids, the same measurement is repeated, allowing determination of the force–distance profile of this irreversible assembly $F_{ir}(h)$. The force $F_{ir}(h)$ includes two contributions, one being the repulsive background and the other originating from the presence of the linker. In the limit of low grafting densities, these two forces must be additive. Then, by subtracting $F_{bg}(h)$ from $F_{ir}(h)$, they computed the force–extension profile of the linkers $F_l(h)$. Finally, from the average number of links per particle they can deduce the force–extension law of a single complex, assuming biomolecules behave like springs in parallel.

To validate this technique, Koenig et al. [26] considered size-controlled molecules: double-stranded DNA (dsDNA from 76 to 315 base pairs) [27]. A sample of average diameter 176 nm is used. This sample is covalently grafted with streptavidin; 185 available biotin binding sites per grafted bead are measured by inverse titration using fluorescent biotin [28]. As one of the most important requirements, this colloidal dispersion is stable and does not aggregate during a storage period of several months or under a magnetic field. Indeed, in the absence of specific linkers the molecules redisperse immediately once the field is switched off. This property is ensured by the presence of the repulsive colloidal forces, which in turn

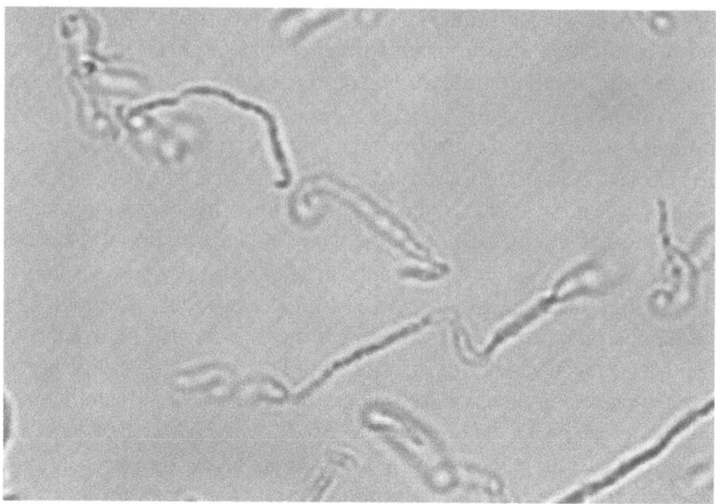

FIGURE 7.4. Optical microscopy image of permanently linked magnetic particles. Each streptavidin coated particle is linked to its neighbors by on average 3 dibiotin dsDNA of 151 base pairs. After the field has been switched off, the chains persist and spontaneously bend under thermal fluctuations. (From [27], with permission.)

guarantees the absence of any unspecific aggregation. Grafted colloidal particles and dsDNA are mixed in a pH 7.2 phosphate buffer at a final particle concentration $C_p = 1.3\ 10^{11}$ particles/ml and $C_{\text{DNA}} = 2.7\ 10^{11}$ to $7.3\ 10^{12}$ molecules/ml. The sample is then incubated at room temperature for 10 min to allow coupling of DNA onto the particles. A magnetic field of 80 mT is finally applied for 240 s. As a first demonstration of the controlled formation of permanent bounds between particles, two distinct experiments, one with dibiotin dsDNA, and one with mono-biotin ds DNA, are performed. In each case, the field is turned off after 240 s, and the chains are directly observed under the microscope (Fig. 7.4). As expected, only the dibiotin dsDNA which has biotin at each end is able to bridge particles.

The force–distance profile is systematically measured by applying five times the same protocol: the field is decreased from 80 to 10 mT, and then increased again from 10 to 80 mT. The 10 different profiles arising from these 10 repeated ramps all superimpose on a single set, which confirms that only a reversible compression/extension of the engaged links is probed and that no additional links are formed during these ramps. In this procedure, averages over all these force–distance profiles are systematically computed. In the very low regime of forces, when magnetic interactions become comparable to $k_B T$, the chains persist because particles are linked together, but their local orientation relative to the direction of observation becomes less pronounced. This leads to a reduction of the average measured particle separation, an experimental artefact that sets the lower limit of the force detection in this technique.

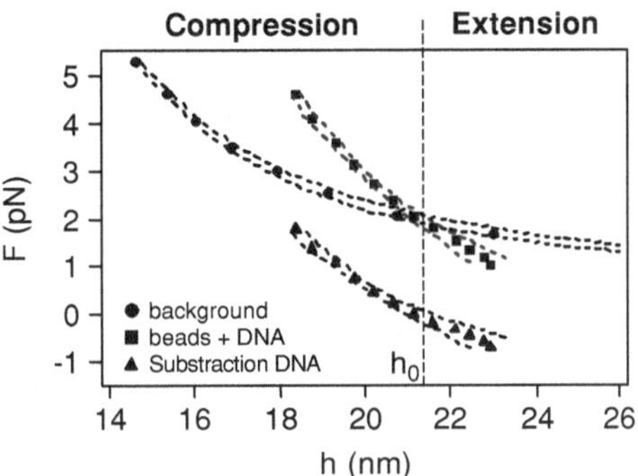

FIGURE 7.5. Comparison between average force–distance profiles in the absence and in the presence of 1.9 DNA on average between adjacent particles. Circles, F_{bg}; squares, F_{ir}; triangles, $F_l = F_{ir} - F_{bg}$. Dashed lines correspond to the error in force–distance profiles, calculated from the standard deviation of the measured profiles. (Adopted from [26].)

The background $F_{bg}(h)$ and the irreversible chain profile $F_{ir}(h)$, as well as the subtraction of one from the other, which represents the "linker contribution" $F_l(h)$, are shown in Fig. 7.5, in the case of a 151-base-pair dsDNA linker. Koenig et al. [27] also find perfect agreement of their data with a theoretical prediction initially derived for actin filaments, which concerns semiflexible biopolymers in the regime in which their contour length L_c is less than or equal to their persistence length L_p [26,29]. As a consequence of the presence of both magnetic attractive and colloidal repulsive forces, $F_{ir}(h)$ crosses $F_{bg}(h)$ at a particular separation h_0 for which these two forces are equal. In other words, at that particular separation these two forces are equal because the complex is not exerting any force; thus this separation gives the natural length of the linker at which it remains at rest. This crossover also demonstrates that this technique can explore both the regimes of extension and compression. Magnetic attractive forces are used to probe the compression regime while the repulsive colloidal ones are used to explore the extension regime. For each case, this technique leads to a value for L_p very close to 50 nm, the expected value which also corresponds to measurements on longer DNA molecules [19–21]. More importantly, a good agreement for L_0 which is predicted to be 86% of the contour length, that is, 44 nm for 151 base pairs, is also found [29]. From these results one can assume that any complex from typically a few nanometers to about 100 nm can be characterized both in terms of its length and rigidity. To capture the limits of this approach, Koenig and co-workers performed the same measurements with various DNA contour lengths, from 76 to 315 base pairs. The results are shown in Fig. 7.6. The same behavior is recognized. But the crossover may disappear for too short or too long chains because its position may not be accessible owing to the limited range and intensity of both the background force and the applied magnetic one. However, in most cases extrapolating F_{ir} would give a rather good estimate of h_0.

These two examples show how emulsion-based magnetic colloids can become the essential elements for two apparently very different issues. In fact, this section introduces the basics of using superparamagnetic emulsion-based colloids, and their spontaneous self-assembling ability under a field, as new biosensors. The

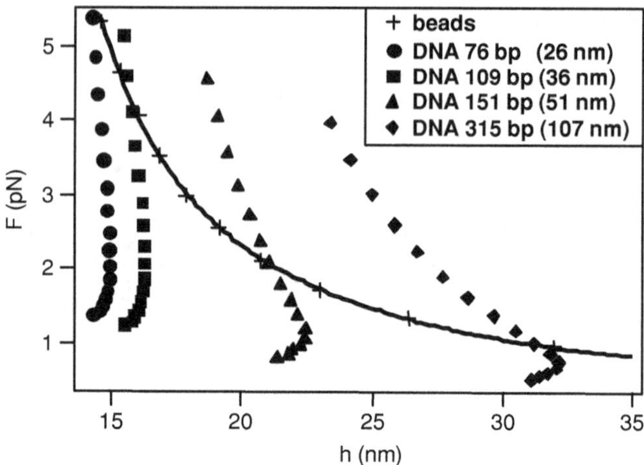

FIGURE 7.6. Comparison between average force–distance profiles in the absence and in the presence of 1.9 DNA molecules on average between adjacent particles, in 10^{-2} mol/l phosphate buffer, pH 7.2. Solid line, F_{bg}; black symbols, F_{ir}, for various dsDNA linker contour lengths from 76 to 315 base pairs. (Adapted from [27]).

possibility to vary magnetic interactions, and therefore the distance between particles within chain-like structures, addresses new possibilities to resolve and assay recognition at a molecular level. However, the particles must fulfill some very specific requirements in order to be used in this strategy: particles must be spherical, monodisperse, strongly magnetic, and Brownian. The last characteristic imposes a maximum diameter of about 500 nm. In addition, their surface must provide very strong colloidal stability in any type of buffer, serum, or plasma. Finally, the particles must also possess functional groups in order to be further grafted with biomolecules of interest. These unique characteristics are obtained from an emulsion-based process that possesses the required versatility for bringing together all of these properties [30,31].

7.3. Emulsions as Nano-reactors

The possibility of entrapping a single gene per water-in-oil droplet within an inverse emulsion, as well as concentrating within the same drop the products of various biochemical processes encoded by a single gene, is an attractive tool in molecular biology. This may be applied to gene libraries and therefore to biological questions. Similarly, single colloids can be entrapped within each minute reservoir made out of inverse emulsion droplets; therefore products from one drop can be concentrated on the surface of the entrapped colloid. After breaking of the emulsion, each droplet that bears multiple copies of the same transcript can be further sorted. In other words, this allows creating many clones encoded by the same single gene. The use of emulsions or double emulsions as micro-reservoirs provides a straightforward way of miniaturizing very large libraries. Indeed, an emulsion (inverse or double) can be viewed as a collection of minute aqueous reservoirs able to entrap separated single molecules originating from a large library. This section particularly focuses on the use of emulsion drops as single gene reservoirs and reactors as a tool to direct in vitro enzyme evolution.

7.3.1. Screening of Large Libraries and Directed Enzyme Evolution

Evolution, as postulated by Darwin in 1880, is a powerful process that is based on natural selection. Genes, which make up the blueprints for living things, are passed from one generation to the next. However, the genes can pick up mistakes, mutations, resulting in variations in the molecules they encode. Genes carrying mutations that encode "better" molecules, which increase the chances that the organism will survive and reproduce, are more likely to be transmitted to the next generation. Over many generations, this continual process of optimization as a result of repeated rounds of mutation and selection has resulted in the vast range of organisms we see today.

Life is sustained by a complex web of chemical reactions. Catalysts, molecules that accelerate the rate of a chemical reaction but that are unchanged by the overall reaction, are essential for life as most reactions would otherwise occur far too slowly. Indeed, it can be argued that the evolution of life is essentially the story of the evolution of catalysis. In nature, most catalysts are proteins and these catalytic proteins, or enzymes, are one of the most remarkable classes of molecules to have been generated during evolution. Enzymes catalyze an enormous range of different reactions and their performances typically far exceed those of man-made catalysts. They can accelerate reactions by anything up to 10^{17}-fold relative to the uncatalyzed reaction, enabling reactions that would otherwise have half-lives of tens of millions of years to be performed in milliseconds.

All evolutionary systems require a link between genotype (a nucleic acid that can be replicated) and phenotype (a functional trait such as binding or catalytic activity), whether in nature or in the laboratory. We recall that the genotype refers to the DNA sequence while the phenotype refers to its function in relation to environment; the same concept can be used at the level of a single gene encoding for a single protein, but in this case the phenotype is restricted to a binding or a particular catalytic activity. Evolution becomes possible if a physical link between genotype and phenotype exists in order for a particular phenotype to further transmit its genotype to the next generation. Completely in vitro selection systems are a highly advantageous means of investigating directed evolution, as they allow selection of larger repertoires and much greater control than in vivo systems. Direct selection for all enzymatic properties can be achieved by compartmentalization in cells (as in nature). Unfortunately, in vivo selections are virtually always restricted to functions that affect the viability of the organism and are often complicated by the complex intracellular environment and the need to transform the gene library. More commonly, 10^3–10^5 clone libraries are screened (rather than selected) in a plate assay using a fluorogenic or chromogenic substrate. However, crossing long evolutionary distances, and evolving completely novel proteins and activities, requires much larger libraries, for which selection rather than screening is preferable.

When subjected to selection pressure, genes evolve through a series of mutations that modify the function of the enzymes they encode. New enzymes can be produced in the laboratory via a process of "directed evolution" [32–47]. Following this strategy, selection pressure can be focused on a single enzyme rather than on a whole organism. In this context, Tawfik and Griffiths have developed a completely in vitro system for the selection of enzymes, based on linking genotype and phenotype, not by compartmentalization of genes in cells as in nature, but by in vitro compartmentalisation (IVC) in aqueous microdroplets in water-in-oil emulsions [39]. IVC can be used to select nucleic acids and proteins and can select for binding [41,42,46] as well as catalysis [39,43–46]. Each microdroplet is ~2 μm in diameter and only ~5 femtolitre volume (about the same size as a bacterial cell), which means that a 50-μl aqueous reaction mixture is dispersed into $\sim 10^{10}$ microdroplets. This is equivalent to more than 10^6 1536-well plates! In a typical IVC experiment, the mutated genes are compartmentalized into aqueous microdroplets dispersed in mineral oil, with the concentration of DNA set such that the majority of droplets contain no more than one gene per droplet. According to the Poisson distribution,

$$P(x) = e^{-m}\left(\frac{m^x}{x!}\right) \tag{7.2}$$

where $m = 0.1$ is the mean number of genes per droplet, and $P(x)$ is the probability of finding x genes per droplet. According to Eq. (7.2), 90.5% of droplets contain no gene, 9.05% contain one gene, 0.45% contain two genes, and 0.016% contain more than two genes. In addition, each droplet contains an in vitro transcription/translation system. Tawfick and Griffiths [39] and Bernath et al. [47] have pioneered various approaches to sort the active enzymes coupled to their genotypes. In all cases, the fluorescence activated cell sorter (FACS) technique is employed. One version consists of breaking the inverse emulsion; however, this requires some manipulation to keep the link between genotypes and phenotypes after the emulsion is broken. One example of this approach is detailed in Fig. 7.7. Another version that avoids any restrictions imposed by the requirement of physically linking the phenotype to the genotype consists of making a water continuous double emulsion of the initial inverse emulsion containing the gene library. This version of IVC indeed requires fluorogenic substrates to allow the selection of enzymes. This approach can directly flow sort fluorescent double microdroplets containing genes that encode active enzymes (at up to 10^5 droplets/s) once conversion of the original water-in-oil emulsion into a water-in-oil-in-water double emulsion [47] is not spoiled by leakage from inner aqueous droplets toward the

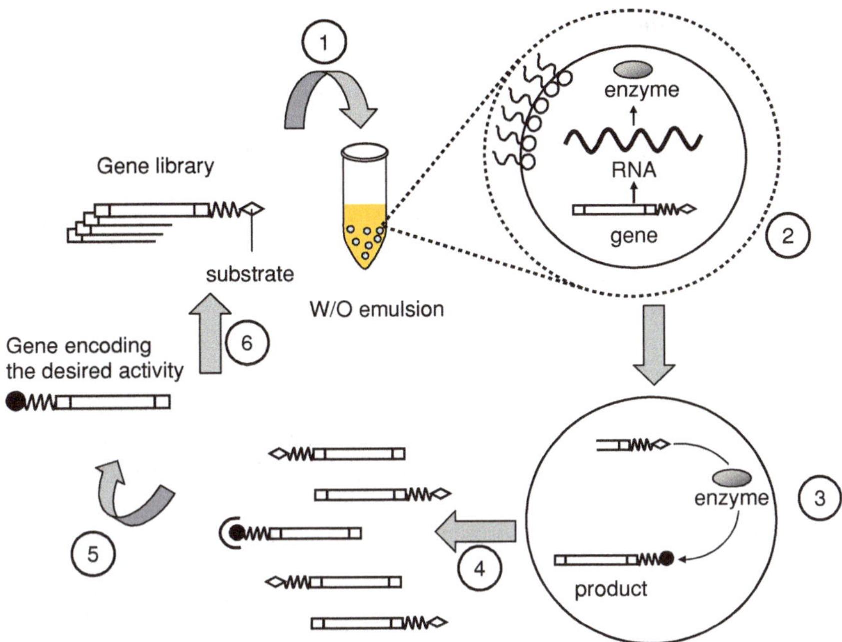

FIGURE 7.7. Schematic representation of gene selection by compartmentalization. Step 1: An in vitro transcription/translation reaction mixture containing a library of genes linked to a substrate for the reaction being selected is dispersed to form a water-in-oil emulsion with typically one gene per aqueous compartment. Step 2: The genes are transcripted and translated within their compartments. Step 3: Proteins (or RNAs) with enzymatic activities convert the substrate into a product that remains linked to the gene. Compartmentalization prevents the modification of genes in other compartments. Step 4: The emulsion is broken; all reactions are stopped and the aqueous compartments are combined. Genes linked to the product are selectively enriched, then amplified, and either characterized (step 5) or linked to the substrate and compartmentalized for further rounds of selection (step 6). (Adapted from [39].)

external phase. Many rounds of evolution are possible because the selected genes can then either be reselected directly, or first remutated randomly using nucleoside analogues [48] or recombined by DNA shuffling [49] before reselection. This allows selections to be performed using a variety of different mutational loads and under an asexual (random mutation alone) or sexual (with recombination) system.

So far the major application of IVC has been to select enzymes. Griffiths has used IVC to select DNA methyltransferases and has successfully generated variants that can efficiently modify novel DNA target sequences [38]. This study provides a rare example of a laboratory-produced enzyme with a catalytic efficiency that surpasses that of the wild-type enzyme with the principle substrate. IVC has also been used for the directed evolution of *Taq* DNA polymerase variants with greater thermostability and expanded substrate range [50,51].

Using microdroplets as micro-reservoirs for various biochemical or biomolecular reactions has many advantages. It allows dramatically reducing the sample volumes and enormously increasing the throughput rate. Therefore the size of libraries that can be screened can easily be increased by orders of magnitude. However, the use of bulk emulsions (simple or even double) restricts the range of IVC applications to screening assays that do not require any droplet feeding after emulsification has set the composition of each droplet. Combining microdroplets and microfluidic technologies opens a new and even more exciting possibility: this would allow processing each droplet reservoir with exquisite control over each drop separately. Each molecule or cell can then be processed independently, and the products of the processing will remain physically confined and associated exclusively with the starting material. This is the only method that enables individual molecules or cells not only to be detected, but also to be processed within completely controlled confinement and controlled feed of any other species at an adjusted chemical potential (drop recombination). If in the future microfluidic technologies turn to be a reliable option to produce, manipulate, merge, mix, split, read, and sort water-in-oil droplets at a throughput rate exceeding 1 kHz, it might become a breakthrough for all applications that require complex screening of large libraries, from genes to chemicals.

7.4. Emulsions as Microtemplates

This section deals with the use of emulsion droplets as microtemplates to direct the self-assembly of a small number of colloidal particles into well controlled aggregates. To some extent, the underlying idea is again to confine species in order to select and direct one particular assembly. In a second step, these colloidal aggregates might be arranged into a larger scale material that is predicted to possess unusual optical properties. Indeed, these nonspherical aggregates could serve as building blocks for more complex assemblies, such as colloidal crystals, which could find applications as photonic materials.

7.4.1. Colloidal Clusters and Micro-optics

Colloidal crystals can be viewed as the mesoscopic counterpart of atomic or molecular crystals. They have been used to explore diverse phenomena such as crystal growth [52–54] and glass transition [55,56], and have many interesting applications for sensors [57], in catalysis [58,59], advanced coatings [60], and for optical/electro-optical devices for information processing and storage [61,62]. In particular, their unusual optical properties, namely the diffraction of visible light and the existence of a photonic stop band, make them ideal candidates for the development of photonic materials [61,63–66]. They may lead to the fabrication of more efficient light sources, miniature waveguides, detectors, modulators, and circuits.

Any study of colloidal crystals requires the preparation of monodisperse colloidal particles that are uniform in size, shape, composition, and surface properties. Monodisperse spherical colloids of various sizes, composition, and surface properties have been prepared via numerous synthetic strategies [67]. However, the direct preparation of crystal phases from spherical particles usually leads to a rather limited set of close-packed structures (hexagonal close packed, face-centered cubic, or body-centered cubic structures). Relatively few studies exist on the preparation of monodisperse nonspherical colloids. In general, direct synthetic methods are restricted to particles with simple shapes such as rods, spheroids, or plates [68]. An alternative route for the preparation of uniform particles with a more complex structure might consist of the formation of discrete uniform aggregates of self-organized spherical particles. The use of colloidal clusters with a given number of particles, with controlled shape and dimension, could lead to colloidal crystals with unusual symmetries [69].

The preparation of colloidal clusters with well-defined structures is difficult because the interactions between the colloids are generally isotropic. Therefore, when suspended in a liquid medium, spherical colloids tend to form discrete aggregates but with a very broad size distribution and no well-defined shape. An approach that has proven to be efficient consists of physically confining the particles using a template [70–80]. This allows the clusters to be isolated from each other by the walls of the templates so that uncontrolled aggregation between the clusters is prevented. The structure of the cluster is essentially controlled by the respective shapes and dimensions of the template and of the particles [74]. Different sorts of templates can be used to direct the self-assembly of spherical colloids into well-defined clusters. Some success has been achieved by patterning surfaces with chemical functions [70,71], or with geometric structures [72–74]. Aggregates with well defined shapes such as polygonal or polyhedral clusters, rings, and chains were obtained. Other authors relate the use of emulsion droplets as templates by adsorbing particles on their surface [75,76], or by confining them into their core [77,78]. However, the majority of this work led to aggregates with a large number of particles.

Manoharan et al. have reported the first method for fabricating small clusters of particles using direct emulsions [79,80]. Microspheres (diameter 844–900–1170 nm) made from polystyrene, silica or PMMA were dispersed in the oil of an oil-in-water emulsion with each oil droplet containing a small number N of spheres. The oil was then selectively evaporated. Removing the liquid from a droplet containing particles bound on its surface generated compressive forces that drew the particles together into compact clusters. Surprisingly, for all $N < 11$, the final particle packings were unique: the observed packings closely corresponded to those previously identified as minimizing the second moment of the mass distribution, $M = \sum_i (r_i - r_0)^2$, where r_0 is the center of mass of the cluster. Separation of the aggregates by density gradient centrifugation resulted in aqueous suspensions containing relatively high amounts of identical stable clusters (10^8 to 10^{10}

clusters). The same results were recently obtained via use of inverse emulsions and silica-in-water particles [81]. However, in both cases lack of control of the emulsification conditions led to polydisperse emulsions and severely limited the yield of each type of cluster. Indeed, encapsulation of particles in polydisperse emulsions led to a wide variation in the number of spheres in different droplets. Consequently, to obtain a relatively high proportion of small aggregates ($N < 11$), they used a small volume fraction of colloids in the oil phase.

On the basis of Manoharan's work, Zerrouki et al. [82] have recently demonstrated that controlling emulsification allows one to make much larger quantities of small clusters, as well as to adjust the mean number of colloids per droplet. More importantly, the statistical distribution of the number of spheres per droplet is significantly narrowed by tightly controlling the droplet diameter and by narrowing its distribution. This leads to an increase of the proportion of small aggregates over a large range of conditions. The preparation of clusters containing fewer than five particles, that is, doublet, triangular, and tetrahedral clusters is particularly suitable because, as mentioned earlier, these kinds of aggregates may lead to interesting colloidal crystals, particularly with a diamond-like structure [69]. As shown by Zerrouki and co-workers [82], and summarized below, the amount of colloids in the dispersed phase, the viscosity of the continuous phase, and the shear rate during emulsification are all important for obtaining a high yield of clusters of a given size.

In Zerrouki's experiments, the preparation of aqueous phases of identical clusters is performed in six steps. First, colloidal particles of silica, 1.2 μm in diameter, are synthesized. Next, the surface of the particles is made hydrophobic by chemical grafting. Then, an oil-in-water premix emulsion is made by adding an octane suspension of the colloids in an aqueous solution. Controlled shear of the premix in a Couette-type apparatus is subsequently performed to obtain a quasi-monodisperse

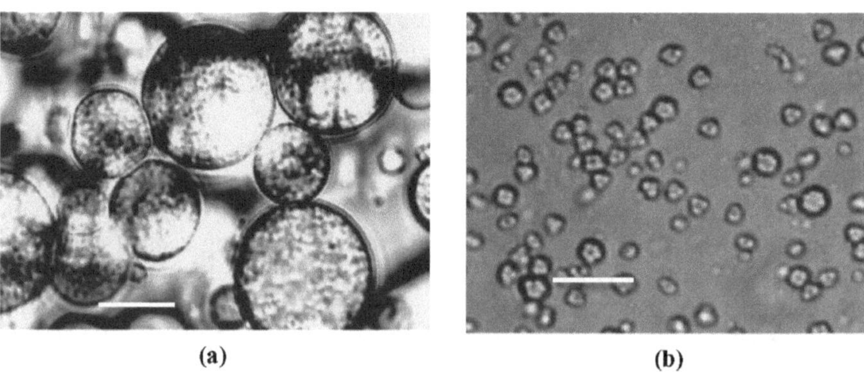

FIGURE 7.8. **(a)** Photomicrograph of a premix silica-in-octane-in-water emulsion. The octane contains 17% vol of silica particles. Compositions are given in the text. **(b)** Same sample after being sheared at 3750 s^{-1} in a Couette geometry device. The scale bar corresponds to 10 μm.

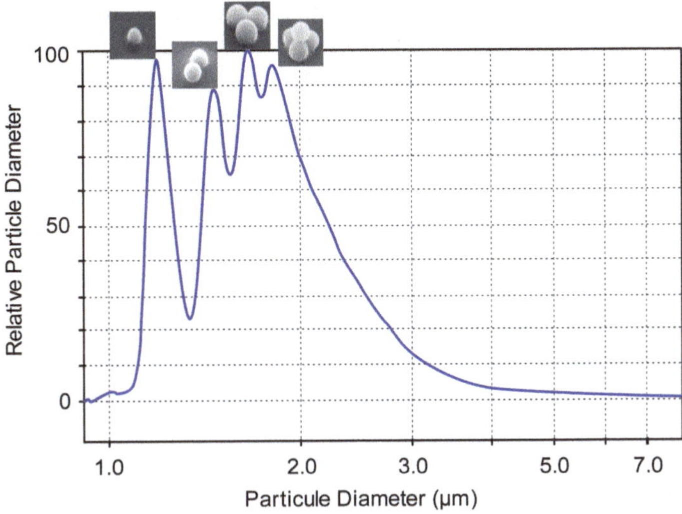

FIGURE 7.9. Histogram from disc centrifugation (see text) of the cluster size distribution obtained from shearing the sample with conditions identical to those in Fig. 7.8. Each discrete peak corresponds to one cluster size from singlets to quadruplets. The larger masses are not resolved. Each type of cluster is indexed by its respective TEM image obtained from the same sample after separation on a gravity column. (Reproduced from [82].)

emulsion. Elimination of the octane from the emulsion droplets leads to an aqueous suspension of the colloidal clusters. Finally, the clusters are separated by sedimentation in a liquid density gradient column. The premix emulsion is actually made by slowly adding the organic colloidal solution (silica particles dispersed in octane), in an aqueous phase containing a hydrophilic surfactant (Synperonic PE/F68) and a thickening agent (PVP-K90). About 30% of PVP in the continuous phase is added to increase the viscosity ratio between the continuous phase and the dispersed phase and thereby improve droplet breakup of the crude emulsion during the subsequent shearing step. The mass fraction of the dispersed phase in the continuous one is fixed at 30%. Figure 7.8a shows an optical micrograph of a premix emulsion ($\phi_{\text{int}} = 17\%$). The premix is highly polydisperse (sizes range from 2 to more than 20 μm) and the droplets generally contain a large number of silica particles. Figure 7.8b shows an optical micrograph of the emulsion after shearing at 3124 s^{-1}. Compared to the premix, the droplets are more monodisperse with a much smaller size and generally contain only a small number of particles. Thus, as previously observed for simple oil-in-water emulsions [83–87], the controlled shear allows a regular fragmentation of the premix droplets. The sheared emulsion is then diluted in a 1% w/w solution of Synperonic PE/F68 in water, with a mass fraction of 10%. When octane is removed by slow evaporation, aggregation of the particles in the emulsion droplets occurs and thus a stable aqueous suspension of colloidal clusters is formed. The proportion of each type of cluster as a function of their particle number N in the colloidal suspension is determined by disc centrifugation. Figure 7.9 shows an example of the cluster size distribution in a suspension, each peak corresponding to a cluster with a given number of particles (%(PVP) = 30%, $\phi_{\text{int}} = 17\%$, shear rate 3124 s^{-1}). From right to left, the four well-defined peaks correspond respectively to single particles, doublets, triplets, and quadruplets. The maximum achieved yield with respect to the initial amount of silica in the crude suspension is about 13% for doublets, 15% for triplets, and 14% for quadruplets.

At this stage, Zerrouki et al. [82] obtained about 100 mg for each type of cluster (about 10^{11} in number); however, since this process can easily be scaled up to several grams of silica, it could be used to prepare several grams of identical clusters. Emulsions as templates are therefore likely to become a propitious strategy to make sufficient amounts of these controlled clusters to further use them as building blocks for performing a larger scale material. It is also likely that compartmentalization into oil or water droplets may offer quite soon the right solution to create the necessary reactive patches onto particles within clusters to promote self-assembly with the correct symmetry.

References

[1] S.K. Mandal, N. Lequeux, B. Rotenberg, M. Tramier, J. Fattaccioli, J. Bibette, and B. Dubertret: "Encapsulation of Magnetic and Fluorescent Nanoparticles in Emulsion Droplets." Langmuir **21**, 4175 (2005).

[2] J. Bibette: "Monodisperse Ferrofluid Emulsions." J. Magn. Magn. Mater. **122**, 37 (1993).

[3] F. Leal-Calderon, T. Stora, O. Mondain Monval, P. Poulin, and J. Bibette: "Direct Measurement of Colloidal Forces." Phys. Rev. Lett. **72**, 2959 (1994).

[4] C.P. Price and D.J. Newman: in "Principles and Practices of Immunoassay." Macmillan, London (1997).

[5] G.U. Lee, S. Metzger, M. M. Natesan, C. Yanavich, and Y.F. Dufrene: "Implementation of Force Differentiation in the Immunoassay." Anal. Biochem. **287**, 261 (2000).

[6] J. Baudry, C. Rouzeau, C. Goubault, C. Robic, L. Cohen-Tannoudji, A. Koenig, E. Bertrand, and J. Bibette: "Tuning the Recognition Rate between Grafted Ligands and Receptors with Magnetic Forces." Proc. Natl. Acad. Sci. USA, vol. 103 pp. 16076–16078.

[7] S. Chesla, P. Selvaraj, and C. Zhu: "Measuring Two-Dimensional Receptor-Ligand Binding Kinetics by Micropipette." Biophys. J. **75**, 1553 (1998).

[8] J.M. Singer and C.M. Plotz: "The Latex Fixation Test : I. Application to the Serologic Diagnosis of Rheumatoid Arthritis." Am. J. Med. **21**, 888 (1956).

[9] B.K. Van Weemen and A.H.W.M. Schuurs: "Immunoassay Using Antigen—Enzyme Conjugates." FEBS Lett. **15**, 232 (1971).

[10] J.M. Singer and C.M. Plotz: "The Latex Fixation Test : II. Results in Rheumatoid Arthritis." Am. J. Med. **21**, 893 (1956).

[11] Immunological Applications 1999, Bangs Labs Technotes, www.Bangslabs.Com.

[12] K. Sato, M. Tokeshi, T. Odake, H. Kimura, T. Ooi, N. M. and T. Kitamori: "Integration of an Immunosorbent Assay System: Analysis of Secretory Human Immunoglobulin a on Polystyrene Beads in a Microchip." Anal. Chem. **72**, 1144 (2000).

[13] J.H.E. Promislow, A.P. Gast, and M. Fermigier: "Aggregation Kinetics of Paramagnetic Colloidal Particles." J. Chem. Phys. **102**, 5492 (1995).
[14] R. Dreyfus: "Magnetic Filaments: Application to the Conception of Force Sensor and Artificial Microscopic Swimmers." Ph.D thesis, Paris VI University (2005).
[15] C. Bustamante, J.C. Macosko, and G.J.L. Wuite: "Grabbing the Cat by the Tail: Manipulating Molecules One by One." Nat. Rev. Mol. Cell. Biol. **1**, (2000).
[16] C. Bustamante, Z. Bryant, and S.B. Smith: "Ten Years of Tension: Single-Molecule DNA Mechanics." Nature **421**, 423427 (2003).
[17] A.F. Oberhauser, P.E. Marszalek, H.P. Erickson, and J.M. Fernandez: "The Molecular Elasticity of the Extracellular Matrix Protein Tenascin." Nature **393**, 181 (1998).
[18] R. Merkel, P. Nassoy, A. Leung, K. Ritchie, and E. Evans: "Energy Landscapes of Receptor-Ligand Bonds Explored with Dynamic Force Spectroscopy." Nature **397**, 50 (1999).
[19] S. Smith, Y. Cui, and C. Bustamante: "Overstretching B-DNA: The Elastic Response of Individual Double-Stranded and Single-Stranded DNA Molecules." Science **271**, 795 (1996).
[20] P. Cluzel, A. Lebrun, C. Heller, R. Lavery, J.-L. Viovy, D. Chatenay, and F. Caron: "DNA: An Extensible Molecule." Science **271**, 792 (1996).
[21] C. Bouchiat, M.D. Wang, J.-F. Allemand, T. Strick, S.M. Block, and V. Croquette: "Estimating the Persistence Length of a Worm-Like Chain Molecule from Force-Extension Measurements." Biophys. J. **76**, 409 (1999).
[22] H. Clausen-Schaumann, M. Rief, C. Tolksdorf, and H.E. Gaub: "Mechanical Stability of Single DNA Molecules." Biophys. J. **78**, 1997 (2000).
[23] H. Dietz and M. Rief: "Exploring the Energy Landscape of GFP by Single-Molecule Mechanical Experiments." Proc. Natl. Acad. Sci. USA **101**, 16192 (2004).
[24] M. Rief, M. Gautel, F. Oesterhelt, M. Fernandez, and H.E. Gaub: "Reversible Unfolding of Individual Titin Immunoglobulin Domains by Afm." Science **276**, 1109 (1997).
[25] M.S.Z. Kellermayer, S.B. Smith, H.L. Granzier, and C. Bustamante: "Elastic Properties of the Titin Molecule Measured Using an Optical Trap." Biophys. J. **72**, A2 (1997).
[26] A. Koenig, P. Hébraud, C. Gosse, R. Dreyfus, J. Baudry, E. Bertrand, and J. Bibette: "Magnetic Force Probe for Nanoscale Biomolecules." Phys. Rev. Lett. **95**, 128301 (2005).
[27] A. Koenig: "Study of a Semi-Flexible Polymer with Magnetic Colloids." Ph.D thesis, Paris VI University (2005).
[28] L. Dorgan, R. Magnotti, J. Hou, T. Engle, K. Ruley, and B. Shull: "Methods to Determine Biotin-Binding Capacity of Streptavidin-Coated Magnetic Particles." J. Magn. Magn. Mater. **194**, 69 (1999).
[29] F. MacKintosh, J.J. Kas, and P. Janmey: "Elasticity of Semiflexible Biopolymer Networks." Phys. Rev. Lett. **75**, 4425 (1995).
[30] J. Baudry, E. Bertrand, N. Lequeux, and J. Bibette: "Colloidal Assemblinfg as Bio-Sensors." J. Phys. Condens. Matter **16**, R469 (2004).
[31] B. Grillet, M. Gaboyard, J. Baudry, and J. Bibette: "A New Magnetic Agglutination Assay for Detection of the Hepatitis B Surface Antigen." Clin. Chem. **50**, A152 (2004).
[32] P. Soumillion and J. Fastrez: "Novel Concepts for Selection of Catalytic Activity." Curr. Opin. Biotechnol. **12**, 387 (2001).

[33] I.P. Petrounia and F.H. Arnold: "Designed Evolution of Enzymatic Properties." Curr. Opin. Biotechnol. **11**, 325 (2000).
[34] G. Georgiou: "Analysis of Large Libraries of Protein Mutants Using Flow Cytometry." Adv. Protein Chem. **55**, 293 (2000).
[35] J.E. Ness, S.B. Del Cardayre, J. Minshull, and W.P. Stemmer: "Molecular Breeding: The Natural Approach to Protein Design." Adv Protein Chem **55**, 261 (2000).
[36] A. Pluckthun, C. Schaffitzel, J. Hanes, and L. Jermutus: "In Vitro Selection and Evolution of Proteins." Adv. Protein Chem. **55**, 367 (2000).
[37] D. Wahler and J.L. Reymond: "Novel Methods for Biocatalyst Screening." Curr. Opin. Chem. Biol. **5**, 152 (2001).
[38] A.D. Griffiths and D.S. Tawfik: "Man-Made Enzymes—from Design to In Vitro Compartmentalisation." Curr. Opin. Biotechnol. **11**, 338 (2000).
[39] D.S. Tawfik and A.D. Griffiths: "Man-Made Cell-Like Compartments for Molecular Evolution." Nat. Biotechnol. **16**, 652 (1998).
[40] N. Doi and H. Yanagawa: "Stable: Protein-DNA Fusion System for Screening of Combinatorial Protein Libraries In Vitro." FEBS Lett. **457**, 227 (1999).
[41] A. Sepp, D.S. Tawfik, and A.D. Griffiths: "Microbead Display by In Vitro Compartmentalisation: Selection for Binding Using Flow Cytometry." FEBS Lett. **532**, 455 (2002).
[42] M. Yonezawa, N. Doi, Y. Kawahashi, T. Higashinakagawa, and H. Yanagawa: "DNA Display for In Vitro Selection of Diverse Peptide Libraries." Nucleic Acids Res. **31**, e118 (2003).
[43] Y.F. Lee, D.S. Tawfik, and A.D. Griffiths: "Investigating the Target Recognition of DNA Cytosine-5 Methyltransferase Hhai by Library Selection Using In Vitro Compartmentalisation." Nucleic Acids Res. **30**, 4937 (2002).
[44] A.D. Griffiths and D.S. Tawfik: "Directed Evolution of an Extremely Fast Phosphotriesterase by In Vitro Compartmentalization." Embo J. **22**, 24 (2003).
[45] H.M. Cohen, D.S. Tawfik, and A.D. Griffiths: "Altering the Sequence Specificity of Haeiii Methyltransferase by Directed Evolution Using In Vitro Compartmentalisation." Prot. Eng. Des. Sel. **17**, 3 (2004).
[46] N. Doi, S. Kumadaki, Y. Oishi, N. Matsumura, and H. Yanagawa: "In Vitro Selection of Restriction Endonucleases by In Vitro Compartmentalization." Nucleic Acids Res. **32**, e95 (2004).
[47] K. Bernath, M. Hai, E. Mastrobattista, A.D. Griffiths, S. Magdassi, and D.S. Tawfik: "In Vitro Compartmentalization by Double Emulsions: Sorting and Gene Enrichment by Fluorescence Activated Cell Sorting." Anal. Biochem. **325**, 151 (2004).
[48] M. Zaccolo, D.M. Williams, D.M. Brown, and E. Gherardi: "An Approach to Random Mutagenesis of DNA Using Mixtures of Triphosphate Derivatives of Nucleoside Analogues." J. Mol. Biol. **255**, 589 (1996).
[49] W.P. Stemmer: "Rapid Evolution of a Protein In Vitro by DNA Shuffling." Nature **370**, 389 (1994).
[50] F.J. Ghadessy, J.L. Ong, and P. Holliger: "Directed Evolution of Polymerase Function by Compartmentalized Self-Replication." Proc. Natl. Acad. Sci. USA **98**, 4552 (2001).
[51] F.J. Ghadessy, N. Ramsay, F.F. Boudsocq, D. Loakes, A. Brown, S. Iwai, A. Vaisman, W.R., and P. Holliger: "Generic Expansion of the Substrate Spectrum of a DNA Polymerase by Directed Evolution." Nat. Biotechnol. **22**, 755 (2004).

[52] P.N. Pusey, W. Van Megen, P. Bartlett, B.J. Ackerson, J.G. Rarity, and S.M. Underwood: "Structure of Crystals of Hard Colloidal Spheres." Phys. Rev. Lett. **63**, 2753 (1989).

[53] U. Gasser, E. Weeks, A. Schofield, P.N. Pusey, and D.A. Weitz: "Real-Space Imaging of Nucleation and Growth in Colloidal Crystallization." Science **292**, 258 (2001).

[54] H. Miguez, F. Meseguer, C. Lopez, A. Mifsud, J.S. Moya, and L. Vazquez: "Evidence of FCC Crystallization of SiO_2 Nanospheres." Langmuir **13**, 6009 (1997).

[55] A. Van Blaaderen and P. Wiltzius: "Real-Space Structure of Colloidal Hard-Sphere Glasses." Science **270**, 1177 (1995).

[56] A. Yetrihaj and A. Van Blaaderen: "A Colloidal Model System with an Interaction Tunable from Hard Sphere to Soft and Dipolar." Nature (2003).

[57] J.H. Holtz and S.A. Asher: "Polymerized Colloidal Crystal Hydrogel Films as Intelligent Chemical Sensing Materials." Nature **389**, 829 (1997).

[58] R. Schroden, C. Blanford, B. Melde, B.J.S. Johnson, and A. Stein: "Direct Synthesis of Ordered Macroporous Silica Materials Functionalized with Polyoxometalate Clusters." Chem. Mater. **13**, 1074 (2001).

[59] B.J.S. Johnson and A. Stein: "Surface Modification of Mesoporous, Macroporous, and Amorphous Silica with Catalytically Active Polyoxometalate Clusters." Inorg. Chem. **40**, 801 (2001).

[60] S. Matsushita, T. Miwa, T. D. and A. Fujishima: "New Mesostructured Porous TiO_2 Surface Prepared Using a Two-Dimensional Array-Based Template of Silica Particles." Langmuir **14**, 6441 (1998).

[61] O. Velev and E. Kaler: "Structured Porous Materials Via Colloidal Crystal Templating: From Inorganic Oxides to Metals." Adv. Mater. **12**, 531 (2000).

[62] J. Veinot, H. Yan, S. Smith, J. Cui, Q. Huang, and T.J. Mark: "Fabrication and Properties of Organic Light-Emitting "Nanodiode" Arrays." Nano Lett. **2**, 333 (2002).

[63] Y. Xia, B. Gates, and Z.-Y. Li: "Self-Assembly Approaches to Three-Dimensional Photonic Crystals." Adv. Mater. **13**, 409 (2001).

[64] D. Norris and Y. Vlasov: "Chemical Approaches to Three-Dimensional Semiconductor Photonic Crystals." Adv. Mater. **13**, 371 (2001).

[65] Y. Vlasov, X. Bo, J. Sturm, and D.J. Norris: "On-Chip Natural Assembly of Silicon Photonic Bandgap Crystals." Nature **414**, 289 (2001).

[66] W. Wang, B.H. Gu, L. Y. Liang, and W.A. Hamilton: "Fabrication of near-Infrared Photonic Crystals Using Highly-Monodispersed Submicrometer SiO_2 Spheres." J. Phys. Chem. B **107**, 12113 (2003).

[67] Y. Xia, B. Gates, Y. Yin, and Y. Lu: "Monodispersed Colloidal Spheres: Old Materials with New Applications." Adv. Mater. **12**, 693 (2000).

[68] Y. Lu, Y. Yin, and Y. Xia: "Three-Dimensional Photonic Crystals with Non-Spherical Colloids as Building Blocks." Adv. Mater. **13**, 415 (2001).

[69] A. Van Blaaderen: "Colloidal Molecules and Beyond." Nature **301**, 470 (2003).

[70] J. Aizenberg, P.V. Braun, and P. Wiltzius: "Patterned Colloidal Deposition Controlled by Electrostatic and Capillary Forces." Phys. Rev. Lett. **84**, 2997–3000 (2000).

[71] K. Chen, X. Jiang, L. Kimerling, and P.T. Hammond: "Selective Self-Organization of Colloids on Patterned Polyelectrolyte Templates." Langmuir **16**, 7825 (2000).

[72] Y. Yin and Y. Xia: "Self-Assembly of Monodispersed Spherical Colloids into Complex Aggregates with Well-Defined Sizes, Shapes, and Structures." Adv. Mater. **13**, 267 (2001).

[73] Y. Yin, Y. Lu, B. Gates, and Y. Xia: "Template-Assisted Self Assembly: A Practical Route to Complex Aggregates of Monodispersed Colloids with Well-Defined Sizes, Shapes and Structures." J. Am. Chem. Soc. **123**, 8718 (2001).

[74] Y. Xia, Y. Yin, Y. Lu, and J. Mc Lellan: "Template-Assisted Self-Assembly of Spherical Colloids into Complex and Controllable Structures." Adv. Funct. Mater. **13**, 907 (2003).

[75] O. Velev, K. Furusawa, and K. Nagayama: "Assembly of Latex Particles by Using Emulsion Droplets as Templates. 1. Microstructured Hollow Spheres." Langmuir **12**, 2374 (1996).

[76] A. Dinsmore, M. Hsu, M. Nikolaides, M. Marquez, A.R. Bausch, and D.A. Weitz: "Colloidosomes: Selectively Permeable Capsules Composed of Colloidal Particles." Science **298**, 1006 (2002).

[77] O.D. Velev, K. Furusawa, and K. Nagayama: "Assembly of Latex Particles by Using Emulsion Droplets as Templates. 2. Ball-Like and Composite Aggregates." Langmuir **12**, 2385 (1996).

[78] O. Velev, A. Lenhoff, and E.W. Kaler: "A Class of Microstructured Particles Through Colloidal Crystallization." Science **287**, 2240 (2000).

[79] V. Manoharan, M. Elsesser, and D.J. Pine: "Dense Packing and Symmetry in Small Clusters of Microspheres." Science **301**, 483 (2003).

[80] G.-R. Yi, V. Manoharan, E. Michel, M.T. Elsesser, S.-M. Yang, and D.J. Pine: "Colloidal Clusters of Silica or Polymer Microspheres." Adv. Mater. **16**, 1204 (2004).

[81] Y.-S. Cho, G.-R. Yi, S.-H. Kim, D.J. Pine, and S.-M. Yang: "Colloidal Clusters of Microspheres from Water-in-Oil Emulsions." Chem. Mater. **17**, 5006 (2005).

[82] D. Zerrouki, B. Rotenberg, S. Abramson, J. Baudry, C. Goubault, F. Leal-Calderon, D.J. Pine, and J. Bibette: "Preparation of Doublet, Triangular, and Tetrahedral Colloidal Clusters by Controlled Emulsification." Langmuir **22**, 57 (2006).

[83] C. Goubault, K. Pays, D. Olea, J. Bibette, V. Schmitt, and F. Leal-Calderon: "Shear Rupturing of Complex Fluids: Application to the Preparation of Quasi-Monodisperse W/O/W Double Emulsions." Langmuir **17**, 5184 (2001).

[84] T.G. Mason and J. Bibette: "Shear Rupturing of Droplets in Complex Fluids." Langmuir **13**, 4600 (1997).

[85] C. Mabille, V. Schmitt, P. Gorria, F. Leal Calderon, V. Faye, and B. Deminière: "Rheological and Shearing Conditions for the Preparation of Monodisperse Emulsions." Langmuir **16**, 422 (2000).

[86] C. Mabille: "Fragmentation in Emulsions Submitted to a Simple Shear." Ph.D thesis, Bordeaux I University (2000).

[87] C. Mabille, F. Leal-Calderon, J. Bibette, and V. Schmitt: "Monodisperse Fragmentation in Emulsions: Mechanisms and Kinetics." Europhys. Lett **61**, 708 (2003).

General Conclusion

Technical progress is driven by the in-depth knowledge and control of material properties at the very smallest scale. While nanosciences are set to change our future, our daily life has already been transformed by the use of micrometric objects such as the droplets present in emulsions. The wide range of processes and products that involve emulsions requires command over the phenomena governing their behavior. The scientific background of emulsions has reached some maturity which allows today to envision more audacious applications. In particular, nano or micro-technologies are more and more benefiting from this background though the relatively new input of emulsions and this adventure is just at its early stage. Indeed, because of the very large versatility offered by fragmentation techniques, emulsions as intermediary materials are providing unique advantages for designing multi-function colloidal particles. For instance as described in this book, this has given access to Brownian and highly super-paramagnetic colloids, which because of their self and reversible assembling ability, can enter into new strategies in diagnostic, and biophysics. Liquid droplets can compartmentalize minute amounts of defined reactants and can modify their concentration from slight diffusion through the continuous phase. As femtoliter reservoirs, emulsions allow screening a large compound library and as semi-permeable reservoirs they allow parallelizing a directed process imposed by confinement. As substrates, reservoirs, reactors, templates, or simply intermediaries, emulsions seem to have successfully met the nano and micro-technologies adventure.

On a scientific point of view, numerous questions relative to emulsions remain unanswered. For instance, the origin of coalescence is not fully understood. This phenomenon has a strong impact on the emulsion lifetime, as discussed in this book and is certainly at the origin of the well known empirical Bancroft rule. It is increasingly clear that there is not just one, but several coalescence mechanisms, each of them involving different sets of microscopic parameters (surface forces, interfacial tension, surface elasticity, surfactant spontaneous curvature, defects, etc.). Understanding the role of the microscopic parameters controlling film rupturing should significantly consolidate the fundamental knowledge of emulsions and should further allow one to monitor their kinetic evolution. In the same vein, the mechanism of destabilization under high shear, though being a very important phenomenon, still raises challenging questions. Protein or solid-stabilized emulsions are known to be very sensitive to shear and would be very interesting candidates for elucidating these questions. Clarifying this matter would certainly make more precise the understanding of the coupling between thin-film metastability and shear. The possibility of making emulsions out of crystallizable oils is reinforcing the range of applications of emulsions and opens interesting perspectives for the processing of solid colloids or the for the preparation of materials with tuneable rheological properties and structures. However, the link between surface properties in both the liquid and the solid states, and stability is still obscure. The knowledge of emulsification methods that lead to concentrated, controlled materials has been progressing, but understanding the phase inversion mechanism, which is very useful in the case of highly viscous dispersed phases, requires further investigations.

To conclude, research, applications and expectations with regard to emulsions are permanently evolving. Changes in the macro-economic context along with various political decisions (e.g., sustainable development, Registration, Evaluation and Authorisation of Chemicals (REACH) directive for the European Union) are factors which influence the conditions in which emulsion-based products are fabricated and used. It is probable that these changes will motivate technological developments and new advances in emulsion science.

Index

SPRINGER NATURE

GPSR Compliance

The European Union's (EU) General Product Safety Regulation (GPSR) is a set of rules that requires consumer products to be safe and our obligations to ensure this.

If you have any concerns about our products, you can contact us on ProductSafety@springernature.com

In case Publisher is established outside the EU, the EU authorized representative is:

Springer Nature Customer Service Center GmbH
Europaplatz 3
69115 Heidelberg, Germany

Zeitfracht Medien GmbH
Ferdinand-Jühlke-Straße 7
99095 Erfurt, Deutschland
produktsicherheit@kolibri360.de